Transferability of Fracture Mechanical Characteristics

NATO Science Series

A Series presenting the results of scientific meetings supported under the NATO Science Programme.

The Series is published by IOS Press, Amsterdam, and Kluwer Academic Publishers in conjunction with the NATO Scientific Affairs Division

Sub-Series

I. **Life and Behavioural Sciences**	IOS Press
II. **Mathematics, Physics and Chemistry**	Kluwer Academic Publishers
III. **Computer and Systems Science**	IOS Press
IV. **Earth and Environmental Sciences**	Kluwer Academic Publishers
V. **Science and Technology Policy**	IOS Press

The NATO Science Series continues the series of books published formerly as the NATO ASI Series.

The NATO Science Programme offers support for collaboration in civil science between scientists of countries of the Euro-Atlantic Partnership Council. The types of scientific meeting generally supported are "Advanced Study Institutes" and "Advanced Research Workshops", although other types of meeting are supported from time to time. The NATO Science Series collects together the results of these meetings. The meetings are co-organized bij scientists from NATO countries and scientists from NATO's Partner countries – countries of the CIS and Central and Eastern Europe.

Advanced Study Institutes are high-level tutorial courses offering in-depth study of latest advances in a field.
Advanced Research Workshops are expert meetings aimed at critical assessment of a field, and identification of directions for future action.

As a consequence of the restructuring of the NATO Science Programme in 1999, the NATO Science Series has been re-organised and there are currently Five Sub-series as noted above. Please consult the following web sites for information on previous volumes published in the Series, as well as details of earlier Sub-series.

http://www.nato.int/science
http://www.wkap.nl
http://www.iospress.nl
http://www.wtv-books.de/nato-pco.htm

Series II: Mathematics, Physics and Chemistry – Vol. 78

Transferability of Fracture Mechanical Characteristics

edited by

Ivo Dlouhý

Institute of Physics of Materials,
Academy of Sciences of the Czech Republic,
Brno, Czech Republic

Kluwer Academic Publishers

Dordrecht / Boston / London

Published in cooperation with NATO Scientific Affairs Division

Proceedings of the NATO final project workshop on
Fracture Resistance of Steels for Containers of Spent Nuclear Fuel arranged within
Science for Peace Program
Brno, Czech Republic
5–6 November 2001

A C.I.P. Catalogue record for this book is available from the Library of Congress.

ISBN 1-4020-0794-9 (HB)
ISBN 1-4020-0795-7 (PB)

Published by Kluwer Academic Publishers,
P.O. Box 17, 3300 AA Dordrecht, The Netherlands.

Sold and distributed in North, Central and South America
by Kluwer Academic Publishers,
101 Philip Drive, Norwell, MA 02061, U.S.A.

In all other countries, sold and distributed
by Kluwer Academic Publishers,
P.O. Box 322, 3300 AH Dordrecht, The Netherlands.

Printed on acid-free paper

CONTENTS

PREFACE

Five laboratories from France, Hungary and the Czech Republic have solved a Project supported financially by NATO within the Science for Peace Program (under Nr. 972655) for three years. The project, titled *Fracture Resistance of Steels for Containers of Spent Nuclear Fuel*, was focused (i) on the generation of data needed for the qualification procedure of a new container introduced by Škoda Nuclear Machinery and (ii) on a number of topics of scientific nature associated with the interesting field of transferability of fracture mechanical data. It has been found during numerous conference presentations of project results that the knowledge developed within the project would be more attractive when published in a more comprehensive form. This was the reason why the final project workshop was arranged as a meeting of project collaborators and contributing invited experts working in very similar field.

The main scope of the final project workshop, titled *Transferability of Fracture Mechanical Data* and held in Brno from 5 to 6 November 2001, was to bring together project collaborators with a number of invited international experts, both covering the spectrum of topics solved within the project and reviewing the project results in the presence of these specialists. A total of 34 colleagues from 7 European countries and the USA participated in the workshop.

The contents of this book contains: (i) contributions of project partners setting forth the main project results after they were evaluated and discussed at the workshop, (ii) contributions of the experts participating in the workshop, and finally (iii) invited contributions of experts working in the same field that could not participate personally on the workshop but were interested in the presentation of their results commonly with our project results. It was decided at the workshop that two reviewers will check each contribution in the volume except for standard editorial comments.

A great deal of effort has been exerted on co-ordination of the content of contributions from the project collaborators in order not to present the same results in more contributions of the book. All knowledge either of experimental or theoretical nature, once it has been developed within the common NATO SfP project, is thus here presented only in one of contributions.

The workshop proceedings should be interesting to engineers and scientists, both in academia and industry, working in the field of transferability of fracture mechanical data from smaller specimens to larger ones. This includes the associated transfer from the laboratory measurement to the real component. The topics included in the volume should be attractive for experts not only concerned with nuclear and industrial machinery but also from all other areas dealing with component integrity, prediction of fracture behaviour etc.

We are indebted to all reviewers who have supported this volume preparation by careful reading of the manuscripts, remarks or comments to the concept of separate contributions or book itself.

Our warm thanks belong to project consultant J. Leonard, USA for his numerous discussions and for introducing the personal dimension to our three-year collaboration.

We wish to acknowledge our sponsors, without whom the project and workshop, and the present volume would have never existed: NATO Science for Peace Steering Committee for financial support within the project Nr. 972655 and authorisation of this Volume, Ministry of Education of the Czech Republic (project ME 303) and Hungarian fund (OTKA T030057) for additional supports during the project solution, and, finally to the Institute of Physics of Materials of the Academy of Sciences for the organisational and financial background for the workshop arrangement.

I am particularly indebted to members of the Brittle Fracture Group for their great help in organising the workshop and considerable effort to put everything together for the elaboration of the present volume.

Ivo Dlouhý

ADDRESS LIST OF AUTHORS

Prof. Zitouni AZARI
L. F. M.
Université de Metz – ENIM
Ile du Saulcy
57045 Metz cedex 01
FRANCE
azari@sciences.univ-metz.fr

Dr. Wolfram BAER
Department of Materials Engineering
Federal Institute for Materials Research
and Testing
Unter der Eichen 87
DE-12205 Berlin
GERMANY
wolfgang.bayer@bam.de

Dr. Zsolt BALOGH
Bay Zoltán Foundation for Applied Research
Institute for Logistic and Production Systems
P. O. B. 59
H-3519 Miskolctapolca
HUNGARY
bazso@bzlogi.hu

Dr. Clotilde BERDIN
École Centrale Paris, LMSS-Mat
Grande Voie des Vignes
F-92295 Châtenay-Malabry
FRANCE
berdin@mssmat.ecp.fr

Dr. Petr BERTSCHINGER
Mat-Tec SA
Unter der Graben 27
CH-8401 Winterthur
SWITZERLAND
bertschinger@mat-tec.ch

Dr. Jiří BRYNDA
ŠKODA Nuclear Machinery plc
Orlík 266
316 06 Plzeň
Czech Republic
jbrynda@jad.in.skoda.cz

Dr. Ivo DLOUHÝ
Institute of Physics of Materials AS CR
Žižkova 22
616 62 Brno
CZECH REPUBLIC
idlouhy@ipm.cz

Dr. Zdeněk FIALA
Institute of Theoretical and Applied
Mechanics AS CR
Prosecká 76
190 00 Praha
CZECH REPUBLIC
Fiala@itam.cas.cz

Mr. Petr HAUŠILD
École Centrale Paris, LMSS-Mat,
Grande Voie des Vignes
92295 Châtenay-Malabry
FRANCE
hausild@mssmat.ecp.fr

Czech Technical University
Faculty of Nucl. Sci.&Phys. Eng.
Dept. of Materials
Trojanova 13
120 00 Praha 2
CZECH REPUBLIC

Dr. Norbert HEGMAN
Bay Zoltán Foundation for Applied Research
Institute for Logistic and Production Systems
P. O. B. 59
H-3519 Miskolctapolca
HUNGARY
lenkey@bzlogi.hu

Ass. Prof. Miloslav HOLZMANN
Institute of Physics of Materials AS CR
Žižkova 22
616 62 Brno
CZECH REPUBLIC
idlouhy@ipm.cz

Dr. Pavel HOSNEDL
ŠKODA Nuclear Machinery plc
Orlík 266
316 06 Plzeň
CZECH REPUBLIC
phosnedl@jad.in.skoda.cz

Dr. Zdeněk CHLUP
Institute of Physics of Materials AS CR
Žižkova 22
616 62 Brno
CZECH REPUBLIC
chlup@ipm.cz

Dr. Aleš JANÍK
Institute of Physics of Materials AS CR
Žižkova 22
616 62 Brno
CZECH REPUBLIC
kozak@ipm.cz

Mr. Ladislav JURÁŠEK
Institute of Physics of Materials AS CR
Žižkova 22
616 62 Brno
CZECH REPUBLIC
jurasek@ipm.cz

Dr. Josef KASL
SKODA Research Ltd.
Tylova 57
316 00 Plzeň
CZECH REPUBLIC
jkasl@vyz.in.skoda.cz

Dr. Zbyněk KERŠNER
Brno University of Technology
Faculty of Civil Engineering
Institute of Structural Mechanics
Veveří 95
662 37 Brno
CZECH REPUBLIC
kersner.z@fce.vutbr.cz

Prof. Amar KIFANI
L. M. M.
Ufr MFS
Faculté des sciences
Université Mohammed V, Avenue Ibn
Batouta, Bp 1014,
Agdal – Rabat
MOROCCO
kifani@fsr.ac.ma

Dr. Jan KOHOUT
The Brno Military Academy
Kounicova 65
612 00 Brno
CZECH REPUBLIC
Jan.Kohout@vabo.cz

Prof. Sergey KOTRECHKO
G.V. Kurdymov Institute for Metal
Physics NA SU
36 Vernadsky Blvd.
03142 Kyiv 142
UKRAINE
kotr@d24.imp.kiev.ua

Dr. Vladislav KOZÁK
Institute of Physics of Materials AS CR
Žižkova 22
616 62 Brno
CZECH REPUBLIC
kozak@ipm.cz

Dr. Libor KRAUS
COMTES FHT Ltd.
Borská 47
320 13 Plzeň
CZECH REPUBLIC
lkraus@comtesfht.cz

Dr. Dana LAUEROVÁ
Nuclear Research Institute Řež, plc.
250 68 Řež
CZECH REPUBLIC
lau@nri.cz

Dr. Gyöngyvér B. LENKEY
Bay Zoltán Foundation for Applied Research
Institute for Logistic and Production Systems
P. O. B. 59
H-3519 Miskolctapolca
HUNGARY
lenkey@bzlogi.hu

Prof. M. LOUAH
LaMAT
E.N.S.E.T.
BP 6207 R.I
MOROCCO
louah@caramail.com

Dr H. EL MINOR
LaMAT
E.N.S.E.T.
BP 6207 R.I
MOROCCO
elminor@caramail.com

Dr. Karsten MÜLLER
Department of Materials Engineering
Federal Institute for Materials Research
and Testing
D-12205 Berlin
GERMANY
karsten.mueller@bam.de

Dr. Stanislav NĚMEČEK
NTC, University of West Bohemia
Univerzitní ul.
Plzeň
CZECH REPUBLIC

Ass. Prof. Dr. Drahomír NOVÁK
Brno University of Technology
Faculty of Civil Engineering
Institute of Structural Mechanics
Veveří 95
662 37 Brno
CZECH REPUBLIC
novak.d@fce.vutbr.cz

Dr. Jiří NOVÁK
Nuclear Research Institute Řež, plc.
250 68 Řež
CZECH REPUBLIC
novak.j@ujv.cz

Mr. Miroslav PICEK
ŠKODA Nuclear Machinery plc
Orlík 266
316 06 Plzeň
CZECH REPUBLIC
mpicek@jad.in.skoda.cz

Prof. Guy PLUVINAGE
L. F. M
Université de Metz – ENIM
Ile du Saulcy
57045 Metz cedex 01
FRANCE
pluvina@sciences.univ-metz.fr

Dr. Hans-Jakob SCHINDLER
Mat-Tec SA
Unter der Graben 27
CH-8401 Winterthur
SWITZERLAND
schindler@mat-tec.ch

Prof. Bohumír STRNADEL
Technical University of Ostrava
Department of Materials Engineering
17. listopadu 15
708 33 Ostrava
CZECH REPUBLIC
bohumir.strnadel@vsb.cz

Prof. Břetislav TEPLÝ
Brno University of Technology
Fakulty of Civil Engineering
Institute of Structureal Mechanics
Veveří 95
662 37 Brno
CZECH REPUBLIC
teply.b@fce.vutbr.cz

Mr. Tibor THOMÁZY
Bay Zoltán Foundation for Applied Research
Institute for Logistic and Production Systems
P. O. B. 59
H-3519 Miskolctapolca
HUNGARY
lenkey@bzlogi.hu

Prof. Lázsló TÓTH
Bay Zoltán Foundation for Applied Research
Institute for Logistic and Production Systems
P. O. B. 59
H-3519 Miskolctapolca
HUNGARY
tlaszlo@bzlogi.hu

Ass. Prof. Stanislav VEJVODA
Institute of Applied Mechanics Brno, Ltd.
Veveří 95
611 39 Brno
CZECH REPUBLIC
uam@telecom.cz

Mr. Libor VLČEK
Institute of Physics of Materials AS CR
Žižkova 22
616 62 Brno
CZECH REPUBLIC
vlcek@ipm.cz

Dr. Hans-Peter WINKLER
GNB Gesellschaft für Nuklear-Behälter
Hollestr. 7A
D-45127 Essen
GERMANY
Hans-Peter_Winkler@gns-gnb.de

Dr. Peter WOSSIDLO
Department of Materials Engineering
Federal Institute for Materials Research
D-12205 Berlin
GERMANY
peter.wossidlo@bam.de

ŠKODA CASK TESTING AND LICENSING REQUIREMENTS

J. BRYNDA, P. HOSNÉDL, M. PICEK
ŠKODA Nuclear Machinery plc, Orlík 266, 31606 Plzeň, Czech Republic

Abstract: Last activities in field of licensing cask from cast ferritic steels are summarised with special attention to aspects of material fracture resistance. Main principles of container design are shortly described focusing on the role of thick walled cask made of cast ferritic steel in integrity of whole container. Licensing requirements prescribed by IAEA rules are roughly analysed leading to Certificate of Accreditation from the Czech authority in this field. Testing program including the drop test facility development has been described. As important part of container operational life the monitoring program has been introduced and ways for evaluation of cask material residual lifetime introduced.

Keywords: cask, storage, transport, spent nuclear fuel, cast steel, ferritic steel, safety requirements

1. Introduction

Each application of a new material for container production has (a) some general guidelines and (b) specific problems.

(a) The general aspects and long term experiences of other laboratories have been summarised in "Guidelines for safe design on shipping packages against brittle fracture" which are regulating present assessment procedures from the point of view of IAEA [1]. Three methods have been included in this document:

(i) The evaluation and use of materials, which remain ductile and tough throughout the required service temperatures down to – 40 °C.

(ii) The evaluation of material having ferrite matrix by means of nil ductility transition temperature measurements correlated to fracture resistance.

(iii) The assessment of fracture resistance based on a design evaluation using fracture mechanics.

(b) The specific problems of a material application to manufactured products, such as container for spent fuel, have to show the advantages and disadvantages of that particular solution. These include such material characteristics assessment like the inherent scatter of material fracture toughness, susceptibility (predisposition) of material to embrittlement (radiation, low temperature), the effect of loading rate, etc. Every approach able to predict these characteristics of fracture resistance will enhance the quality of pure experimental evaluation of materials for production and confidence in material integrity. Recently, several concepts have been followed in the world.

1

I. Dlouhý (ed.), Transferability of Fracture Mechanical Characteristics, 1–14.
© 2002 *Kluwer Academic Publishers. Printed in the Netherlands.*

The cask design conception based on ductile cast iron (DCI) has been followed for a long time and has found extensive application in case of CASTOR® containers mainly [2-6 etc.]. Ductile cast iron can be characterised as a heterogeneous material the properties of which are significantly affected by size, quantity and shape of nodular graphite particles. Chemical composition, particularly the carbon and silicon content, has a major effect here. Increase of silicon content (usually being in range of 1.2 to 2.4 wt. %) leads to eutectic temperature increase, stabilises ferrite in the microstructure and during heat treatment suppresses formation of pearlite. Strength properties increase and at the same time ductility decreases. In addition to that, the properties of DCI are significantly dependent on impurities content, which even at low concentrations can lead to changes in graphite morphology and reduction of ductility and fracture toughness. Standard DCI constitution allows ratio value of yield stress to tensile strength to be on the level of 0.8 or more. Minimum ductility values needed is 6 %, recently after careful production technology development and monitoring are satisfied being on level of up to 15%. What is important and demanding to assure the required properties in entire thick walled vessel with respect to the technological procedures, which have to be followed during the whole period of fabrication.

ŠKODA Nuclear Machinery currently produces CASTOR® casks for transport and storage of 84 spent fuel assemblies from VVER 440 reactor. The CASTOR® casks are made under the licence and in collaboration with GNB mbH, Essen Company [7].

Procedures for brittle fracture evaluation of ductile cast iron used for packaging production in Japan have been investigated [6]. For development of brittle fracture assessment procedures methodologies accepted for reactor pressure vessel steel were applied. The use of a small specimen for fracture toughness lower bound prediction is analysed for ductile cast iron. Carbon steels SA 350 Gr. LF1 are used in Japan [10]. To achieve uniform properties in high thickness (300 mm) hollow ingot technique used for the manufacture of rings.

In United Kingdom Nirex has developed re-usable shielded transport containers. They are testing three candidate materials [e.g. 11]. It has been shown that it is possible to satisfy the safe design criterion against brittle fracture for cast martensitic stainless steel. For complex design and integrity calculations dynamic J_{ld} integral values have been found more convenient.

Various aspects of methodologies for evaluation of the risk of brittle fracture for thick walled cask made from cast iron or forged ferritic steels have been analysed by French authors [e.g. 13]. Dynamic fracture toughness determined using the pre-cracked Charpy type specimen and dynamic analysis is preferred for fracture resistance of followed materials assessment

The use of non-austenitic materials in containers requires consideration of the brittle fracture risk under severe loading conditions. There is motivation for utilising ferritic steels or titanium alloys for container application in USA [14-15].

Brittle fracture prevention criteria for casks based upon fracture mechanics principles remain uncodified and further regulatory work is needed to develop guidance for the evaluation of brittle fracture. According to Westinghouse and recently also TEPCO (Japan) projects, cask bodies are made of forged and welded parts. This design, with respect to material and technology, is very similar to the manufacture of reactor pressure vessel. E.g. ring forgings of steels SA 508 Cl.4b and Cl.5 are used for cask

MC-10. It is a high-strength steel with 3.5 % of Ni and 1.5 % Cr for which a special welding method under preheating shall be applied. Required properties are determined by ASME Code and particularly the brittle fracture characteristics are on high level.

A different solution of the material issue for cask has been proposed and investigated in ŠKODA Nuclear Machinery. It was enabled by the cask design itself, which differs significantly form the previous designs by its double barrier cover.

Aim of the contribution can be seen in summarisation of results of development and recent possibilities of the application of cast ferritic steel for casks of containers of spent nuclear fuel. The topics connected with the problem of evaluation of resistance against brittle fracture and container cask integrity is addressed as a main topic where the transferability of data from laboratory experiments to the container cask plays key role.

2. Cask Body Design

The ŠKODA 440/84, dual-purpose container is designed for application and use as a transport and storage container of VVER 440 type pressurized water reactor spent fuel. It is designed applying a cast steel cask with a removable stainless steel hermetic canister. Spent fuel is held within the hermetic canister at spacing to maintain subcriticality under all conditions. The whole cask is encased in a shell with neutron shielding material. For intermediate term storage the shell can be replaced by storage of cask(s) within a concrete silo. Some technical data of the container are introduced in TABLE I.

TABLE I. Technical data of the container ŠKODA 440/84

Number of stored fuel assemblies	84
Cask body, bottom and lid material	cast steel
Empty cask weight	111 320 kg
Total weight of filled cask	129 820 kg
Cask outside diameter	2 800 mm
Transport chock absorber outside dimension	3090 x 3090 mm
Inside cask diameter	1 846 mm
Total height (without shock absorber)	4 745 mm
Inside pressure in the canister	0.07 MPa
Inside pressure in the cask	0.5 MPa
Radiated thermal output	max. 24 kW
Heat exchanging medium	Helium
Cask life time	min. 60 years

Basic design elements of the container are as follows (see also *Figure 1* for the design overview):

Cask body, bottom and lid are fabricated using manganese cast steel (Nr. 42 2707 according to Czech Standards, equivalent to international identification W. Nr. 1.1120, DIN 17182, SA 352 Gr. Lcc) components bolted and welded together. Helicoflex

4

sealing provides stability of helium filled gap between canister and their interior cask surface.

Canister is produced from austenitic cylindrical material and represents the first barrier of protection for the stored nuclear fuel. The canister body and lid are connected with both seal and structural welds. The seal weld is created with automated welding equipment after the insertion of spent fuel into the canister. After drying and vacuum degassing, the canister is filled with helium.

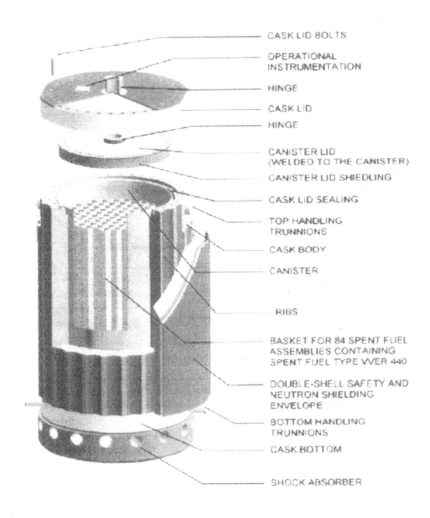

Figure 1. Main design characteristics of container ŠKODA 440/84 nuclear fuel transport and storage cask

Basket provides the cask internal matrix for proper alignment and spacing of stored fuel assemblies. The basket tubes are precision extruded AlMg material arranged in hexagonal geometry. Shielding plates of ATABOR material are inserted between individual dovetails of the hexagonal tubes.

Safety and shielding envelope is comprised of triple–shell steel cylinder filled with BISCO shielding material and air. The safety envelope is fastened to the cask.

Instrumentation consists of a helium pressure detector in the gap between the canister and cask body. Temperature monitoring of the cask exterior surface is also provided.

3. Licensing Requirements (Licence)

The degradation of material properties due to operational conditions and particularly in the radiation environment is in many cases limiting for the lifetime of structures. Transport and storage casks of type B according to IAEA classification are used for the storage of spent nuclear fuel. As mentioned in introduction these casks shall comply with the TECDOC 717 regulations.

A sufficient reserve of plasticity in operational and emergency conditions is a limit criterion for safe operating of a cask which implicitly includes brittle fracture characteristics of material such as T_{K0} – critical temperature of material brittleness in initial (non-degradated) condition, T_K – critical temperature of material brittleness in various stages of material degradation including the radiation embrittlement, temperature degradation and other effects.

3.1. TESTING PROGRAM BEFORE LICENSING

The basic brittle fracture parameters include K_{IC} – critical value of fracture toughness in the field of linear fracture mechanics and/or J_{IC} – critical value of fracture toughness for the area of elastic-plastic fracture mechanics. These values have been established as initial and are compared with calculation codes and actual value K_I - stress intensity coefficient. The limit brittleness curves for the cask design lifetime have been prepared based on these determinations and calculations. For the evidential condition of the cask material property degradation the manufacturer proposes to apply a system of methodologies for monitoring and evaluation of material properties.

Casting material of the cask body was tested first of all from the point of view of achieving more advantageous brittle fracture and plastic properties in comparison to other materials used in this application and with respect to optimal properties throughout the wall thickness. It has been found that mechanical properties reached high values also in the half of the wall thickness; yield stress $R_p0.2$ being between 276 to 290 MPa and tensile ductility being between 40 to 43 % at the test temperature 20 C. Ductility remains on a relatively high level even at the test temperature of -40 C (ranging from 24 to 28 %).

Overview of main mechanical properties followed on the several melts and castings is introduced in *Figure 2*.

6

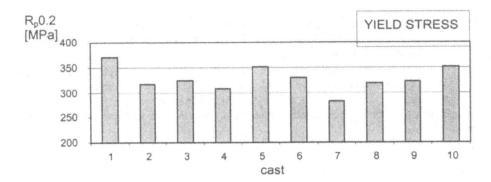

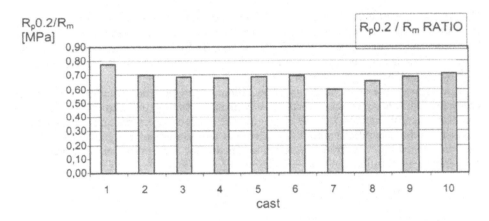

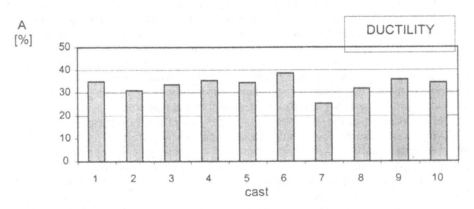

Figure 2a. Basic mechanical properties of different steel melts and locations tested

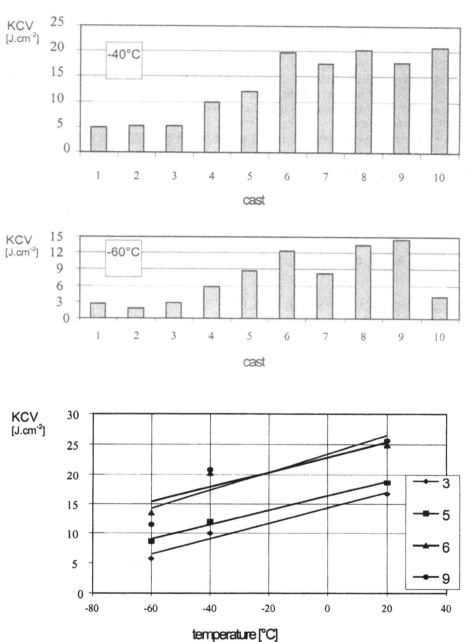

Figure 2b. Basic mechanical properties of different steel melts and locations tested

There are data from tensile and Charpy impact tests corresponding to four different melts of the investigated steel. The different casts have been labelled by the following way:

a) Melt Nr. 83936 produced as 2 : 1 model of the cask body with added proofs, the wall thickness of the cask was about 123 mm, the thickness of proofs about 80 mm. As *cast 1 and 2* in *Figure 2* the specimens cut from the cap are introduced. *Cast 3* represents the properties of the model cask body. As *cast 4* the properties of proofs (PU 282) are given.

b) For Melt Nr. 92459 the properties of added proofs (PU 282) are only introduced under designation *cast 5*.

c) Melt Nr. 06331 suppose 1 : 1 cask body, the initial wall thickness was about 322 mm. The mechanical properties were followed either as average values for the cask body (under *cast 6*) or as properties in midthickness (bores) of the wall thickness (designated as *cast 7*). In addition, properties of added proofs (PU 282) are also supplied (*cast 8*).

d) Melt Nr. 07289 suppose mainly the bottom casting. The properties are given as average data for this container part (*cast 9*) and as obtained from added proofs (PU 282) - *cast 10*.

3.2. DROP TEST FACILITY AND RESULTS

A necessary condition to obtain the licence for the manufacture and use of the cask is, besides other, to perform drop tests. Cask properties in possible extreme conditions, which may occur during the transport of spent nuclear fuel, have been tested by such tests.

Drop tests are performed according to requirements stipulated in IAEA document - "Rules for safety transport of radioactive material (Safety Series No. 6) [1]. A cask is dropped during such tests from a defined height 9m on a massive base plate, and preservation of leak- proof is the main criterion.

In order to detect leakage the cask is filled with helium the possible leakage is monitored using special devices (see as example *Figure 3*). Besides the verification of sealing properties, also tensometric measurements are performed during the drop tests. These measurements can detect to what mechanical loading the cask material is subjected during the drop test. The data from measurements are then used for verification of calculations which in addition documents the reliability and safety of produced casks.

To perform the tests a drop test facility has been built in ŠKODA, Plzeň, It consist of special steel plate of weight 14 tons under which there is a 450t Ferro concrete mono damping the drop reaction. In 1997 drop tests of ŠKODA 440/84 cask model were performed in this facility; the model was a replica of an actual cask, only reduced in size in scale 1 : 2. A drop from 9 m height on the edge, and also a flat drop on a spine protruding from the basis plate from 1 m height were performed there. During the tests and after them all the above-mentioned measurements have been performed and it could be stated that there was no helium leakage and no plastic deformations present in the critical components of the cask.

Successful drop test results of the cask were included in the safety documentation file which was submitted to Regulation authority (SONS) for licensing.

Figure 3. Cask model before the drop test from 9m height according to IAEA requirements (furnished with shock absorbers)

3.3. POSSIBLE RADIATION AND OPERATIONAL DAMAGE

We are aware of the fact the operational fluencies of spent nuclear fuel are in case of a cask lower - order fluencies in comparison with a reactor pressure vessel (total analysis is out of scope of this contribution).

According to designed values and verification calculations it has been found that radiation loading of the cask is 5 times lower than of a reactor pressure vessel (considering the initial value of fast neutron fluency having more detrimental effect in degradation mechanism of the VVER 440 RPV material) in comparison to designed and calculated value at the end of the cask life (that is assumed for 50 years). Evaluation from the point of view of thermal mechanical loading and mainly by the character of loading would be similar. The cask can be considered as a stationary vessel without dynamic effects of loading while during the reactor operation the RPV is subject to thermal mechanical dynamic regimes of loading with incomparably higher inside overpressure.

A critical point in the assessment of the cask life is unstable condition during the storage but possible emergency situation during the cask transport. For this reason all safety regulations emphasize verification of material properties in the initial conditions first of all from the point view of static and dynamic brittle fracture characteristics in comparison with the critical embrittlement value. These values were verified by calculation codes modelling mainly dynamic loading and/or crash test with factual drop and fire test on a model or possibly on actual cask bodies where the cask leak proof is the main aspect.

The TECDOC 717 requires keeping cask resistance under emergency conditions, which are simulated by a drop test of a cask body with a reference postulated crack at a material temperature of -40°C from the height of 9 m, or from the height of 1 m in to a spine, respectively. Therefore, also a relatively lower degree of material embrittlement can be limiting for the cask material resistance against a brittle fracture and thus also for its lifetime. The integrity of the cask shell is the most important criterion.

Figure 4. Leakage measuring equipment (model cask with helium indicator)

3.4. CERTIFICATE OF ACREDITATION

The performed calculation of ŠKODA 440/84 cask of brittle failure resistance met the required criteria of IAEA TECDOC 717 (3) for a design defect equal to 1/4 of the body wall thickness, i.e. 65 mm with the safety coefficient $K_B = 1.463$.

Based on documentation of the experimental results obtained and calculations carried out the Czech Authority (State Office for Nuclear Safety) issued permission for the commercial production and the use of the ŠKODA 440/84 containers was mentioned.

4. Monitoring Program and Evaluation of Cask Material Residual Life Time

The competitiveness of a cask is limited not only by the price but also by safety guarantees with minimum risks for the environment. The monitoring of cask properties can be positively projected also to a decision-making process for the extension of a cask design and safe lifetime. For the monitoring of container operational life the following general scheme was developed enabling a complex approach to the solution of prediction of cask material residual life time including the cask integrity assessment based on experimentally determined characteristics.

4.1. GENERAL SCHEME OF SOLUTION AND APPLIED METHODOLOGIES

The philosophy is based on the monitoring of material degradation with the utilization of a combination of non-destructive means and methods and the application of semi-destructive methods. The solution for the given design results from the following step sequence:

4.2. MAIN STEPS

4.2.1. Stress determination
Determination of maximum values of stress intensity coefficient for a reference postulated crack in a cask body for the most significant operational and emergency modes of a cask. In this respect we focus on the question of prescribed drop tests. This point is in relation with already performed calculations and drop tests of the cask body dummy.

4.2.2. One point bend test
The performance of One Point Bend Tests methodology with the aim to determine the dynamic fracture toughness of bodies with different degree of simulated degradation in conditions conservatively representing the load (stress) of material on the face of a crack in case of an actual cask body drop test, or a cask dummy with a postulated crack, respectively. It is not possible to perform drop tests of cask dummies with different material degradation degree. The results may be compared with the fracture toughness results performed on an instrumented hammer and with the static fracture toughness K_{IC}, J_{IC}, values.

4.2.3. Non-destructive methods
Non-destructive methods hardness measurement methods were chosen (HB and dynamic instrumented hardness's) and then methods for determination of complex electro magnetic hysteresis (ELMH) properties of material and through them detection of microstructure changes in material.

4.2.4. Semi destructive methods and punch test
Semi destructive methods are based on cautious sampling of operated cask material with Rolls Royce device using a so-called Boat Samples method. Such disc-shaped

sample not causing a notch effect is used for the manufacture of test microspecimens for examinations by punch test type and microspecimens for determination of conventional mechanical properties of a material sample, respectively.

Among others the punch tests allow to determine brittle fracture characteristics K_{IC} and J_{IC}, respectively, on actual material of an operated structure.

The application scheme of evaluation of cask resistance against brittle fracture in the course of its operation results consequently from the above mentioned general solution scheme (2.1) and the above project philosophy.

The goal is to determine correlation relations between the dynamic fracture toughness values determined within One Point Bend Tests, standard dynamic fracture toughness and static fracture toughness on one hand and other mechanical properties of material on the other hand and their transfer into computation codes - both as an experiment and based on the calculation within the fracture mechanics regularities.

4.2.5. Correlation characteristics

Determination of correlation characteristics between fracture toughness values determined in point 4.2.2. and methods of non-destructive and/or semi-destructive determination of characteristics in actually operated material (in this case it is a simulated degradation).

The stress is laid on finding correlation characteristics between the fracture toughness values determined using destructive methods and material characteristics determined within non-destructive and semi destructive examinations (see Chapter 2).

The results of successful project realization will be projected also to the increase of culture and technical level of nuclear equipment lifetime evaluation and prediction of safe lifetime.

5. Summary

All countries operating the nuclear power plants have to deal with the question of spent nuclear fuel storage. The application of cheaper universal casks designed for transport and permanent storage seems to be the most effective solution otherwise other treatment of spent nuclear fuel will be developed on usable level. The casks have to meet a number of requirements resulting from international regulations for equipment working with radioactive materials. Materials used for the manufacture of casks are chosen and tested so that they meet all these demanding regulations assuring a long-term safety operation. A critical point when assessing the cask and container safety and life have to be seen not in a standard conditions kept during the spent fuel storage in the storage facility but in possible extreme emergency situations during the transport. For this reason all safety regulations strongly emphasize verification of material properties in the initial state, first of all from the point of view of static and dynamic brittle-fracture characteristics. ŠKODA Nuclear Machinery plc paid a special attention to this question and all performed tests including drop tests met the requirements of IAEA TECDOC-717 regulation.

There are some advantages of ŠKODA design (ŠKODA approach) that may be taken into account: Number of fuel assemblies – 84 with average burn-up 43 500 MWd/tU. Double shell cask consisting of massive design, manganese cast steel body with built-in removable stainless steel canister with a special basket for spacing of spent fuel assemblies. This design provides total protection against the leakage of radioactive materials. Double-purpose cask, meaning for both the transport and storage, acceptable with respect to all considered regimes (severe accidents during transport, failure of cask cooling due to accident when the cask is buried by debris etc.).

Another important aspect of the container tested is its compatibility, possibility and monitoring. It means that the container is fully compatible with NPP equipment – the cask is designed with respect to the existing equipment in the transport container shaft. Possibility to handle spent fuel placed in the canister – it is designed so that it enables handling in reprocessing plant in Sellafield (Great Britain), or La Hague (France). Possibility to monitor cask leakage having a system of pressure and temperature measurements that is designed so that it provides actual data for the operator about sealing barriers condition for the whole lifetime of the cask within new monitoring methods (non destructive and semi-destructive methods).

Acknowledgements

Main part of fracture resistance assessment of the container cask material was carried out by Brittle Fracture Group at the Institute of Physics of Materials, Brno and within the common NATO SfP project also by collaborating groups from the University of Metz, France, the Bay Zoltán Institute for Logistic and Production Systems, and the University of Miskolc, Hungary. The results obtained substantially contributed to successful container accreditation procedure. Having in mind that NATO has supplied financial support to this NATO project our thanks belongs also to this program support.

References

1. IAEA TECDOC - 717, (1993), Guidelines for safe design of shipping packages against brittle fracture, revision from June 1996, IAEA Vienna.
2. H. P. Winkler, (2002), IAEA requirements for cask testing and short summary of developments at GNB, Contribution in this Volume.
3. Helms R., and Ziebs J., (1987), Nodular cast iron containers - results of materials evaluations 1981-1987, Proc. of Seminar held at BAM On Containers of Radioactive Materials Made from Nodular Cast Iron.
4. Baer, W., Putsch G., Michael, A., (1995), Investigations on the fracture mechanical behaviour of ferritic nodular cast iron in dependence on the applied load and temperature, RAMTRANS, Vol. 6, No. 2/3, pp. 149-154.
5. Sappok, M. (1987) Heavy weight ductile iron castings for the nuclear and machine building industry in the range up to 200 tons, 54th Int. foundry congress, New Delhi, 1987, p. 24.
6. Warnke, E.P., Bounin D. (1997) Fracture mechanics considerations concerning the revised IAEA-TECDOC-717 guidelines, Trans. of 14th Int. Conference on Structural Mechanics in Reactor Technology, SMiRT 14, Lyon, GMW/6, pp. 571-578.
7. Brynda, J., Hosnedl, P., Jilek, M., Picek, M., (1999), Material issues in manufacturing and operation of transport and storage spent fuel casks, Trans. of 15th Int. Conference on Structural Mechanics in Reactor Technology, SMiRT 15, Seoul, D07/4, pp. III- 247-253.

14

8. Saegusa, T. et al., (1995) Application of IAEA TECDOC 717 to packagings and comparison with reactor vessels, RAMTRANS, Vol. 6, No. 2/3. pp. 127-131.
9. Manufacture of Cask Forgings for Spent Nuclear Fuel Transport, (1985), Kawasaki Steel Technical Report No.13.
10. Urabe N., Arai, T. (1995), Fracture toughness of DCI casks and prediction by small specimen test, RAMTRANS, Vol. 6, No 2/3, pp. 171-174.
11. Gray, I. L. S. et al, (1995), Application of IAEA TECDOC 717 to the assessment of brittle fracture in transport containers with plastic flow shock absorbers, RAMTRANS, Vol. 6, No. 2/3, pp. 183-189.
12. Holman, W.R., Langland R.T. Recommendations for protecting against failure by Brittle fracture in ferritic steel shipping containers up to four inches thick.
13. Moulin, D., Yuritzin, T. and Sert, G. (1995), RAMTRANS, Vol. 6, No. 2/3, pp. 145-148.
14. Sorenson, K.B. et al (1993) results of sandia national laboratories MOSAIK cask drop test program, PATRAM 92, Proc. of Symp., Yokohama, Tokyo.
15. Sorenson, S.K. & McConnel, P., (1995), A survey of codes and standards in the USA relevant to the prevention of brittle fracture in cask containment boundaries, RAMTRANS, Vol. 6, No. 2/3, pp. 133-136.

MECHANICAL PROPERTIES OF CAST MN STEEL AFTER INTERCRITICAL HEAT TREATMENT AND MICROALLOYING

L. KRAUS[1], S. NĚMEČEK[2], J. KASL[3]

1 – COMTES FHT s.r.o., Borská 47, 320 13, Plzeň, Czech Republic
2 – NTC, University of West Bohemia, Univerzitní ul., Plzeň, Czech Republic
3 – ŠKODA RESEARCH Ltd., Tylova 57, 316 00, Plzeň, Czech Republic

Abstract: This study deals with intercritical heat treatment optimization of cast low-carbon manganese steel. The steel has been intended for a spent nuclear fuel container and low temperature applications. Proposed austenitizing treatment has been examined by means of thermal expansion measurements. Furthermore, standard heat treatment has been compared with intercritical heat treatment in terms of proper microstructure and optimum transition temperature. An influence of microadditions of Ti and V on properties of as-cast Mn steel (Czech standard CSN 422707 grade) was also analysed. Relationships among individual addition-contents, values of mechanical properties, and the real microstructure of samples are described here.

Keywords: Mn cast steel, mechanical properties, austenitization kinetics, heat treatment, microalloying.

1. Introduction

A key problem in production of thick-walled components, such as pressure vessels and container casks, is achievement of optimum combination of mechanical properties and fracture resistance throughout the wall thickness. It is generally known that in wall midthickness fracture toughness drops to about one half of its value in surface locations, despite optimum ultimate strength and yield strength of the nuclear pressure vessel steel. This is partly a consequence of typically poorer metallurgical quality of central parts of thick-walled castings and partly due to different cooling rates observed in these locations.

The first problem, i.e. the presence of critical non-metallic inclusions and atoms of impurities dissolved in matrix, can be effectively minimized by pure steel production processes, including secondary metallurgy. However, undesirable effects can still be observed, such as different contents of impurities and alloying elements (chromium) due to segregation, and different volume fractions of pores (i.e. different density of material), which cause differences in material's behaviour during heat treatment and under mechanical loading.

I. Dlouhý (ed.), Transferability of Fracture Mechanical Characteristics, 15–32.
© 2002 *Kluwer Academic Publishers. Printed in the Netherlands.*

The latter problem, i.e. the difference in cooling rates of more than one order between surface and central locations, results in different conditions for austenite decomposition. Consequently, there is variation in attained microstructures, from predominantly bainitic surface layer to the mainly pearlitic region of plate midthickness. Obviously, fracture resistance differs in these locations, possibly reaching critical values in midthickness, if coarse lamellar pearlite forms in this central part of a thick-walled component.

This problem may become significant to spent nuclear fuel container cask production if low alloy steel with predominantly ferritic matrix is applied. Since the container cask must exhibit high structural integrity in practical operations, such as transport and storage of spent nuclear fuel, it has been suggested to modify the heat treatment processing with the aim of elimination of this problem.

The only way to avoid the negative effect of different cooling rates is to achieve ferrite stabilization, and have only a small fraction thereof transformed to austenite.

Since 1985, cast C-Mn steel (according to Czech Standard designated as 422707) has been considered as a perspective material for nuclear waste casks in SKODA HOLDING company. Development of suitable steel for large castings (up to 60 tonnes), which would possess high values of CVN notch toughness at low temperatures (down to −60 °C), has been one of our main tasks. The required parametres (both mechanical and economic) have been preliminarily satisfied by the mentioned CSN 422707 steel (similar to the steels C23-45 BL according to ISO, GS-CK16 in Germany and Gr LCC in the USA). Low-alloyed ferritic-pearlitic manganese steel (1.00 − 1.60 % Mn, max. 0.12 % C) with binding value of Charpy-V notch toughness at low temperatures is concerned. The development is focused on Charpy V-notch toughness (KV) enhancement in lower shelf region and reproducible achievement of the needed cask high lifetime. Two methods were selected to improve the quality of the original base material: microalloying with V, Ti and Ni, and optimisation of intercritical heat treatment [1].

The study aims at summarizing the knowledge obtained during optimization of microstructure of cast ferritic steel as a material for production of spent nuclear fuel container casks. The optimization involved development of intercritical heat treatment and optimization of chemical composition of the steel by means of microalloying. Since microstructure of the steel has a key importance to resulting fracture resistance of the cask, it has become a focus of the study rather than description of transformation kinetics and ensuing microstructural changes.

In order to suggest the heat treatment programme it was necessary to firstly measure the steel transformation characteristics. For these purposes, casting and intercritical heat treatment of the testing plate from the base material, i.e. the CSN 422707 steel were carried out. Then small castings (1.5 kg) with different microalloying additions (V, Ni, Ti) were cast [3, 4].

Following the transformation kinetics analyses, considerable ferrite grain refinement, yield strength increase from 300 to 400 MPa, yield-strength/ultimate-tensile-strength ratio enhancement, and notch Charpy V-notch toughness KV (-20°C) improvement were achieved. On this small sample series mainly the relationship between microalloying and mechanical properties was then examined.

2. Experimental Methods and Results Obtained

2.1 KINETICS OF AUSTENITIZING

Kinetics of the ferrite → austenite phase transformation were measured by means of dilatometric technique. Heating at a constant rate of 10°C min^{-1} within the temperature range from 20 up to 600°C followed by different rates from 1 up to 10°C min^{-1} within the temperature range from 650 up to 930°C was applied. Kinetics of austenite decomposition during cooling were evaluated too. During the heating and cooling the dilatation curves had standard characters (heating dLh, cooling dLc).

In order to evaluate the kinetics of the α→γ phase transformation it was necessary to determine those parts of dilatation characteristics of ferrite and austenite that could be approximated by linear segments. In case of ferrite it is the range of approx. 600 – 700°C (heating curve dLh$_{(lin)}$), and in case of austenite the temperature range of approx. 930 – 785°C (cooling curve dLc$_{(lin)}$).

The instant volume fraction of austenite is given by the relationship:

$$V_\gamma = \frac{(dLh_{(lin)} - dLh)}{(dLh_{(n)} - dLc_{(lin)})} \tag{1}$$

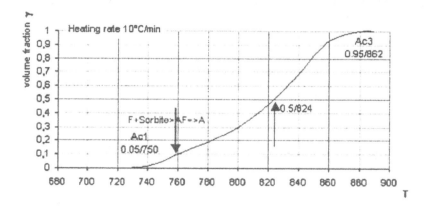

Figure 1. Ac$_1$ and Ac$_3$ temperatures for the heating rate of 10°C/min

18

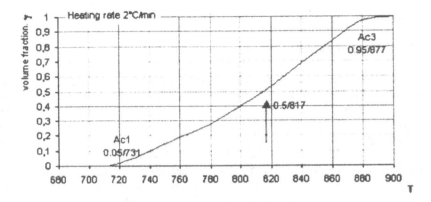

Figure 2. Ac₁ and Ac₃ temperatures for the heating rate of 2°C/min

It is possible to read Ac_1 and Ac_3 temperatures from the given dependencies for the selected limit volumes Vγ. Usually the values of 5% and 95% are used. When the measurement is sensitive enough, it is possible to use the values of 1% and 99%. The measured data is shown in graphic dependencies in the *figures 1 – 4*. In *figure 3*, the transformation kinetics are compared for the selected range of heating rates. The differences between the rates of 1°C min⁻¹ and 2°C min⁻¹ were not important considering the measurement accuracy. We assume the heating rate of a casting body just in the field of the mentioned rates, i.e. 1-2 °C min⁻¹.

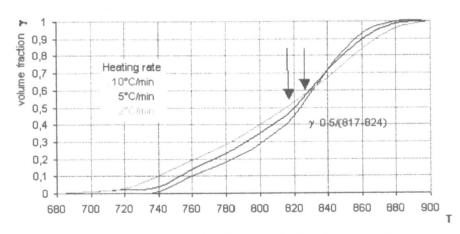

Figure 3. Comparison of the course of transformation for various heating rates

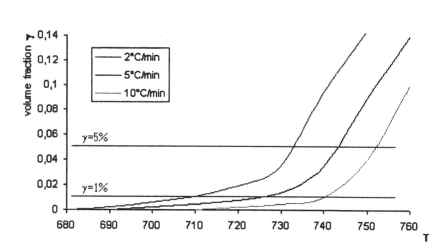

Figure 4. Beginning of transformation for various heating rates

The dependence of the "ferrite + pearlite → austenite" transformation kinetics on the heating rate is controlled by several factors. It is especially the character of the original microstructure (grain size, size and spatial distribution and morphology of M_3C carbides and their chemical composition) and, last but not least, the general chemical composition and corresponding diffusion characteristics. Although there exist general theoretical descriptions of austenitizing kinetics, their current numerical form is usually determined experimentally by means of an adequate measurement method.

The transformation starts at a certain level of overheating above the equilibrium temperature when certain magnitude of a thermodynamic driving force is achieved, which can be expressed by the difference between free energies $\Delta G_{\alpha \to \gamma}$. At the same time sufficient number of nuclei of a new phase must be available. *Figures 3* and *4* show that Ac_1 temperature decreases, when the heating rate declines, while the end of transformation is shifted toward higher temperatures (possible control by means of the structure and chemical composition). These facts can be explained by lower number of initial growth nuclei, hence the growth of larger grains and resulting longer diffusion paths, which is, on the whole, in accord with the fact that the transformation rate $(dV_\gamma/dT)_{dT/d\tau=const}$ is less dependent on the heating rate in the transformation range of $0<V_\gamma<0.5$. When the transformation rate is greater, approximately above $V_\gamma>0.6$, the differences in transformation rates are well visible (see *figure 3*).

Determination of the $V_\gamma=f(T)_{dT/d\tau=const}$ dependence is important for the practical use. It is possible to choose various model functions, most easily the polynomial of the n-th grade. For the complete description of the kinetic curve the polynomial of the 5th and 6th grade is sufficient, while for the reduced description ($0.2<V_\gamma<0.80$), the one of the 3rd grade suffices. The last mentioned model was used for the description of transformation in the evaluated steel. The measured values fit the calculated austenitizing curve very well.

20

Diagram of non-isothermal austenite decomposition (see *figure 5*) was obtained by measurements of cooling curves for the given steel. According to thus built CCT diagram, the heating temperatures in the regimes of heat treatment were determined.

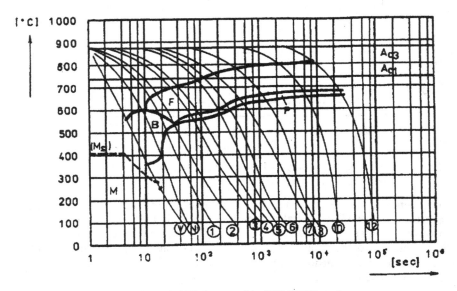

Figure 5. CCT diagram of the CSN 422707 steel

2.2 PROPERTIES OF BASIC STEEL

Testing plate of the size of $550 \times 350 \times 750$ mm (height × width × length) was cast into a sand mould. Chemical composition (wt %) of the plate is given in table I. Prisms of dimensions $60 \times 85 \times 120$ mm were cut from the plate.

TABLE I. Chemical composition of the test melt

	C	Mn	Si	P	S	Cr	Ni	Cu	Mo	Al
wt%	0.090	1.180	0.370	0.010	0.025	0.120	0.290	0.290	0.030	0.028

The optimum microstructure for achieving high values of toughness is formation of homogeneous sorbitic microstructure within fine former austenite grains. Since the steel is not capable (even at dimensions of used specimens) of achieving a hardened matrix throughout the volume, owing to its chemical composition, intercritical heat treatment (ICHT) was proposed as one of the heat treatment alternatives.

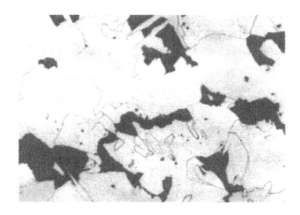

Figure 6. Typical microstructure of the investigated steel after casting

Heat treatment applied to experimental blocks simulated the treatment of large castings. According to equations found in the literature, the temperature of ICHT was 780°C for achieving about 25% recrystallization of the studied steel matrix. As another alternative even higher IC temperature (820°C) was chosen, more than 50% of matrix volume has transformed. Cooling rates of samples, given in table II, were calculated for various cooling media (air, water, oil) and the casting dimensions by means of simulation programme.

TABLE II. Overview of heat treatment of specimens

Sample	Heat treatment condition
I	930°C/4h/2000°C.hr^{-1}
	930°C/4h/29000°C.hr^{-1}
	650°C/12h/200°C.hr^{-1}
II	930°C/4h/2000°C.hr^{-1}
	930°C/4h/29000°C.hr^{-1}
	780°C/3h/18000°C.hr^{-1}
	650°C/12h/200°C.hr^{-1}
III	930°C/4h/2000°C.hr^{-1}
	930°C/4h/29000°C.hr^{-1}
	820°C/3h/25000°C.hr^{-1}
	650°C/12h/200°C.hr^{-1}
IV	880°C/4h/2000°C.hr^{-1}
	880°C/4h/27000°C.hr^{-1}
	650°C/12h/200°C.hr^{-1}

TABLE III. Basic mechanical properties values after the heat treatment

Sample	Yield strength (R_e) [MPa]	UTS (R_m) [MPa]	Ductility (A_5) [%]	Red. of area (Z) [%]	CVN impact energy (KCV) +20 C [J.cm^{-2}]	CVN impact energy (KCV) -60 C [J.cm^{-2}]	FATT (t_{50}) [°C]
Values prescr. by CSN standard	min. 270	420 570	min. 25	min. 50	min. 80	min. 30	-
I	347	474	31.7	71.6			
	378	499	29.3	69.8	159; 161	18; 35	-45
	375	478	32.3	71.6			
II	336	474	36.7	69.8			
	340	478	38.3	75.0	186; 176	50; 96	-52
	340	478	38.0	78.2			
III	333	467	33.3	66.0			
	343	474	32.7	66.0	141; 194	58; 53	-35
	325	470	36.7	71.6			
IV	343	467	35.0	76.6			
	357	481	31.7	75.0	188; 184	54; 45	-25
	347	474	30.0	75.0			

On ICHT-processed specimens impact tests were performed at room temperature. The fracture surfaces were analysed by scanning electron microscopy. In all of the rods transcrystalline cleavage with little fraction of transgranular ductile bridges was found. Both the type of damage and the size of facets were similar in all samples. Only the size of shear lip zone below the notch was slightly different in separate specimens. Sulphidic inclusions of the type II, distributed as eutectic form, were found on the fracture surfaces of several specimens (see *figure 7*).

Great difference in terms of microstructure between the microstructure of steel after casting and after heat treatment has been observed. Finer grain size and better homogeneity of distribution of carbide particles (cementite) had been achieved by heat treatment, especially the ICHT. The apparent difference in the resulting character of microstructure was also discovered between the standard heat treatment (sample I) and the ICHT (samples II and III) (*figures 8, 9*).

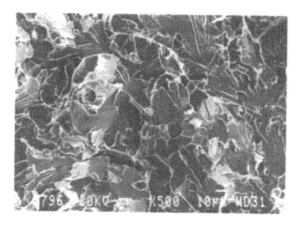

Figure 7. Typical fracture surface, scanning electron micrograph of the CVN specimen

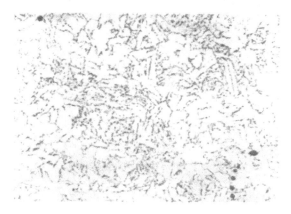

Figure 8. Microstructure of sample I (standard heat treatment)

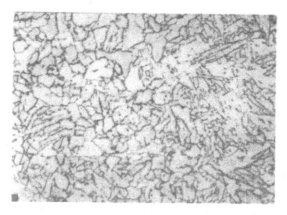

Figure 9. Microstructure of sample II (ICHT)

24

The main apparent difference between the fully austenitizing heat treatment (sample I) and the intercritical heat treatment consisted only in the cementite distribution. In the sample II (less in the sample III) its globular particles were concentrated into narrow strips along the ferritic grain boundaries. This phenomenon can be observed both in the micrographs from the light microscope, and the transmission electron microscope (extraction replicas) in *figures 10* and *11*.

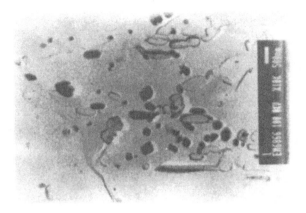

Figure 10. Sample I (standard heat treatment), electron micrograph

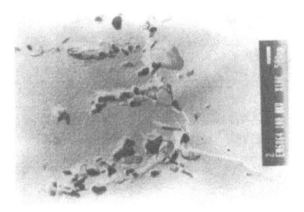

Figure 11. Sample II (ICHT), electron micrograph

In order to find more distinctive differences between microstructures resulting from individual heat treatment procedures, the metallographic samples were analysed by means of scanning electron microscopy in the backscattered electron mode. This method is suitable for revealing differences between crystallographic subgrain orientations inside the ferritic grain documented by light microscopy. The size of ferritic grains found in samples I and II ranged from 5 up to 30 μm. Sample II showed ferritic grains of a smaller size – from 3 up to 20 μm. The size of crystallographic blocks – subgrains – inside the ferritic grains was identical (1-5 μm) in all samples but the fraction of their occurrence differed. While in the sample II virtually all of the grains were divided into subgrains, in the sample I more than one half of the grains was divided in this manner and in the sample III only a small portion of grains exhibited this feature. Identical results were obtained from the analysis of thin foils by means of TEM.

2.3. MICROALLOYING OF MN BASIC STEEL

Blooms of dimensions $400 \times 400 \times 250$ mm (the third dimension corresponds to cask container wall thickness) were cast in order to verify and improve the achieved properties. The contents of microalloying elements were optimized (two levels were selected) and the number of cast blooms was reduced in contrast to small testing castings mentioned in introduction.

The blooms were divided into four groups, with each group including two pieces with two levels of carbon contents of 0.12% or 0.25% (blooms with carbon 0.25% contents are designated with the letter "A"). The first group (blooms "1" and "1A") contains materials without microalloying elements to simulate the cast state of CSN 422707 steel. Blooms "2, 2A" contain vanadium microalloying addition only, while blooms "3, 3A" contain only titanium addition. Material of blooms in the fourth group has higher manganese content, but no further microalloying additions. Fourth group contains newly made representatives of the material, which had been formerly evaluated as sample No. 11, which – as the only one – possessed bainitic matrix. Niobium was not included in the experimental plan, since its low content influence is very similar to that of titanium, in that higher content of niobium (above 0.2%) causes strong low-temperature notch-toughness decrease.

The results of chemical composition analysis are shown in the table IV. The temperature cooling cycle was measured during the casting of blooms no. 1 and 1A by means of thermocouples placed on selected points. The goal of the procedure was to record the time – temperature dependence in the points, for which numerical simulation of solidification had been performed. After cutting the blooms to samples of dimensions of $100 \times 100 \times 250$ mm, special sets of samples were assembled. Each set contained one sample taken from each bloom. The sets were processed by different heat treatments, which were intended to simulate the final heat treatments of the cask casting. The four modified hea treatment processes (with the exception of the standard HT 1) were as follows:

- **HT1** – normalizing + tempering: $930°C/3h/100°C.h^{-1} + 650°C/12h/100°C.h^{-1}$
- **HT2** – quenching + tempering: $930°C/3h/100°C.h^{-1} + 900°C/3h/furnace\ 100°C + 650°C/12h/100°C.h^{-1}$
- **HT3** – intercritical quench hardening without tempering: $930°C/3h/100°C.h^{-1} + 900°C/3h/furnace\ 100°C + 780°C/3h/furnace\ 100°C.h^{-1}$
- **HT4** – intercritical hardening + tempering: $930°C/3h/100°C.h^{-1} + 900°C/3h/furnace\ 100°C + 780°C/3h/furnace\ 100°C + 650°C/12h/100°C.h^{-1}$
- **HT5** – homogenization annealing + normalizing + tempering: $1030°C/5h/furnace\ 930°C/3h/100°C.h^{-1} + 650°C/12h/100°C.h^{-1}$

TABLE IV. Chemical composition of blooms with different
microalloying additions (in wt%)

Sample	C	Mn	Si	P	S	V	Ti
1	0.16	1.08	0.41	0.018	0.011		
1A	0.27	1.01	0.38	0.018	0.010		
2	0.16	1.16	0.31	0.019	0.010	0.13	
2A	0.27	1.20	0.30	0.020	0.013	0.13	
3	0.12	1.12	0.28	0.016	0.009		0.017
3A	0.26	1.12	0.27	0.018	0.010		0.025
4	0.14	2.00	0.28	0.017	0.010		
4A	0.28	2.00	0.29	0.019	0.010		

The metallographic analysis, tensile test and measurements of CVN impact energy with the complete evaluation of the transition curve were performed for each combination of cast bloom and heat treatment.

2.4. MECHANICAL AND METALLOGRAPHIC ANALYSIS OF MICROALLOYED MATERIALS

The microalloying has no substantial influence on yield strength value as observed. In this respect, vanadium is more suitable than titanium. However, a much more important effect is caused by changing carbon content. Increasing the carbon content from 0.12 to 0.25% yields an enhancement of at least 10 per cent in yield strength R_e. Similar phenomenon is exhibited by tensile strength: a 70 MPa enhancement (15%) was found after doubling carbon content. The ductility, however, remains essentially unchanged. Different situation can be observed in case of CVN impact energy. Upon increasing the carbon content the value of impact energy (in HT1- and HT2-processed specimens) decreased to almost 50 % of original value in the low carbon material. Adding vanadium to the low carbon material causes impact energy to reach almost twice as high value, while titanium addition results in slight decrease in this parameter. Both vanadium and titanium are suitable additions with respect to CVN impact energy. After raising the carbon content in the material, even stronger effect is caused by vanadium. Both V and Ti decrease the transition temperature with vanadium yielding more suitable results again. Increasing carbon content brings negative results in all mentioned cases.

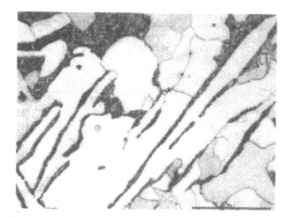

Figure 12: Sample 1.1, base material, 0.16% C, normalizing (HT1)

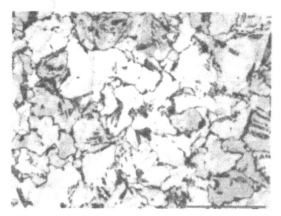

Figure 13: Sample 1.4, base material, 0.16% C, intercritical HT4 treatment

The heat treatment differences influence neither the yield strength R_e, nor ultimate tensile strength R_m. The only exception has been observed in HT3, where ultimate strength – R_m increase after Ti addition. With regard to transition temperature, the intercritical heat treatment appeared to be better (HT3, HT4). Probable reasons include grain refinement and evolution of more uniform microstructure. The most favourable CVN impact energy value of both microalloyed and unalloyed samples has been achieved by HT4 processing. In this case the positive influence of tempering is apparent, although some literature sources claim that its effect is weak.

The presence of 2% Mn increases hardenability of the steel considerably (thus having a positive influence on strength properties). Resulting transition temperature is undesirably high, which also means low toughness. This material was included in the experimental plan in order to evaluate influence of manganese on stabilization of retained austenite after different heat treatments. Its application to cask casting is not currently favoured. Optimization (by microalloying) and utilization of the approved steel CSN 422707 is more probable.

TABLE V. Mechanical properties of HT1-processed samples

Sample	Yield strength (R_e) [MPa]	Ultimate tensile strength (R_m) [MPa]	Ratio R_e/R_m [MPa]	Ductility (A) [%]	CVN impact energy +20°C [J]	FATT (T_t) [°C]
1.1	297	480	0.618	33	54;35	50
1A.1	333	585	0.569	26	24;21	70
2.1	320	502	0.637	34	104;112	15
2A.1	360	583	0.617	32	57;52	40
3.1	300	471	0.637	30	53;48	45
3A.1	330	560	0.589	25	32;28	50
4.1		551		29	92;78	40
4A.1	404	612	0.660	23	46;44	40

TABLE VI. Mechanical properties of HT2-processed samples

Sample	Yield strength (R_e) [MPa]	Ultimate tensile strength (R_m) [MPa]	Ratio R_e/R_m [MPa]	Ductility (A) [%]	CVN impact energy +20°C [J]	FATT (T_t) [°C]
1.2	333	501	0.664	31	83;96	50
1A.2	384	622	0.617	28	28;33	70
2.2	355	502	0.707	32	123;127	15
2A.2	404	599	0.674	28	53;56	40
3.2	325	468	0.694	30	95;90	10
3A.2	355	552	0.643	27	45;37	40
4.2	384	537	0.715	27	82;81	30
4A.2	464	622	0.746	24	38;48	50

TABLE VII. Mechanical properties of HT3-processed samples

Sample	Yield strength (R_e) [MPa]	Ultimate tensile strength (R_m) [MPa]	Ratio R_e/R_m [MPa]	Ductility (A) [%]	CVN impact energy +20°C [J]	FATT (T_t) [°C]
1.3	316	557	0.567	32	83;79	40
1A.3	375	652	0.673	26	46;45	40
2.3	351	563	0.623	34	122;116	0
2A.3	407	649	0.627	31	117;89	10
3.3	333	559	0.595	31	60;65	30
3A.3	401	631	0.635	24	54;53	20
4.3		870		20	15;11	150
4A.3	Samples with failure				7;7	105

TABLE VIII. Mechanical properties of HT4-processed samples

Sample	Yield strength (R_e) [MPa]	Ultimate tensile strength (R_m) [MPa]	Ratio R_e/R_m [MPa]	Ductility (A) [%]	CVN impact energy +20°C [J]	FATT (T_t) [°C]
1.4	310	502	0.617	35	149;123	20
1A.4	354	590	0.600	32	59;60	30
2.4	325	485	0.670	37	182;219	-10
2A.4	380	565	0.672	32	104;102	5
3.4	305	470	0.649	37	108;102	10
3A.4	330	539	0.612	33	49;60	25
4.4		619		31	64;65	20
4A.4	450	612	0.735	26	45;43	40

TABLE IX. Mechanical properties of HT5-processed samples

Sample	Yield strength (R_e) [MPa]	Ultimate tensile strength (R_m) [MPa]	Ratio R_e/R_m [MPa]	Ductility (A) [%]	CVN impact energy +20°C [J]	FATT (T_t) [°C]
1.5	311	502	0.620	35	49	60
1A.5	354	590	0.600	32	23	120
2.5	325	485	0.670	37	153	0
2A.5	380	565	0.673	32	61	40
3.5	305	471	0.648	37	78	15
3A.5	331	539	0.614	33	34	65
4.5	373	619	0.603	31	24	110
4A.5	417	578	0.721	26	40	110

Metallographic specimens were cut from broken samples perpendicularly to the fracture surface after impact bending test. It was observed that there was a portion of pearlitic component in ferritic matrix. The grain size was determined by means of the "Lucia G" digital image analysis system. The results are summed up in table X and in the *figures 12 – 16*.

TABLE X. Volume fraction of pearlite and measured grain size in samples processed with HT1 procedure

Sample	Volume Fraction Pearlite	Grain Size Dia. [μm]	Sample	Volume Fraction Pearlite	Grain Size Dia. [μm]
1.1	22.7	18.4	1.4	25.7	8.1
1A.1	49.9	24.3	1A.4	50.5	8.0
2.1	14.8	13.5	2.4	5.8	6.1
2A.1	50.5	13.7	2A.4	50.1	5.0
3.1	18.8	17.6	3.4	10.8	7.9
3A.1	40.1	11.7	3A.4	20.7	6.1
4.1		Bainite	4.4		Bainite

30

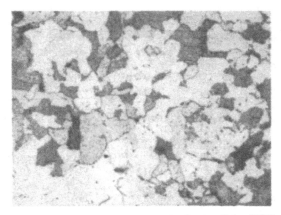

Figure 14. Sample 2.1, 0.13% V, 0.16% C, normalizing (HT1)

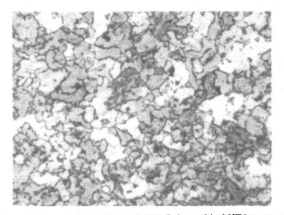

Figure15. Sample 2.4, 0.13% V, 0.16% C, intercritical HT4 treatment

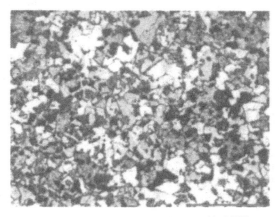

Figure 16: Sample 2A.4, 0.13% V, 0.27% C, intercritical HT4 treatment

3. Conclusions

Experimental programme has proved the suitability of intercritical heat treatment (ICHT) for processing of CSN 422707 low-carbon manganese steel. By application of ICHT, fine-grained matrix and suitable distribution and morphology of carbidic particles were achieved as a result of partial recrystallization and subsequent quench-hardening. When ICHT-treated, the steel exhibits higher toughness at low temperatures in comparison with standard treatments at higher temperature (i.e. temperature recommended in CSN standards) or slightly above Ac_3. The lower temperature (780°C) has been shown to be much more suitable for the intercritical hardening (resulting in reaustenitization of about 25% of matrix volume). At higher intercritical temperature, i.e. at 820°C, more than 50% of matrix volume transformed, thus significantly lowering the progressive effect of ICHT. When the fraction of newly-formed austenite is low, compared to the still untransformed surrounding matrix, it contains relatively high concentrations of carbon and alloying elements (in particular manganese). During subsequent cooling this leads to formation of acicular structures, which improve the strength of the surrounding ferritic matrix upon quench hardening. Large castings, which practically cannot be quench-hardened to obtain microstructures of martensite or martensite + bainite (as optimum initial states for achieving low transition temperature), require non-standard heat treatment procedures. Intercritical heat treatment (ICHT) has been identified as one of them.

Among samples of base material without microalloying additions, better properties have been achieved in those of the steel with higher carbon content. Highest tensile strength and, at the same time, lowest ductility have been exhibited by the samples of melt no. 4 with 2% Mn and with microstructures corresponding to hardened state. In terms of transition temperature, which is one of most critical characteristics of container casks, samples of melt 4 also exhibit high values, by which they resemble the unalloyed samples of melt 1 (tables V – IX).

Microalloying additions affect mechanical properties in a positive way. After normalizing (HT1) and quench-hardening + tempering (HT2), the strength parameters of microalloyed melts with higher carbon contents (designated "A" in table IV) increased. Highest impact resistance is observed in samples microalloyed with vanadium and with low carbon content (designated "2" in table IV), regardless of heat treatment applied. Raising the carbon content improves yield strength and tensile strength, but lowers impact resistance and increases the transition temperature at the same time.

Results of analyses foster the assumption that doubling carbon content causes an approximately twofold increase in pearlite volume fraction. Higher manganese content enhances hardenability and causes pearlitic colonies in matrix to be replaced with bainitic regions. Intercritical heat treatment (HT4 – section 2.3) has positive influence on grain size in terms of grain refinement. Compared with the HT1 procedure, it produces half grain size, which proves the influence of heat treatment more significant than that of microalloying. Addition of vanadium produces smaller grain in cast steel than titanium.

In general, it is possible to conclude that most valuable improvements in impact resistance, transition temperature and yield strength/ultimate strength ratio (R_e/R_m) were achieved by addition of 0.13% V to material with low carbon content. The HT4 processing was selected as the best among heat treatment procedures evaluated, as it lowered the transition temperature by 5°C (in the sample 2.4). It is necessary to emphasize that processing and resulting microstructures of samples simulated the state of the center of container cask wall thickness, i. e. the most unfavorable location. With this in mind it is possible to conclude that the achieved values are very suitable. Utilizing this type of heat treatment procedure, combined with optimized microalloying concept, provides realistic assumption that reproducible desirable properties in large castings of nuclear waste casks will be achieved.

Acknowledgements

The investigations had been carried out with a support of grant No. 106/99/0643 of the Grant Agency of the Czech Republic and the NATO SfP project No 972655.

References

1. Němeček S. and Kraus L. (1999) Materials and Technological Properties of ŠKODA Container, *Research report ŠKODA RESEARCH Ltd. VYZ 0357/99*, Plzeň, Czech Republic.
2. Kraus L., Kasl J. and Němeček S. (2000), Kinetics of Cast Mn Steel Austenitization during Intercritical Heat Treatment, *Proceedings of Metal 2000 conference*, Ostrava, Czech Republic.
3. Němeček S., Kasl J. and Kraus L. (2000), Microstructure and Properties of V, Nb, Ti – Microalloyed Steels after Normalized Annealing, *Proceedings of Metal 2000 conference*, Ostrava, Czech Republic.
4. Kraus L., Kasl J., Němeček S. and Nový Z. (2000), Optimization of Cast Mn Steel Properties with Help of Intercritical Heat Treatment, *Proceedings of SHMD 2000*, Opatija, Croatia.
5. Kraus L. (1996), Cast Materials for the SKODA Container, *Research Report ŠKODA RESEARCH Ltd. VZVU 1141*, Plzeň, Czech Republic.
6. Zeman, J. (1996), Kinetics of Steel Austenitizing, *Research Report CONMET*, Brno, Czech Republic.
7. Němeček S., Kraus L., Macek K. and Cejp J. (2001), Influence of Heat Treatment on Microstructure and Properties of Cast Microalloyed Steel, *Proceedings of EUROMAT 2001*, Rimini, Italy.
8. Kraus L., Němeček S. and Kasl J. (2001), Kinetics of Austenitization of Cast Mn Steel and its Properties after Intercritical Heat Treatment, *Proceedings of EUROMAT 2001*, Rimini, Italy.

IAEA REQUIREMENTS FOR CASK TESTING AND SHORT SUMMARY OF DEVELOPMENTS AT GNB

H. P. WINKLER
GNB Gesellschaft für Nuklear-Behälter, Hollestr. 7A, Essen, D-45127 Germany

Abstract: For high-level radioactive materials, especially for spent fuel from nuclear power stations, the CASTOR® casks were developed and continue to be designed and produced by GNB to meet the necessary high safety requirements economically. The inherently safe concept of dry storage in transport casks was selected and supported by a lot of nuclear power station owners worldwide.

Keywords: ductile cast iron, storage cask, transport cask, CASTOR®, spent nuclear fuel, IAEA safety requirements, drop test,

1. Introduction

"GNB Gesellschaft für Nuklear-Behälter", with headquarters in Essen, Germany, was founded in 1991 and is a daughter company of "GNS Gesellschaft für Nuklear-Service" who has an annual turnover of approx. 80 to 90 million Deutsch marks. GNB's field of activities include: development, design, engineering, licensing, and manufacturing of casks for transport, interim storage and final disposal of radioactive waste.

As a base material for the cask body, ductile cast iron (DCI) was chosen, which is good for shielding, easy to machine, relatively inexpensive, and has been developed to a reproducible superior quality in cooperation with renowned foundries. To fulfil the IAEA requirements more than 70 drop tests and a multitute of fire tests Airplane crash simulations and Terrorist attak simulations had been done. In 1999 a real explosion of a liquid tank in front of a CASTOR® -cask was realised. For the extensive test programs for licensing, numerous model test casks and prototypes were fabricated according to this method. The tests were the basis for the licensing of the CASTOR®-family as a Type B (U) transport cask design and for the long-term interim storage as can be seen in *Figure 1*. Thus, the family of CASTOR® casks (Cast Iron Casks for Storage and Transport of Radioactive Materials) was born. The system is the reference solution for interim fuel storage both in Germany and abroad.

In Switzerland, the CASTOR® Ic was loaded in 1983 and is presently stored at the Paul Scherrer Institute in Würenlingen. In the USA, the CASTOR®s became the first dry storage casks licensed by the NRC in 1985 and are currently being used at the Surry

I. Dlouhý (ed.), Transferability of Fracture Mechanical Characteristics, 33–46.

34

Power Station facility owned by Virginia Power. Even these storage casks are fabricated in compliance with the IAEA Regulations for Transportation. The competent authority in France subjected ductile cast iron to an unusually intensive analysis for licensing as a material for spent fuel transport casks. This process, finished in 1987,

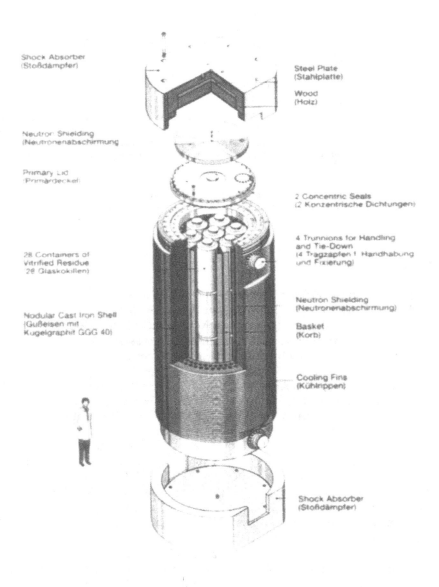

Figure 1. Details of CASTOR cask

led to the unrestricted approval of the corresponding CASTOR® for both domestic transports in France and international ones.

To date a large number of cask types in the CASTOR® family have been licensed for long-term interim storage in Germany and three types in the USA (CASTOR® Ic, V/21 and CASTOR® X). In France, interim storage of spent fuel from the Super-Phénix Fast Breeder Reactor on site has received basic approval. In its family of casks, GNB designs and delivers:

- casks for spent fuel from all types of reactors as well as for MTR's
- casks for high-level vitried and liquid wastes
- casks for unirradiated nuclear fuel and other radioactive materials
- the auxiliary handling equipment for these casks.

The cask material is not limited to ductile cast iron, but can be made of steel/lead (as yet for transport only in USA, FRG, and France). Forged steel designs are also possible. However as far as ductile cast iron casks are concerned, because the CASTOR® casks for spent nuclear fuel are dual purpose, i.e. for both the storage and transport of nuclear waste, this offers a tremendous advantage in the long-term aspects of handling and economics.

GNB has delivered over 700 CASTOR® casks for spent fuel and vitrified high active wastes for storage and transport. More than 550 of these are currently in interim storage (for example see the *Figure 2*), loaded with pebble bed reactor fuel, with vitrified high active waste, and a large number of casks have been loaded with more than 700 t of heavy metal in the form of LWR fuel.

Figure 2. Storage facilities of nuclear power station Dukovany, Czech Republic

Activities have branched out into the development of transfer systems for spent fuel, i.e. for the dry loading of spent fuel assemblies into a DCI cask. Such systems have already been delivered to Würenlingen, Switzerland and to the Super-Phénix Fast Breeder in Creys-Malville, France.

GNB has also full access to the storage operations know-how of GNS, who operates the storage facilities of Gorleben and Ahaus in Germany; furthermore GNB is always informed about the experiences by other storage facilities. Also, for the design and construction of storage facilities, GNB works together with GNS.

2. Advantages of the Cask Dry Storage Concept in CASTOR® Casks

The principle of the dry cask concept is based on the containment of the fuel assemblies in casks suitable for both transport and storage. The main characteristics of the CASTOR® cask design are (*Figure 3*):

- The cask body is made of ductile cast iron (DCI) to guarantee gamma shielding, leak tightness and protection against mechanical and thermal loads under both normal operational and hypothetical accident and test conditions.
- The monolithic cask body is cast in one piece and has no welds. This means that the casks are less expensive than those made of forged steel for comparable cask sizes.

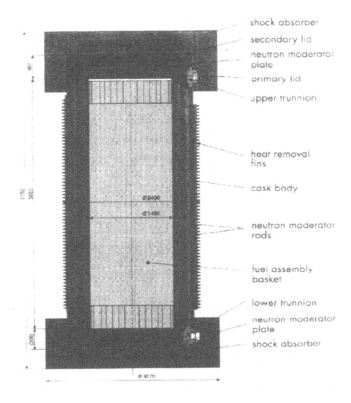

Figure 3. Axial section of CASTOR cask (transport configuration)

- The casting technology gives flexibility in design (cask types can be easily adapted to various requirements).
- DCI is easy to machine so that external cooling fins for heat removal can be directly integrated in the cask body.
- Neutron shielding is provided by polyethylene rods integrated in the cask body wall.
- Double barrier system realized by the lid systems consisting of primary and secondary lids sealed with metallic gaskets.
- Permanent leak tightness control is realized by a special pressure monitoring system.
- Qualification as a type B(U) package according IAEA regulations.
- The cask is independent of an energy supply, ducted cooling or other systems.
- The suitability of the cask has been demonstrated in a large number of tests. The cask is able to withstand the following accident conditions among others: aircraft crash (for some details see *Figure. 4*), drop of heavy loads, gas cloud explosion (details are shown in *Figure. 5*), fire, temporary burial, earthquake, drop (some examples are in *Figure. 6 a-c*) etc.

Figure 4. Airplane crash simulations

3. The CONSTOR® Family of Casks

The CONSTOR® cask, a modification of the CASTOR® cask technology from GNB, is a steel-concrete cask for spent nuclear fuel which uses the advantages of steel given by our experience in the past with metal casks as well as the advantages of heavy concrete for shielding purposes. It is a multipurpose system for storage, transport and possibly final disposal which keeps all options open for the future and minimizes the loading and unloading procedures necessary for handling – see *Figure 7*.

The cask itself has been designed to provide a high level of safety at a very moderate price, while at the same time being able to be manufactured in all countries with suitable industrial capabilities. The basis for the development was the large experience from the CASTOR®-casks. The cask itself consists of a steel-concrete base body with a double lid system. The "sandwich" cask body is made of an inner and outer steel liner, each 40 mm thick, welded to a head piece and filled in with heavy concrete. The double lid system can either be bolted leak-tight to the cask body or bolted/welded. A series of drop tests with the cask performed in May 1997. The design of the CONSTOR® was initially licensed for RBMK fuel. GNB has already delivered 40 CONSTOR® casks to the nuclear power plant IGNALINA in Lithuania.

Figure 5. Explosion test

A compared to other systems available, the CONSTOR® cask has the following bε lefits and advantages:
- inherent safety against a large number of accidents, including: type B(U) transport accident conditions, airplane crash, burial under rubble in case of collapse of storage building or for other reasons and gas cloud explosion;

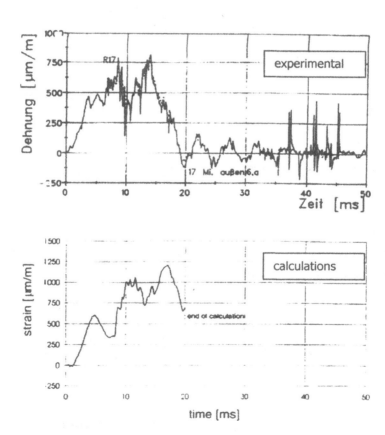

Figure 6. Drop test of the cask: (a) Real test -container in 9m position; (b) Comparison of experimental and calculated loading strain - time curves

40

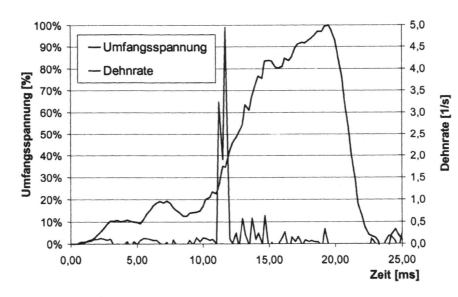

9-m Mantellinienfall

Figure 6 (c) Drop test – FEM model and calculation results

- double barrier system for zero release concept;
- passive cooling by means of natural air convection or heat radiation along the surface;
- the fabrication concept is clear and simple - the manufacturing can be performed by any qualified welding company;
- proven fabrication methods are used;
- all components are readily available on the market;
- no sensitive sealing surfaces and metallic seals have to been used, because welded barriers can be used for containment;
- overall costs of fabrication and for storage are low.

This last advantage is of extreme importance. The price for a CONSTOR® cask is considerably lower than that of a CASTOR® cask, making it more competitive than its predecessors.

4. Experience with Spent Fuel Storage Projects

4.1 GENERAL

GNB is currently delivering or has delivered transport and storage casks for spent fuel to the interim storage facilities, see the TABLE I on the following page. More than 550 casks are loaded. Having delivered a total of approx. 700 casks for high-level radioactive material, GNB has extensive experience in the design, licensing, testing, manufacture and operation of casks for spent fuel.

4.2 REFERENCE SPENT FUEL STORAGE PROJECTS

In order to briefly describe the experience of GNB in the design, licensing and manufacture of spent fuel casks, a few of the above mentioned projects are described in the following parts.

4.2.1. Dry Storage of Spent Fuel at Surry Nuclear Power Station, Virginia/USA

For the storage of spent from PWR reactors, GNB has designed and licensed two casks types in the USA: the CASTOR® V/21 for 21 assemblies (Westinghouse type) and the CASTOR® X/33 for 33 assemblies. The CASTOR® V/21 was among the first dry storage casks licensed by the Nuclear Regulatory Commission and has received a Certificate of Compliance in accordance with 10 CFR Part 72, which is valid until the year 2010.

As part of the licensing process, cask performance tests were conducted on the first serial cask at INEL between 1984 and 1991. These tests included examination of the behavior and condition of the spent fuel during long-term storage. During the period from 1986 to 1996, a total of 25 CASTOR® V/21 casks and one CASTOR® X/33 were delivered. All are currently in outdoor storage at the independent spent fuel storage installation at Surry Power Station.

TABLE I. Loaded CASTOR cask over the world (as per January 2001)

Location	Cask Type	No. of loaded casks
■ Paul Scherrer Institut (PSI), Switzerland	CASTOR Ic-Diorit	1
■ Department of Energy (DOE), Idaho Falls, USA	CASTOR V/21	1
■ Gundremmingen Power Station, Germany unloaded in Clab, Sweden	CASTOR KRB-MOX	4
■ Surry Power Station, Virginia/USA	CASTOR V/21	25
	CASTOR X/33	1
■ Ahaus Storage Facility, Germany	CASTOR THTR	305
	CASTOR V/19	3
	CASTOR V/52	3
■ Jülich Research Center, Germany	CASTOR THTR	106
■ Gorleben Interim	CASTOR Iia	1
Storage Facility, Germany	TS28V	1
	CASTOR Ic	1
	CASTOR V/19	3
	CASTOR HAW 20/28 CG	2
■ Dukovany Power Station, Czech Republic	CASTOR 440/84	38
■ Department of Energy (DOE), Hanford,	CASTOR GSF	6
Washington, USA	GNS-12	2
■ Greifswald Power Station, Germany	CASTOR 440/84	17
■ Ignalina Nuclear Power Plant, Lithuania	CASTOR RBMK	20
■ Neckarwestheim NPP	CASTOR V/19	6
■ Biblis NPP	CASTOR V/19	2
■ La Hague Reprocessing Plant	CASTOR HAW 20/28 CG	18

4.2.2. Dry Storage of Spent Fuel at Dukovany NPP in the Czech Republic

In 1989, the nuclear fuel services offered by the former Soviet Union changed, and it became urgent in the Czech Republic to build an interim spent fuel storage facility. The owner of the Dukovany NPP CEZ, a. s. had to ensure additional spent fuel storage capacity starting in 1995, beyond the existing wet storage pools at the reactor. The tendering process began in February 1991 and was finished in 1992. Preliminary bids were submitted by 10 bidders from 7 countries. The proposed types of storage systems covered nearly all spent fuel storage methods. The technical concept of a dry cask storage offered by GNB fulfilled all requirements and is very similar to the existing reference plants Ahaus and Gorleben in Germany. Therefore a short licensing and realization time was expected. For the fuel type VVER 440 a new cask, the CASTOR® 440/84, based on the design and safety principles of the CASTOR® family had to be developed. With the preparation of the Site Permit Documentation including a Preliminary Safety Case, the race against time started.

The CASTOR® 440/84 has been specially designed for the transport and the long-term interim storage of fuel assemblies from reactors of the type VVER 440 in operation in Eastern Europe. Cask dry storage using the CASTOR® 440/84 is currently being realized at the Nuclear Power Plant Dukovany and Greifswald in eastern part of Germany. The CASTOR® 440/84 can accommodate 84 fuel assemblies with an average initial enrichment of 3,5 % U235, an average burn-up of 33000 MWd/tHM and a cooling time of 5 years.

The hexagonal arrangement of this type of fuel assemblies and their large number required a special constriction for the fuel basket. The chosen design guarantees heat removal and criticality safety under normal and testing conditions. The hexagonal tubes made of borated steel in the fuel basket have a honeycombed arrangement. Between the fuel positions of this arrangement, aluminum plates are adapted to the contours. In the radial direction, the plates have different lengths and assure sufficient heat removal. The residual free spaces are filled with small aluminum plates to guarantee a sufficient strength.

The casks are stored in a building which protects them from the weather, provides handling capabilities and minimizes the impact of the radioactivity on the environment. It is built in a purely conventional way inside the area of Dukovany Nuclear Power Plant. The features of the storage building itself are as follows: length 55 m, width 26 m, height 20 m, capacity 60 casks with a total of 5040 fuel elements, overhead crane with 130 tons lifting capacity, central control room for cask tightness surveillance and radiation control, cask maintenance room, road and rail connection, safety related data are transferred directly to the NPP control room.

The close cooperation between Czech and German Authorities, along with the very safe concept, provided the best conditions for a smooth licensing procedure of the plant in the Czech Republic. The erection period was limited due to the small storage capacities remaining in the wet pools. For the storage building, erected by domestic subsuppliers of the consortium, a completion time of about one year was needed.

By November 1995, the old C30 cask type holding 30 irradiated fuel assemblies was the only cask type which could be used by the NPP Dukovany. The new cask CASTOR® 440/84 providing higher capacity for 84 fuel assemblies of VVER 440 type and the appropriate handling equipment had to be introduced into the spent fuel system management.

One month prior to the start of loading activities, cold handling trials were performed with the new cask design. Staffs of both the reactor and the supplier of the loading machine were involved to check the loading of the basket by means of a dummy. Additionally the exact coordinates of the cask in the pool were transferred to the software of the loading machine. In accordance with a sequence plan, the reactor staff

was trained in all handling steps which are necessary to prepare the cask for dry storage of irradiated fuel, such as the following:
- assembly of lid seals
- checking the cleanliness of sealing surfaces
- connecting and disconnecting of the lifting yoke arms at the pond
- draining of water inside the cask and sealing gaps
- screwing the lid bolts
- vacuum drying of the cavity
- performing the helium leak test on all gaskets
- back filling of the cavity with helium
- assembly of the pressure monitoring device
- transfer of the cask to the dry storage facility

After successfully testing, the Czech licensing authority SONS gave their permission for the hot loading of the cask, which was performed in November 1995. In the meantime, approx. 40 CASTOR casks have been loaded, and placed into storage.

4.2.3. Dry Storage of Spent Fuel at the Ignalina NPP

In December 1993, GNB Gesellschaft für Nuklear-Behälter mbH was contracted by the Lithuanian Ministry of Energy to deliver 60 casks of the type CASTOR® RBMK for storage and transport of spent fuel from the reactors at IGNALINA NUCLEAR POWER PLANT. In the contract, GNB also stated that it was going to develop a new type of cask which could be fabricated with conventional manufacturing technology in all countries with nuclear power stations and which would be less expensive than the CASTOR® cask.

In 1995, GNB was informed by the Lithuanian Minister of Energy that beginning with Cask no. 21, Lithuania would agree to purchase casks of the new design. For this reason, GNB concentrated all its development and engineering work on the new cask family made of steel and concrete called CONSTOR®. Initially, the design work was completed to take into account fuel from RBMK and VVER reactors. GNB is also planning to apply for licenses to accommodate Western PWR and BWR fuel types in the CONSTOR® cask as well as high level waste.

Within the scope of the licensing procedure, GNB performed a series of drop tests with a prototype cask from the new CONSTOR® cask family in May and June 1997. The tests serve to qualify the new package design, as well as to benchmark computer codes used for licensing of the casks. The tests were performed with a half-scale model which had a weight of approx. 12 t. Results of the test are to be transferred to a full-scale cask type.

The first test consisted of a 9-m free drop onto the sidewall of the cask with shock absorbers on an unyielding foundation, in accordance with the IAEA requirements Safety Series No. 6 for Type B package designs. No deformations of the cask body resulted from the test. Only the shock absorbers were affected by the drop, as was to be expected. It is, in fact, the purpose of the shock absorbers, to protect the cask during transport by absorbing energy from impacts.

The second test was a 1-m pin drop onto the side-wall of the cask without shock absorbers also in accordance with the IAEA conditions. In this case, the cask exhibited only a minor deformation on the outer surface which corresponded to the anticipated results from analytical methods.

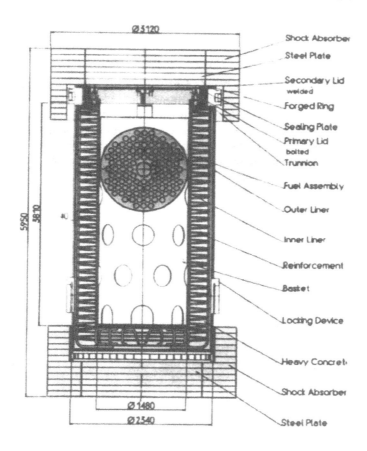

Figure 7. The CONSTOR cask (steel-concrete cask)

The test series also included the following:
- 1-m pin drop onto the bottom of the cask with shock absorber
- 1-m pin drop onto the lid area of the cask with shock absorber
- 1-m handling drop onto the bottom without shock absorber
- 1-m handling drop onto the bottom corner without shock absorber.

The results of the drop tests exceeded all expectations, so that it was clearly demonstrated that the CONSTOR® cask is capable of meeting the requirements for transport and storage, both under normal and accident conditions. In this way, the CONSTOR® has been shown to meet the high standards of quality and safety worldwide. Licensing of the CONSTOR® was completed, and 40 casks have already been delivered. The first casks have been successfully loaded and placed into storage in early 2001.

5. Summary

With a large number of tests and calculations GNB has shown, the casks of the CASTOR nad the CONSTOR family fulfile all the IAEA requirements in all standard and extreme situations. The cask design is based on modern method of calculations, that have been verified with a lot of experiments and large scale tests. In summarry of all experience, it can be concluded, that the casks of both families are a very save and inexpensive solution for the transport and storage of spent nuclear fuel and high active waste.

FRACTURE RESISTANCE OF CAST FERRITIC C-MN STEEL FOR CONTAINER OF SPENT NUCLEAR FUEL

I. DLOUHÝ, M. HOLZMANN, Z. CHLUP
Institute of Physics of Materials ASCR, Žižkova 22, 61662 Brno, Czech Republic

Abstract: Fracture resistance of cast ferritic steel predetermined for containers of spent nuclear fuel has been evaluated based on sets of different fracture mechanical test specimens and assessment procedures. Standard fracture toughness values were determined from 1T SENB specimens and compared with data from pre-cracked CVN specimens (P-CVN). The other parameters that have been in focus of interests were: the effect of metallurgical technology (two melts followed), specimen location (midthickness vs. surface locations) in thick walled plate, the effect of loading rate (here followed on CVN and P-CVN specimens), statistical effects etc.

Key words: ductile to brittle transition, fracture toughness, cast ferritic steel, specimen size effect, strain rate effect

1. Introduction

Given by the evolution of analytic capabilities, test methods, and fabrication processes, confidence in material integrity has increased to a point where certain structural materials until now not considered for radioactive material transport and storage cask construction are being proposed for this application. Brittle fracture can occur under specific combinations of temperature, mechanical and environmental loading conditions. Without the development of new guidance for the evaluation of the brittle fracture potential in ferritic steels (and other similar candidate materials) the cask producers will stay largely limited to constructing containers from austenitic stainless steel and/or similar materials with high safety factors.

For the safe enclosure of the radioactive material during transportation it must be shown that the crack extension will not occur at the tip of the postulated crack like reference flaw. For an acceptable design the applied stress intensity factor even in case of the most severe accident loading (earthquake etc.) has to be smaller than the fracture toughness value at the design conditions (temperature, loading rate) divided by a safety factor [2-4].

For the safe storage of the spent nuclear fuel additional embrittling effects should be taken into account (age hardening due to service temperatures and irradiation). Irradiation of steel (although comparably smaller than that in nuclear pressure vessels) decreases the fracture toughness and increases risk of brittle fracture [3-7]. The

47

I. Dlouhý (ed.), Transferability of Fracture Mechanical Characteristics, 47–64.
© 2002 *Kluwer Academic Publishers. Printed in the Netherlands.*

container lifetime (supposed at minimum about 60 years) is controlled by shift of the transition temperature region resulting in a shift to the completely brittle state at service temperature [5].

Each application of new material for container production has to (a) satisfy some general rules and (b) respect specific problems of the particular solution.

(a) The general aspects and experiences of other laboratories have been summarised in "Guidelines for safe design on shipping packages against brittle fracture" which are regulating present assessment procedures from the point of view of IAEA [2]. Three methods have been included in this material:

(i) The evaluation and use of materials, which remain ductile and tough throughout the required service temperatures down to $-40\,°C$.

(ii) The evaluation of material having ferritic matrix by means of nil ductility transition temperature measurements correlated to fracture resistance.

(iii) The assessment of fracture resistance based on a design evaluation using fracture mechanics.

(b) The specific problems of a material application to manufactured products, such as a container cask for spent nuclear fuel, have to show the advantages and disadvantages of that particular solution. These include such material characteristics assessment as the inherent scatter of material fracture toughness (as a characteristics of material resistance against catastrophic failure), susceptibility (predisposition) of material to embrittlement (radiation, temperature exposition, low temperature), the effect of loading rate, etc. [3-7]. Every approach able to predict these characteristics of fracture resistance will enhance the quality of pure experimental evaluation of materials for production and confidence in material integrity.

When assessing if the material satisfies the demands on container resistance against catastrophic failure the following key problems have to be addressed from the fracture mechanical point of view:

(i) the *transferability of fracture toughness* characteristics measured on small specimens (or a specimen of limited dimensions) to the component of much larger dimensions, and

(ii) the *prediction* with a good probability of brittle fracture in case of the most severe accident loading and in case of radiation embrittlement.

Several approaches may be applied or developed in order to solve the problem of the container integrity from the point of view of material fracture resistance and its prediction. These include the master curve methodology [11-12], local approach [13,14] and several concepts of toughness scaling models represented mainly by original works of Nevalainen and Dodds [15-16] and of Koppenhoeffer and Minami [17-19] etc. [20 and others]. Capability of these approaches to predict the fracture behaviour for any configuration of defect and component should be accepted as a well-defined criterion of the assessment procedure.

For container casks design itself, there are several concepts being developed applicable to the areas mentioned above on commercial level. An alternative being developed by Škoda Nuclear Machinery Ltd. to higher cost steel containers proposes a universal cask (the same design for transport container and for storage ones) and a not as expensive solution. Cast low carbon manganese steel treated by intercritical hat treatment possesses good strength properties and fracture toughness of ferrite microstructure even in the centre of cask walls with their thickness of 270 mm. The high level of the fracture resistance throughout the required service temperature range, including down to –40 °C, and sophisticated design supported by the extensive computer simulation ensures the optimal container integrity.

For quality planning and control during production and for a safe and environmentally acceptable operation of newly developed containers, the producer needed additional fundamental knowledge about material fracture resistance of the different microstructures under the different loading conditions. Recent knowledge and fracture mechanics approaches were intended to be applied and developed. The anticipated knowledge and approaches included:

(i) Statistically meaningful number of material fracture resistance data showing inherent scatter of material properties measured on pre-cracked specimen loaded statically and dynamically.

(ii) Assessment of fracture resistance based on design evaluation using fracture mechanics approach (transferability of laboratory fracture mechanics data, geometry and size effects, influence of loading rate etc.).

(iii) The application of alternative and cheaper methods of material toughness assessment (prediction) such as small Charpy type specimen or notched tensile specimen. The first enabling testing in irradiated state and the second reflecting the effect of (notch) stress triaxility.

Almost all the above-mentioned aspects were subject of intensive investigation within the scientific part of the project NATO SfP 972655 [21]. Substantial part of the experimental results, new knowledge of experimental and/or theoretical nature, respectively, have been exploited in a number of contributions included into this volume. Aim of this paper has been shortly to summarise the main experimental program as a whole and to characterise the fracture resistance of the cast ferritic steel by, more or less, standard approaches and applying those data only that could not be exploited in other specific topics addressed in this proceedings.

2. Material and Experimental Methods

2.1. MATERIAL DESCRIPTION

Standard manganese cast steel has been used in investigation having the chemical composition given in TABLE I.

TABLE I. Chemical compositions of experimental steel melts used (in wt. %)

	C	Mn	Si	P	S	Cr	Ni	Cu	Mo	Al
melt I	0.09	1.18	0.37	0.01	0.025	0.12	0.29	0.29	0.03	0.028
melt II	0.12	1.35	0.44	0.15	0.007	0.06	0.10	0.15	014	0.023

The experimental material was supplied by Skoda Nuclear Machinery in form of two melts; the experimental program thus has been carried out in two stages:

(i) "Melt I" was supplied in form of 250 mm thick plate produced commonly with container model. All the specimens from this plate were oriented in the through thickness orientation (see also *Figure 1*). The test results and data obtained have been mostly used for almost all subtasks of theoretical and computational nature as it has been shown in papers in this volume.

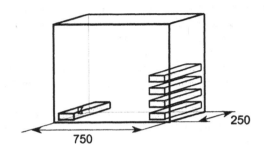

Figure 1. Dimensions of experimental plate and schema of cutting plan (1T SENB specimens) for the Melt I

(ii) "Melt II" was produced as a 270 mm thick plate for purposes of the project. This plate was produced as a casting technology development for the full-scale container. Dimensions selected (1000 x 3100 x 270 mm) corresponded to the dimensions of a segment of a thick walled container cask. Special effort has been paid to quality assurance corresponding to future container cask material. Detailed non-destructive quality testing of all of the plate has been followed by analyses of heterogeneity in chemical composition and microstructure characterisation. The test specimens were cut from the plate in tangential direction (relating to future container cask) with crack propagation in through thickness direction. The schema of cutting plan for standard SENB specimens is shown in *Figure 2*. The mechanical and fracture properties were evaluated in two locations through the plate thickness - very near to the surface (labelled as E) and in the central plane of thick walled plate (labelled as C).

In order to produce homogeneous mechanical properties throughout the plate thickness special heat treatment was developed being based on controlled cooling from intercritical temperature range (930°C/3h/air + 900°C/3h/watter + 780°C/3h/watter + 650°C/12h/air). The treatment resulted in predominantly fine grained ferritic matrix (see also paper [22] in this volume) with regular distribution of small islands of bainite (in surface layers – location E) and fine pearlite (in central parts of the plate – location C).

TABLE II. Basic mechanical properties of both experimental melts

	d_F [μm]	YS [MPa]	UTS [MPa]	FATT [°C]	CVN impact energy [J]
melt I	14	281	419	0	122
melt II / loc E	12	319	465	-45	180
melt II / loc C	14	311	458	-40	182

For both melts, TABLE II shows average value of ferritic grain size d_F as well as the basic room temperature mechanical properties as they have been determined during the experimental investigation. An overview of all these mechanical and fracture mechanical characteristics is also available in [23].

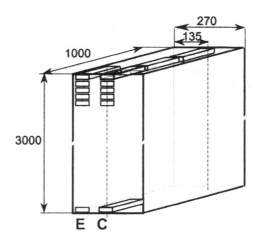

Figure 2. Dimensions of experimental cask segment and cutting plan schematics (for SENB specimens) for the "melt II"

2.2. MECHANICAL TESTING

Tensile properties and true stress-strain curves were determined using cylindrical specimens with a diameter of 6 mm in the temperature range from -180°C to -20°C at a cross-head speed of 2 mm.min⁻¹.

Standard Charpy V notch specimens (CVN) were tested under three types of loading (strain rates):

(i) Temperature dependencies of CVN impact energy were measured by using standard impact tester. The testing was carried out over a temperature range of -196 to 20 °C and fracture appearance transition temperature (FATT) has been determined among others.

(ii) To evaluate the cleavage fracture stress, σ_{CF}, the same CVN specimens were tested at three loading rates: quasistatically in three point bending on standard testing machine and dynamically at two loading rates on instrumented impact tester (according to procedures prescribed in [24]). In the first case the loading rate was 2 mm.min⁻¹, in the second case the full impact (about 5 m.s⁻¹) and/or lowered hammer drop rate (1 ms⁻¹, by low blow method) was applied. Load - displacement traces were recorded, and the maximum force, F_m, fracture force, F_{fr}, and general yield force, F_{gy}, respectively were read from the records. For one selected temperature in lower shelf region (at general yield temperature, t_{gy}, - see later) a range of bend tests was performed to obtain data for statistical analyses.

Fracture toughness was determined using three test specimen geometries. The first one was the standard three-point bend (1T SENB) specimen 25x50x220 mm³. The other two ones were selected to receive in specimen shallow cracks (with crack length to specimen width ratio, a/W, of about 0.2 and 0.1) and, at the same time, comparable ligament area under the crack tip. All specimens were tested in three point bending at 1mm.min⁻¹ cross-head speed in the temperature range from -180 to + 20 °C. For a/W ≈ 0.5 the fracture toughness values have been determined according to ASTM E1820-99a and/or similar standards (ISO standard [25], ESIS P2/92 procedure etc.). The results of the crack length effect analysis are presented in other contribution in this book [26].

Pre-cracked Charpy type specimens (P-CVN) have been static rate tested, and in addition, by low blow method on instrumented impact machine in a wide temperature range. Experimental methods applied included recommended procedures [27].

TABLE III. Complete test matrix carried out with the cast C-Mn steel,
last two columns show the minimum number of tested specimens

Specimen type	Test conditions	Temperature range	MELT I	MELT II
Standard tensile	6 mm in diameter	(-196 – 20) °C	30	25 + 25
with U - notch	R ≈ 0.3 mm	(-180 – -100) °C	10 + 30	
with U - notch	R ≈ 0.7 mm	(-180 – -100) °C	10 + 30	
with U - notch	R ≈ 1.0 mm	(-180 – -100) °C	10 + 30	
Charpy type tested dynamically	different notch radii	(-180 – 20) °C	60	
CVN dynamically	12 different microstructures	(-180 – 20) °C		160
P-CVN	8 different microstructures	(-180 – 20) °C	10 + 30	80
1T SENB - static	a/w ≈ 0.5	(-180 – 20) °C	20 + 30	60 + 60
1T SENB - static	a/w ≈ 0.2	(-180 – 20) °C	20 + 20	
1T SENB - static	a/w ≈ 0.1	(-180 – 20) °C	20 + 20	
1T SENB - drop weight	at ≈ 2-3 m/s	(-120 – -20) °C		30 + 30
CVN – static		(-180 – -100) °C	10 + 30	40 + 40
CVN – instr. pendulum	at ≈ 1.0 m/s	(-160 – 20) °C	20 + 20	30 + 30
	at ≈ 5.0 m/s		20 + 20	30 + 30
P-CVN - static		(-180 – 20) °C	10 + 30	30 + 30
P-CVN - instr. pendulum	at ≈ 1.0 m/s	(-160 – 20) °C	20 + 20	30 + 30
	at ≈ 5.0 m/s		30 + 30	30 + 30

Overview of all tests carried out with both steel melts is shown in the TABLE III. In this contribution only those results have been presented which were not been applied in other results descriptions included in this volume, the papers connected in some way with the investigated cast ferritic steel.

Extensive fractographic observations by using scanning electron microscope (SEM) were carried out being focused on identification of initiation origins and stretch zone geometry. The experimental details have been presented elsewhere [23].

3. Results description

3.1. LABORATORY COMPARISON

A part of experimental works was shared between two laboratories – the Institute of Physics of Materials (IPM) and the University of Miskolc (UM). Common experimental works were started by comparing data obtained in both collaborating laboratories.

As example, some data obtained from CVN specimens tested by instrumented impact machine are shown in *Figure 3*. Characteristic forces are here plotted against temperature; F_{gy} represents the value of force at general yield point, in this particular case determined for 0.1 mm of plastic deflection, F_{ff} is the onset of unstable fracture and F_{max} is maximum force on load deflection traces. It is obvious that the material is characteristic by large scatter of the mentioned fracture characteristics. This is also evident from the set of more than thirty specimens tested at -70 °C. The data obtained at UM fit together with data from IPM. Also the temperature t_{gy} (the temperature at which the general yield and fracture load are coincident on their temperature dependences) have been found to be the same from both compared laboratories.

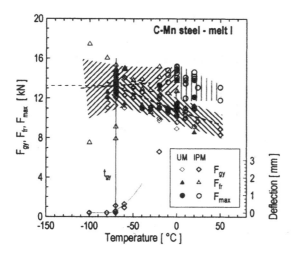

Figure 3. Characteristic forces from CVN impact test
on instrumented impact tester.

Another sets of results from instrumented impact tester are shown in *Figures 4 and 5*. Temperature dependence of CVN impact energy is shown in *Figure 4*. Filled round points represent data obtained at UM. Good correlation of both data sets obtained from

54

cooperating laboratories is obvious. The morphology of fracture surfaces has been evaluated quantitatively and fracture appearance transition temperature, FATT, was determined. The results are plotted in *Figure 5*. Here some difference (the shift by about 10 °C) in FATT between both laboratories can be observed (compare the FATT(hu) in comparison to FATT(cz)).

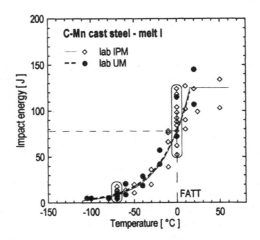

Figure 4. Comparison of data on CVN impact energy obtained in Institute of Physics of Materials (lab IPM) and University of Miskolc (lab UM)

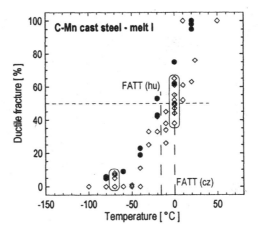

Figure 5. Fracture appearance transition temperature (FATT) determined for data sets shown in previous Figure 4 in both laboratories

After detailed analyses the reason for these differences has been assigned to the conjunction of standard subjective errors and large scatter in material behaviour. This scatter in CVN impact energy, inherent to the applied steel, is evident from the *Figure 3* where except for temperature dependence also larger set of data obtained at one temperature (-100 ºC) is shown.

Nevertheless the tests by using instrumented impact pendulum and Charpy V-notch specimens have shown very good agreement of obtained data. This was a very good initial position for the following experiments on other larger sets of test specimens in collaborating institutions.

3.2. FRACTURE BEHAVIOUR OF MELT I AND MELT II (THE ROLE OF METALLURGICAL QUALITY)

Extensive investigation was carried out during the time interval between Melt I and Melt II production with aim to increase the fracture resistance of the cast steel and, at the same time, to enlarge the metallurgical and microstructural homogeneity of the thick walled plate. It has no importance to compare here in detail all the microstructural, mechanical and fracture characteristics obtained, two representative examples have been selected however – one connected with an effect of wall thickness on fracture behaviour and the other comparing the fracture behaviour of Melt I and Melt II.

Direct comparison of characteristic forces determined from load - deflection traces of CVN specimens tested by means of instrumented impact tester (ipm) and quasistatically (qst) are shown in *Figure 6*. The characteristic forces have been selected as one of the most susceptible characteristics reflecting small (subtle) differences in fracture behaviour. There are introduced data for both specimen locations, E and C. No difference has been found between the data from both locations.

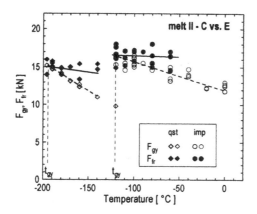

Figure 6. Temperature dependence of characteristic forces obtained from CVN specimens tested statically and dynamically for both locations of melt II

Applying statistical calculations when fitting the exponential equations throughout the data, total agreement of curves have been obtained these curves are shown in figure by dashed (temperature dependence of general yield force) and full line (fracture and/or maximum forces). This is a result of, partly, actually very small differences in fracture

56

resistance due to good metallurgical quality and small differences in microstructural state, and, partly, small susceptibility of Charpy V type geometry of the specimen for these purposes (and it will be shown later that only some fracture characterictics reflect these differences).

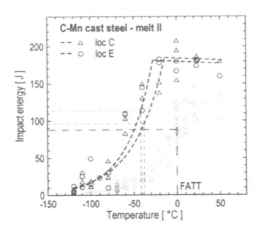

Figure 7. Comparison of data on CVN impact energy obtained with melt II (both locations) with characteristics of melt I.

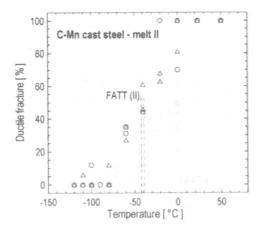

Figure 8. Temperature dependence of ductile fracture ratio (corresponding to data in Fig. 7) for both locations of melt II

The CVN impact energy is shown in *Figure 7* for melt II of the steel and compared to data obtained with melt I. For melt II characteristics of two specimen location are included – midthickness of the plate, labelled as loc C, and surface locations, labelled as loc E. Exponential curves have been fitted through all the data in transition region. The following important findings could be drawn from the figure. Evidently better fracture resistance of melt II when comparing to melt I. This can be seen also from the shift of transition temperature FATT. In upper shelf region, better behaviour, the higher CVN impact energy values, can be observed for the melt II. Finally, very subtle differences in fracture behaviour between midthickness and surface layers are evident showing very comparable properties throughout the wall thickness of the experimental plate. Comparing to other similar component geometries used either for reactor pressure vessel steel or for container casks this is one of key aspects of the steel used in this study. The level of fracture characteristics in midsection of thick walled plates is usually nearly one half of toughness observed in surface locations [28].

The other *Figure 8* shows the quite small differences (but measurable) in FATT between both specimens locations in the plate of melt II.

The data for melt II can thus be taken as a good result of metallurgical technology and heat treatment procedures development. The results have also shown the key importance of heat treatment for fracture properties of thick walled components. Further development in this field is expected in combining the cast steel microalloying and intercritical heat treatment as described in detail in paper [22] in this volume.

3.3. FRACTURE TOUGHNESS TEMPERATURE DIAGRAM

For pre-cracked Charpy type specimens, P-CVN, the fracture toughness values are introduced in *Figure 9*. The correlation of values from standard specimen geometry and those from PCVN specimens is analysed in other contribution in this volume, here very close fracture behaviour between surface location and central plane is documented. Only a few specimens supplied valid fracture toughness data (lying below the K_{Jc} limit line) and for this material the P-CVN cannot be generally used without any other corrections (for statistical size and/or constraint loss effects).

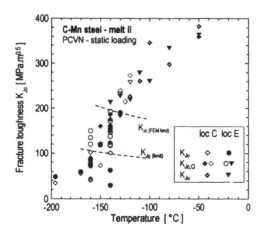

Figure 9. Fracture toughness temperature dependence obtained with pre-cracked Charpy type specimens and both specimens locations

For the standard SENB specimen geometry (crack length a/W ≈ 0.5) and the Melt II, the temperature dependences of fracture toughness values are shown in *Figure 10 and 11*; in the first one for surface location (E) and in the latter for the central part (C) of the thick walled plate. (For the Melt I, similar fracture toughness temperature diagrams have been presented in other contribution in this volume [26]).

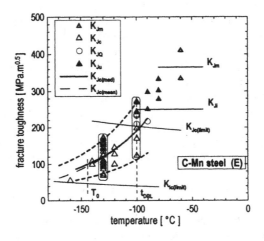

Figure10. Fracture toughness temperature dependence
for specimens from surface location of the melt II

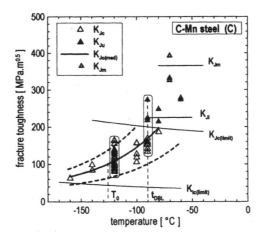

Figure 11. Fracture toughness temperature dependence
for specimens from central parts of the melt II

Specimen sets were tested at two selected temperatures and data obtained are also shown in the figures. The temperature -130°C has been selected in lower shelf region (cleavage initiation only was proved by SEM). The other temperature -100 °C has been selected very close to upper shelf region but still having predominantly cleavage initiation mechanism in specimen fracture surfaces.

Only some specimen kept the condition of LEFM for standard K_{Ic} toughness values and mostly K_{Jc} values have been obtained in transition region and lower shelf regime. From the application point of view one very important conclusion follows from this findings: the material is typical by large crack tip plasticity. In the transition and near the upper shelf region the fracture behaviour has been characterised by K_{Jc} values determined for specimen without detectable ductile tearing before fracture and K_{Ju} for specimen with ductile crack extension preceding the unstable brittle fracture. Curves representing the limit (i) between the validity of LEFM and EPFM and (ii) between the constraint dependent and independent K_{Jc} values are shown in both *Figures 10 and 11*. K_{Ji} represents here the value for specimen with 0.2 mm ductile crack extension and temperature t_{DBL} the lowest temperature of this initiation micromechanism occurrence.

3.4. LOADING RATE EFFECTS (CVN AND PCVN)

The effect of loading rate has been evaluated by using three specimen geometries, standard three point bend specimens, standard Charpy V notched specimens (CVN) and Charpy specimens geometry fatigue pre-cracked (PCVN). The standard specimen geometry (SENB) tested dynamically has been followed elsewhere [29]. Here the results of CVN standard Charpy V notch specimens tested at three different loading rates and the results of pre-cracked CVN specimen will be presented.

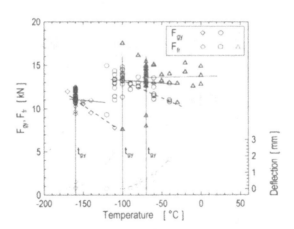

Figure 12. Characteristic forces determined for CVN specimens
tested at three loading rates - Melt I

In the first case, CVN specimens were tested statically at 1 mm.min^{-1}, and dynamically at two loading rates, at about 1 ms^{-1}, corresponding to so called low blow conditions and at about 5 ms^{-1} corresponding to the standard impact test conditions. For

60

purposes of this presentation the temperature dependences of characteristic forces are shown in *Figures 12 to 14*.

The data in *Figure 12* represent the behaviour of steel Melt I. There are presented general yield forces, F_{gy}, and fracture forces, F_{fr}, only in the figure. Then the general yield temperature, t_{gy}, corresponds to the coincidence point of both mentioned characteristic forces on their temperature dependences and is also shown in the figure. Sets of specimens have been followed at this temperature in order to get data for critical (cleavage) fracture stress determination. As it has been supposed the general yield temperature, similarly as any other property characterising the steel transition behaviour, increases with increasing loading rate.

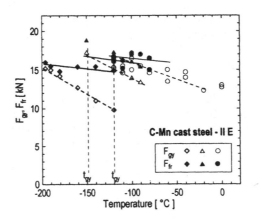

Figure 13. Characteristic forces determined for CVN specimens tested at three loading rates - Melt II – location E

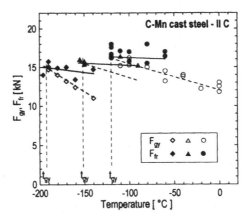

Figure 14. Characteristic forces determined for CVN specimens tested at three loading rates - Melt II – location C

For Melt II, similar observations regarding the susceptibility of general yield temperature to loading rate can be also seen from the *Figures 13 and 14*.

For both melts, in addition to the above mentioned change an increase of fracture forces, F_{fr}, with increasing loading rate can be observed. This is more evident for the Melt I (*Figure 12*) while for the Melt II slighter extent of this susceptibility can be observed, in particular for specimen location E (*Figures 13 and 14*).

To evaluate this effect of loading rate on critical conditions of cleavage fracture initiation more exactly cleavage fracture stress has been determined for data above presented. For CVN specimens loaded statically and by low blow technique, the cleavage fracture stress was evaluated as local peak of maximum principal stress ahead of the notch at fracture. For condition acting at general yield temperatures, t_{gy}, the same procedure of local maximum tensile stress calculation for CVN specimen as used by Brozzo et al. was applied. The value of 2,24 of plastic stress concentration factor, $k_{\sigma p}$, was taken. The relation of measured static and dynamic yield strength and general yield force was determined for the melt I. Arising from the linear relation of static and/or dynamic yield stress σ_y, σ_y^d and general yield forces F_{gy}, F_{gy}^d the simplified approach was possible to use for direct calculation of σ_{CF} from general yield force at t_{gy} thus, $\sigma_{CF} = 112 . F_{gy}$. The cleavage fracture stresses estimated by this way are compared in TABLE IV. The values of σ_{CF} determined from CVN specimens are in very good correlation with principal stress distributions exactly calculated by FEM for different crack length in SENB specimens [31]. The values obtained from CVN specimens correspond about to distance of 0.1 mm from notch, which is the distance of initiation origin from crack tip.

TABLE IV. Cleavage stress, σ_{CF} (in MPa), estimated for both melts and three loading rates applied

	loading rate	Melt I	Melt II / loc E	Melt II / loc C
static	1 mm.min^{-1}	1232	1736	1680
low blow	1 m.s^{-1}	1467	1848	1758
impact	5 m.s^{-1}	1557	1848	1893

There are two important findings that follow from comparisons of cleavage fracture stresses. Firstly, evident difference in resistance of compared steels against cleavage fracture initiation showing poorer properties of Melt I can be seen. Secondly, evident change of σ_{CF} in melt I whereas practically no difference for location E and very small dependence for location C can be observed when assessing the loading rate effects. According to the basic definition the cleavage fracture stress should be independent of strain rate. On the other side this phenomenon is probably connected with the microstructural effects on cleavage failure nucleation - the change occurrence is in some relation to metallurgical quality (higher for Melt I when comparing to Melt II, higher for location C when comparing to location E). This phenomenon needs further investigation.

Another important observation is obvious from comparison of melts I and II. The general yield temperatures are lying at comparably lower temperature when comparing to melt I. In addition to this very small difference can be observed between both specimen locations. It should be noted that this is the only characteristic that is capable to discern the differences between fracture behaviour in both locations of the thick walled plate.

62

The results from pre-cracked CVN specimens (PCVN) tested statically (qst) and dynamically by low blow method (dyn) are shown in *Figures 15 and 16*.

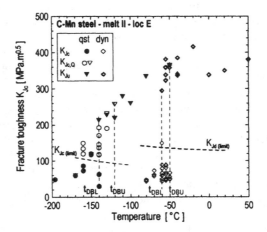

Figure 15. Fracture toughness temperature diagrams obtained at two loading rates from PCVN specimens - location E

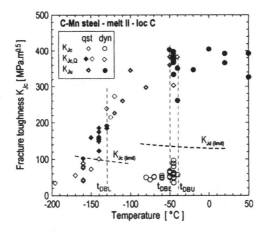

Figure 16. Fracture toughness temperature diagrams obtained at two loading rates from PCVN specimens - location C

The following findings can be summarised from the FTTD plotted in above shown figures:

(i) Shift of transition temperature, here the t_{DBL} (ductile to brittle transition temperature representing the lowest temperature of ductile fracture occurrence preceding the brittle fracture initiations), similarly as in case of t_{gy} for CVN specimens with increasing loading rate. Extremely high susceptibility to strain rate effect has been confirmed for the investigated steel.

(ii) From the application point of view, at dynamic loading the brittle fracture behaviour is taking place at temperatures as low as –50 °C and lower ones.

(iii) For dynamically loaded specimen a gap caused by absent data (and corresponding material fracture behaviour) is possible to observe between the specimens with brittle (cleavage) initiation and specimens displaying the ductile initiation. This is usually prescribed to the effect of adiabatic heating occurring during high strain rates.

(iii) Temperature interval corresponding to contemporary occurrence (overlapping) of both the brittle and the ductile fracture (temperatures between t_{DBL} and t_{DBU}) has been found to be on level of about 10 °C at dynamic loading and 20 °C at static loading

(iv) Validity K_{Jc} limits introduced into figures show that almost all data obtained at static loading are invalid, i.e. in constraint loss regime. No data are invalid at dynamic loading supplying evidence of "additional constraint" due to loading rate effect on yield stress etc.

(vii) Very small differences between properties of both specimen locations of the Melt II, loc E and loc C, have been found out confirming the results of CVN specimens and standard SENB specimens presented in previous chapters.

4. Concluding Remarks

Data set of more than 1200 fracture mechanical values was obtained for the C-Mn cast ferritic steel intended for container of spent nuclear fuel. The complex assessment of steel fracture behaviour has been carried out

Very good agreement of measured quantities (CVN impact energy, characteristic forces from pre-cracked Charpy specimens etc.) has been obtained when comparing data from two collaborating laboratories from Brno and Miskolc.

Based on analyses of Charpy V notch impact energy, fracture toughness obtained from pre-cracked CVN specimens, and standard fracture toughness values, respectively, very small differences in fracture resistance have been observed when comparing the surface and central part of thick walled plate. Comparably better fracture resistance has been found for steel Melt II in comparison to Melt I.

The dynamic fracture toughness values have been the only characteristics reflecting the (very) small differences in fracture behaviour between the surface and central locations of the thick walled plate. High susceptibility of investigated steel to loading rate effects has been proved. Corresponding peculiarities in fracture behaviour have been summarised including the increase of general yield temperature and unexpected increase of fracture load at this temperature with increasing loading rate.

Acknowledgements

The investigations and work on contribution has been financially supported by grant No. A2041003 of the Grant Agency of the Academy of Sciences of the Czech Republic and project No. 972655 within NATO Science for Peace Program.

64

References

1. Cotterell, D.: The past, present, and future of fracture mechanics, *Engng. Fract Mech.*, **69** (2002), pp. 533-553.
2. Guidelines for safe design of shipping packages against brittle fracture, *IAEA-TECDOC-717*, (1993).
3. Saegusa, T. et al: Application of IAEA TECDOC 717 to packagings and comparison with reactor vessels, *RAMTRANS*, **6** (1995), pp. 127-131.
4. Warnke, E.P. and Bounin, D.: Fracture mechanics considerations concerning the revised IAEA/TECDO/717 Guidelines, *Trans. of the SMiRT 14*, Lyon, (1997), pp. 571-578.
5. Bradley, W.L., McKinney, K.E. and Gerhard, P.C. Jr: Fracture toughness of ductile iron and cast steel, *ASTM STP 905*, J.H.Underwood et. al eds, ASTM, (1986), pp. 75-94.
6. Stranghöner, N. et al: Design against brittle fracture of structural steels, *Nordic Steel Construction Conference*, Bergen, (1998), pp. 751-762.
7. Moulin, D., Yuritzin, T., Sert, G.: An Overview of R&D work performed in France concerning the risk of brittle fracture of transport cask, RAMTRANS, Vol. 6, (1995), No. 2/3, pp. 145-248.
8. Winkler, H.P.: IAEA requirements for cask testing and short summary of developments at GNR, *The Transferability of Fracture Mechanical Characteristics*, Contribution in this Volume, (2002).
9. Brynda J., Hosnédl P., Picek M.: ŠKODA cask testing and licensing, Contribution in this Vol. (2002).
10. Heerens J., Zerbst U., and Schwalbe K.H.: Strategy for characterising fracture toughness in the ductile to brittle transition regime, *Fatigue Fract. Engng. Mater. Struct.*, **16** (1993), 11, pp. 1213-1220.
11. Wallin K., Rintamaa, Ehrnstén U.: Transferability of fracture mechanical parameters – philosophy or physics, *26th MPA Seminar*, Stuttgart, (2000), pp. 6.1-6.20.
12. Standard Test Method For the Determination of Reference Temperature T_0 for Ferritic Steels in the Transition range, (1997) ASTM, E1921-97.
13. Beremin. F. M.: A local riterion for cleavage fracture of a nuclear pressure vessel steels, *Metallurgical Transaction*, Vol. **14A**, (1983) pp. 2277-2287.
14. Kozák V. and Janík, A. The use of the local approach for the brittle fracture prediction, *The Transferability of Fracture Mechanical Characteristics*, Contribution in this Volume, (2002).
15. Anderson T.L., Dodds. R.H., Jr.: Simple constraint corrections for subsized fracture toughness specimens, *ASTM STP 1204, Small Spec. Test Techniques Applied to NRPV …*, (1993), pp. 93-105.
16. Nevalainen M., Dodds R.H.: Numerical investigation of 3-D constraint effects on brittle fracture, SE(B) and C(T) specimens, *Int. Journal of Fracture*, Vol. **74**, (1995), pp. 131-161.
17. Koppenhoefer K. C., Dodds R. H.: Loading rate effect on cleavage fracture of pre-cracked CVN specimens: 3-D studies, *Engineering Fracture Mechanics*, Vol. **58**, (1997) pp. 249-270
18. Minami F., Brückner-Foit A., Munz D., Trolldenier B.: Estimation procedure for the Weibull parameters used in the local approach, *Int. Journal of Fracture*, Vol. **54**, (1992), pp. 197-210.
19. Ruggieri C., Dodds R. D.: A transferability model for brittle fracture including constraint and ductile tearing effects: a probabilistic approach, *International Journal of Fracture*, Vol. **79**, (1996), pp. 309-340.
20. Wiesner, C.S., Andrews, R.M., A review of Micromechanical failure models for cleavage and ductile fracture, TWI, (1997), rpt. 592/1997.
21. Pluvinage G., Dlouhý I., Lenkey G.: Fracture resistance of steels for containers of spent nuclear fuel, *The project plan NATO SfP 972655*, (1998).
22. Kraus L., Němeček, S., and Kasl J.: (2002) Mechanical properties after intercritical heat treatment and microalloying, *The Transferability of Fracture Mechanical Characteristics*, Contribution in this Volume,
23. Chlup Z (2002).: Micromechanical aspects of brittle fracture initiation, PhD thesis, Brno University of Technology, Institute of Physics of Materials, Brno,
24. EN ISO 14556: 1999 Steel - Charpy V notch pendulum impact test Instrumented test method, (1999).
25. ISO/DSI 12135 Unified method of test for the determination of quasistatic fracture toughness, (1998).
26. Chlup Z, Dlouhý I. (2002). Micromechanical aspects of constraint effect at brittle fracture initiation, *The Transferability of Fracture Mechanical Characteristics*, Contribution in this Volume,
28. Rosinski S.T., Corwin W.R. ASTM cross-comparison exercise on determination of material properties through miniature sample testing, *Small Specimen Test Techniques, ASTM STP 1329*, W.R.Corvin et al Eds, (1998), pp. 3-14
29. Lenkey G.: Dynamic fracture toughness determination of large SENB specimens, *The Transferability of Fracture Mechanical Characteristics*, Contribution in this Volume, (2002).
30. Vlček L., Chlup Z., Kozák V.: Problems of Q-parameter calculations, Contribution in this Volume, (2002).

MICROMECHANICAL ASPECTS OF CONSTRAINT EFFECT AT BRITTLE FRACTURE INITIATION

Z. CHLUP, I. DLOUHÝ
Institute of Physics of Materials, ASCR, Žižkova 22, 616 62 Brno, Czech Republic

Abstract: Applying the two-parameter fracture mechanics approach to the analysis of failure initiation condition of the three point bend specimens with shallow and deep cracks were tested at various temperatures. Low carbon manganese cast steel was used for the analysis. This steel is tested as one of several candidate materials for containers of spent nuclear fuel. The effect of crack length on the fracture toughness-temperature diagram has been analysed. Although a strong dependence of measured fracture toughness on crack tip constraint was observed no evident differences in fracture morphology have been identified except for quantitative ones. Peculiarities of fracture behaviour in the transition and lower shelf regions of the steel investigated have been explained. The effect of crack tip constraint on brittle fracture characteristics has been quantified by means of the Q-parameter. The role of critical (cleavage) fracture stress in brittle fracture initiation under the influence of crack tip constraint has been analysed.

Keywords: two-parameter fracture mechanics, fracture toughness, cast ferritic steel, crack tip constraint, Q-parameter

1. Introduction

The knowledge in assessment methods of fracture mechanics has increased to a point where certain structural materials, until now not considered for radioactive transport and storage casks design, are being proposed for these applications [1]. For the safe enclosure of the radioactive material during transportation it must be proved that the extension of non-detected cracks after fabrication will not occur even in case of most severe accident loading. For safe storage additional embrittling effects should be taken into account. Brittle fracture can occur under specific combination of temperature, mechanical and environmental loading conditions [2,3]. When assessing if the material satisfies the demand on container resistance against catastrophic failure the following key problems have to be addressed from the fracture mechanical point of view: (i) the transferability of fracture toughness data measured on laboratory specimens to the component of much larger thickness; (ii) the prediction of the probability of brittle fracture in case of the most severe accident loading and in case of radiation embrittlement.

I. Dlouhý (ed.), Transferability of Fracture Mechanical Characteristics, 65–78.
© 2002 *Kluwer Academic Publishers. Printed in the Netherlands.*

Skoda Nuclear Machinery Ltd. has introduced new design of a container for spent nuclear fuel. The cask design is based on thick-walled cask with bolted lids, both fabricated from cast low-alloyed steel with ferritic microstructure. Transferability of material toughness data remains a key issue in applications of fracture mechanics to assess the integrity of structural components. For structures constructed of feritic steels, brittle fracture triggered by transgranular cleavage in the ductile-to-brittle transition region represents a potentially catastrophic failure mode. Several approaches might be applied or developed in order to solve the problem of the container integrity from the point of view of material fracture resistance and its prediction including the local approach [4-6], toughness scaling models [7,8] and master curve methodology [9-12]. The capability to predict the fracture behaviour for any configuration of defect and component could be accepted as a very hard criterion of the assessment procedure.

The aim of the present contribution is to characterise the fracture resistance of cast ferritic steel predetermined for radwaste casks applying two-parameter fracture mechanics approach. This objectives was solved by J-Q characterization of the elastic-plastic crack tip fields in statically loaded three point bend specimen.

2.   Material and Experimental Procedures

2.1.   MATERIAL CHARACTERISATION

Ferritic low carbon manganese cast steel was used as experimental material. Skoda Nuclear Machinery Ltd. produced this experimental material as one of suitable materials for cask of transport and storage container for spent nuclear fuel. Manganese cast steel has been used having the following chemical composition in wt %: 0.09C, 1.18 Mn, 0.37 Si, 0.01 P, 0.025 S, 0.12 Cr, 0.29 Ni, 0.29 Cu, 0.03 Mo and 0.028 Al. The material has been supplied in form of 270 mm thick plate produced commonly with model cask. Special heat treatment based on intercritical austenitisation was developed in order to produce homogeneous properties throughout the plate thickness. The material has been used having the following mechanical properties after heat treatment measured at room temperature: yield stress is equal to 280 MPa, strength is equal to 430 MPa and ductility is equal to 30%.

2.2.   MECHANICAL TESTING AND CALCULATIONS

True stress-strain curves have been determined from load-extension curve obtained by tensile test using cylindrical specimens with a diameter of 6 mm in the temperature range -196°C to -60°C at a cross-head speed of 2 mm.min^{-1}. Measured stress strain curve includes wide Lüders' plastic deformation region. The cast steel examined exhibits relatively low values of lower and upper yield stress and with decreasing temperature these characteristics increase very slowly (e.g. at −100 °C the yield stress is equal to only 380 MPa).

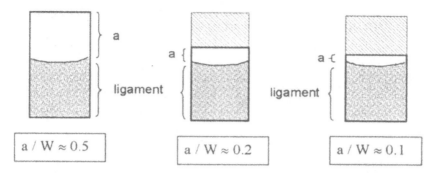

Figure 1. Three point bend specimen geometry

Three-point bend test was used for fracture behaviour determination. Fracture toughness was determined using three test specimen geometries. The first one was the standard three-point bend (1T) specimen 25x50x220 mm (with a/W ratio 0.5 - standard crack length). The other ones were selected to receive in specimen shallow cracks (a/W of about 0.2 and 0.1) and, at the same time, comparable ligament area under the crack tip. In these two specimen types the crack was introduced under condition close to standard one, then the specimens were cut and ground to final dimensions of 25x30x130 mm (a/W ≈ 0.2) and 25x27x120 mm (a/W ≈ 0.1) as is outline in *Figure 1*. All specimens were tested at 1mm/min crosshead speed in the temperature range from -198 to -20°C. The sets of specimen were tested at temperature -100°C for following statistical analyses of fracture behaviour with respect to crack length (influence of constraint). Temperature -100°C was determined with respect to all tested specimens that have to be failure with the same fracture initiation mechanism (cleavage). Fracture toughness values for a/W ≈ 0.5 have been determined according to ASTM E813-89 and similar standards. For calculation of stress intensity factor or J-integral value for specimens with shallow crack the general equation [13, 14] was applied [15].

The standard FEM code ABAQUS 5.8 was used to model elastic-plastic behaviour for all test specimen geometries investigated. Three-dimensional model was used in most of cases due to precise state of stress description. For modelling were used two symmetry axes. Calculations under SSY conditions were accomplished using MBL method in 2D.

Extensive fractographic observations were carried out. The same fracture initiation mechanism by sets of specimen with various crack length designed for three-point bend test tested at -100°C was observed. There are shown for example two micrographs obtained from fracture surface (initiation region) of specimen tested at -100°C with a/W ≈ 0.1 on *Figure 2a)* and a/W ≈ 0.5 on *Figure 2b)*. Both were taken from the middle part of experimentally obtained fracture toughness values scatter band at -100°C. Some differences were discovered in crack blunting and topography in dependence on fracture toughness values and crack length. These differences may be documented by micrographs made on specimen tested at -100°C with a/W = 0.5 from upper and lower section of fracture toughness values scatter band as is evident from

Figure 3a) (lower section) and *Figure 3b)* (upper section). These differences may be explained by size of plastic deformation before unstable fracture.

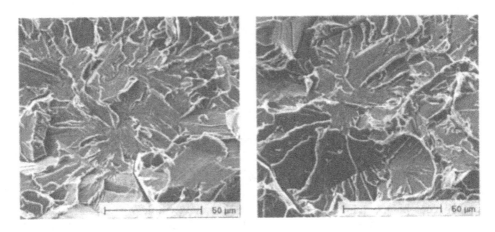

Figure 2. Cleavage fracture morphology for specimen with a) short b) standard crack

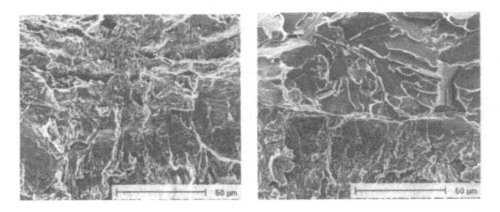

Figure 3. Crack tip blunting for specimen from a) upper b) lower section of scatter, a/W ≈ 0.5

3. Results and Discussion

3.1. FRACTURE TOUGHNESS AND CRACK LENGTH EFFECT

The temperature dependence of fracture toughness determined on specimens with standard crack length (a/W ≈ 0.5) is shown in *Figure 4*.

Two sets of specimens were tested at two selected temperatures and the data obtained are also shown in *Figure 4*. The first one (-100°C) has been selected as test temperature in the lower shelf region common for all crack lengths tested supposing the same - cleavage - fracture initiation mechanism acting on the specimens. The other temperature has been selected very close to the upper shelf region (-70 °C) but still showing cleavage initiation mechanism on fracture surfaces.

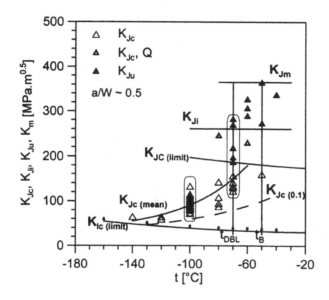

Figure 4: Fracture toughness - temperature diagram for standard specimens

Only some specimens met the condition of LEFM for standard K_{Ic} toughness values and the K_{Jc} values have been obtained in almost all cases. In the transition and near the upper shelf region the fracture behaviour has been characterised by K_{Jc} values determined for specimen without detectable ductile tearing before fracture and by K_{Ju} values for specimens with ductile crack extension preceding unstable fracture. Curves representing the limit between the validity of LEFM and EPFM (eq. 1) and between the constraint dependent and independent K_{Jc} values (eq. 2) are shown in the figure.

$$K_{Ic(\text{limit})} = R_p 0,2 * \sqrt{\frac{\min[a, B, (W - a)]}{2,5}} , \qquad (1)$$

$$K_{Jc(\text{limit})} = \left(\frac{E * B * R_p 0.2}{50}\right)^{0.5} , \qquad (2)$$

70

where a, B, W are dimensions of specimen, E is Young's modulus and $R_p0.2$ is yield strength.

The K_{Jc} (Q) values thus represent the K_J fracture toughness in the constraint dependent regime. K_{Ji} represents the value for specimen with 0.2 mm ductile crack extension and temperature t_{DBL} the lowest temperature for the occurrence of this initiation micromechanism.

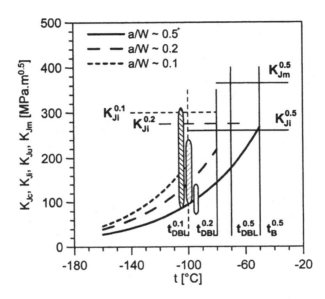

Figure 5. Crack length effect on fracture toughness data (K_{Jc}) in the transition region

Similar fracture toughness - temperature dependencies for bend specimens with shallow cracks (a/W ≈ 0.2 and a/W ≈ 0.1) have been obtained including the larger sets of data generated at -100 °C (common to all specimen geometries) and at the other selected temperature (at -80 and -85 °C, respectively). Exponential curves have been fitted through all three specimen configurations and sets of fracture toughness data (K_{Jc}) in transition and lower shelf region. These fits are introduced in *Figure 5*. It can be seen from this figure that very similar fracture behaviour as in the case of standard crack length (full curve) can be identified in the transition region of specimens with shallow cracks, and only quantitative differences in fracture characteristics could be observed.

The distribution of principal tensile stresses was calculated (3D) for all specimen geometries tested at -100°C and all fracture loads measured. An attempt has been made to estimate the local cleavage fracture stress for all specimen geometries tested. This stress was calculated at the distance of the fracture initiation point as obtained from quantitative fractographic analyses. Average values of about 1255 MPa have been estimated for specimens having standard crack length. Similarly, the local cleavage fracture stress of about 1340 MPa was obtained for specimens having a/W ≈ 0.2.

It is possible to draw an important conclusion from the analysis of stress distributions below the crack tip. For transparent view of dependence distributions of maximum principal stress for all types of specimens on crack length was constructed dependence of sigma peak values on load. This dependence is shown on *Figure 6*. The sigma peak value is maximum (peak) value on stress – distance below crack tip trace for corresponding load step. From this dependence it is possible make the same findings. For a/W ratio from 0.5 to 0.2 stress slowly increase and then quickly drops. It follows that cleavage fracture stress value probably dropping from a/W ratio equal 0.2 due to size of plastic deformation below crack tip respectively due to dimensions of highly stressed region below crack tip.

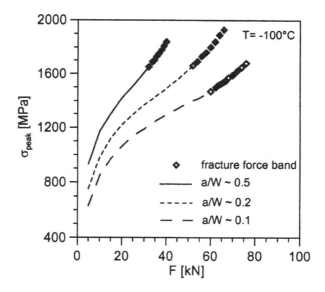

Figure 6. Dependence of sigma peak on load

From comparison of the main trends in fracture toughness temperature dependencies and from the FEM calculation finished until now the following important characteristics of crack length effect (and corresponding crack tip constraint effect) were found:
- A shift of the transition region (the transition temperature t_B) to lower temperatures can be observed for shallower cracks (lower constraint).
- The mean fracture toughness obtained at one test temperature (the same initiation mechanism was proved by SEM for all specimens) is comparably higher for shorter cracks when compared with standard crack length.
- A larger scatter in fracture toughness values can be observed for shorter cracks being strongly influenced by larger crack tip plastic zone (smoother maximum on principal tensile stress distributions below crack tip).

- With decreasing crack length the fracture load and the dimensions of the highly stressed region below the crack tip increase and, in addition, the principal stress corresponding to this region is higher.
- With decreasing crack length the cleavage fracture stress appears to be slightly increased, probably due to larger plastic deformation preceding the fracture initiation.

Using the experimental data obtained on specimen sets tested at -100°C, the dependence of fracture toughness on crack length was obtained as shown in *Figure 7*.

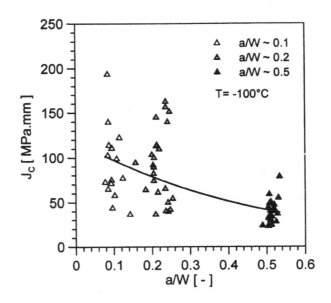

Figure 7: Crack length effect vs. fracture toughness

3.2. STANDARD APPROACH TO Q-PARAMETER DETERMINATION

To explain the obtained dependence of fracture toughness data on crack length (*Figure 7*) it is necessary use two-parameter fracture mechanics approach. The J-Q theory according to Shih and O'Dowd is suitable for constraint phenomena description in this case. Approach is based on describing differences between distributions of stress field under SSY conditions and large scale yielding state. Stress distribution using for standard method for Q-parameter determination is computed by FEM on load level corresponding measured fracture force. Basic definition of distribution difference stress field is given in form

$$\left(\sigma_{ij}\right)_{diff} = \sigma_{ij} - \left(\sigma_{ij}\right)_{SSY, T=0} \qquad (3)$$

Boundary method was used to obtaining distribution of stresses $(\sigma_{ij})_{SSY,\ T=0}$ under SSY condition. For Q-parameter determination is necessary normalised difference stress field by yield stress σ_0 (4). Right value of Q-parameter is evaluated in normalised distance below crack tip given (5) and angle $\theta = 0$. Sufficient conditions for acceptance Q-parameter as parameter describing constraint effect below crack tip is following: the gradient of Q-parameter value evaluated at normalised distance between 1 and 5 is less than 0.1.

$$Q = \frac{\left(\sigma_{ij}\right)_{diff}}{\sigma_0} = \frac{\sigma_{ij} - \left(\sigma_{ij}\right)_{SSY,\ T=0}}{\sigma_0}, \qquad (4)$$

$$r = \frac{2J}{\sigma_0} \qquad (5)$$

The difference stress field characterized by constraint parameter Q is physically interpreted as hydrostatic pressure. Negative (positive) value of Q-parameter means that hydrostatic pressure below crack tip decreasing (increasing) with respect to SSY state.

J-Q dependence determined by standard approach (from basic definition) is in *Figure 8*.

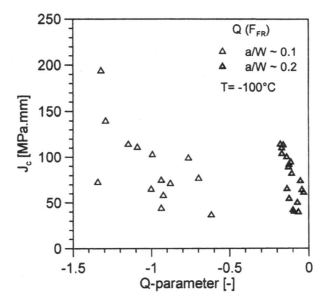

Figure 8. J-Q dependence determined from fracture force - standard approach

3.3. MODIFIED APPROACH TO Q-PARAMETER DETERMINATION

In addition to above-mentioned peculiarities (the slight increase of cleavage fracture stress for shallow cracks in particular) the steel fracture behaviour characterised by high scatter of inherent material properties and extent of plastic deformation preceding the fracture. This behaviour is schematically described in *Figure 9*. The full line represents loading curve calculated by FEM for average input values that characterize the experimental material. Dashed lines define schematically the scatter band of the measured force-deflection trace. For one fracture force is marked the fracture force scatter band ($F_{FEM}\pm2.5\%F_{FEM}$). Relatively small change of measured critical load leads to significant change (about 50% in this case) of fracture toughness value represented by J-integral. That means for the same value of critical load (affected only by inherent scatter) quite different values of J_c-integral have been determined from FEM analyses in comparison with experiment due to feature describing above. In order to overcome the problems two approaches for the Q-parameter determination have been tested: (i) standard calculation arising from the basic definition and fracture loads measured for particular specimens (as was remarked above) and (ii) modified procedure based on more realistic critical parameter of loading - the J integral value.

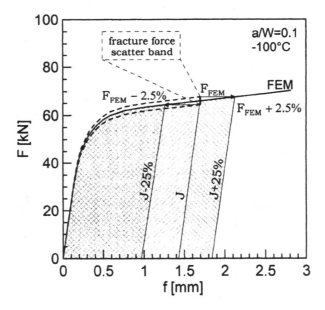

Figure 9. Schema of loading curves

Loading parameter including influence of fracture force value and size of plastic deformation accordant with moment of fracture and can be written in following form

$$Loading\ parameter = \frac{J}{\sigma_0 b}, \tag{6}$$

where J is value of J-integral corresponding to fracture point, b is dimension of unbroken ligament and σ_0 is yield stress. This modification of Q-parameter determination is applicable for similar material behaviour.

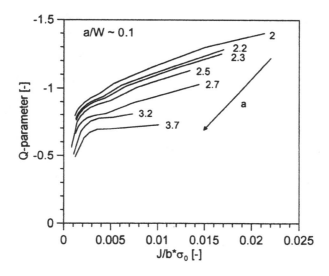

Figure 10. Diagram used for determination of Q parameter from J-integral

Selected curves used for the determination of Q-parameter for particular specimen crack length are shown as example in *Figure 10*. This figure showing easy way to determine Q-parameter value (for specimen of any crack length) if fracture toughness is known. For specimen geometry with shallow cracks (having a/W around 0.1) and given crack length the Q-parameter was read directly from the J values corresponding to critical loads. Results from mentioned modified approach to determination of Q-parameter describing crack tip constraint is shown in *Figure 11*. In this figure were carried out data obtained from both test specimen geometry with shallow cracks at temperature −100°C. The bold triangles are data determined on specimen with a/W   0.2 and the other ones are for a/W   0.1.

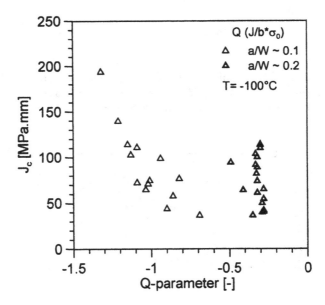

Figure 11. J-Q dependence determined from fracture force - modified approach

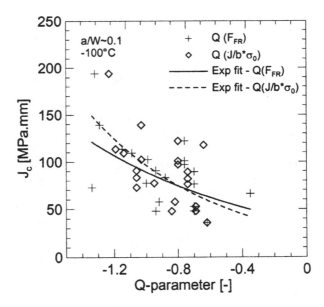

Figure 12. J-Q relation for specimen with shallow cracks and both calculation procedures

Comparison of both used approaches (standard and modified) is shown in *Figure 12* for the crack length a/W ≈ 0.1. The cross points represent the values of Q-parameter determined by standard calculation, the rhombi represent the modified approach. Quite good agreement was found, but further analyses are necessary. By incorporating a much higher number of experimental data the statistical aspect will be necessary to be taken into account before making final conclusions.

4. Concluding remarks

In the presented investigation the fracture resistance of ferritic steel intended for casks for spent nuclear fuels has been analysed combining several approaches.

The characteristics of crack length effect (and corresponding crack tip constraint effect) have been summarised for the steel investigated.

The shift of transition region (towards standard cracks) up to lower temperatures was observed for shallower cracks (loss of constraint).

At one test temperature (-100°C), the mean fracture toughness values comparably higher for specimen with shorter cracks when compared to standard ones (the same initiation mechanism was proved by fractographic analysis).

The larger scatter in fracture toughness values observed for shorter cracks being strongly affected by larger crack tip plastic zone (smoother maximum on principal tensile stress distributions below crack tip). The most important finding is the increase in cleavage fracture stress with decreasing crack length of bend specimens having the same ligament.

A modified approach has been developed for Q-parameter determination taking into account large scatter of inherent material toughness and plastic deformation preceding the fracture.

For these purposes the J-integral has been found to give more consistent data than standard calculation based on critical loads.

The modified approach is useful for Q-parameter determination for materials with similar fracture behaviour as investigated material.

Acknowledgements

The research was financially supported by the grant No. 106/02/0745 of the Grant Agency of the Czech Republic and the project No. 972655 supported in the frame NATO Science for Peace program.

References

1. *Guidelines for Safe Design of Shipping Packages Against Brittle Fracture*, (1993), IAEA -TECDOC-717.

78

2. Warnke, E.P., Bounin, D., (1997) Fracture Mechanics Considerations Concerning the Revised IAEA-TECDOC-717 Guidelines, *Transaction of of 14th Int. Conf. On Structural Mechanics in Reactor Technology*, GMW/6, 571-578.

3. Moulin, D., Yuritzin, T., Sert, G., (1995) An Overview of R&D Work Performed in France Concerning the Risk of Brittle Fracture of Transport Cask, *RAMTRANS*, Vol. 6, No. 2/3, 145-248.

4. Beremin F.M., (1983) *Metal. Trans. A*, Vol. 14A, 2277-2287.

5. Wiesner, C.S., Andrews, R.M., (1997), *A Review of Micromechanical Failure Models for Cleavage and Ductile Fracture*, TWI, rpt. 592/1997.

6. Kozák, V., Novák, A., (2002), The Use of the Local Approach for Brittle Fracture Prediction, Contribution in this Book, 95-112.

7. Ruggieri, C., Dodds, R.H., Wallin, K., (1998) *Eng. Fract. Mech.* Vol. 60, 14-36.

8. Koppenhoefer, K.C., Dodds, R. H., (1997) *Eng. Fracture Mech.*, Vol 58, 224 – 270.

9. *Standard Test Method For the Determination of Reference Temperature T_0 for Ferritic Steels in the Transition Range*, ASTM, E1921-97.

10. Wallin, K, (1997) *In Advances in Fracture Research, ICF 9*, Eds. B. L. KariHaloo, Y. W. Mai, M.I. Ritchie, Pergamon, Amsterdam, 2333-2344.

11. Chlup, Z., Holzmann, M., Dlouhý, I., (1999) Micromechanical Aspects and Methods of Constraint Effect Assessment, *Proc. of Conf. Engineering Mechanics*, Svratka, 315-320.

12. McCabe, D. E., Sokolov, M. A., Nansland, R. K., (1997) In *Struct. Mechanics in Reactor Technology, 14 Int. Conference*, Vol. 4, division G, Lyon, France, 349-356.

13. Al-ani A.M., Hancock J.W., (1991), J-dominance of short cracks in tension and bending, *J. Mech. Phys. Solids*, 39, 23-43.

14. Anderson, T.H., Dodds, R.H., (1991), Specimen size requirements for fracture toughness testing in the transition region, *Inter J. Fracture* 48, 1-22.

15. Chlup, Z.,(2001), *Micromechanical Aspects of ns int Effect*, PhD theses, IPM AS CR.

PROBLEMS IN Q-PARAMETER CALCULATIONS

L. VLČEK, Z. CHLUP, V. KOZÁK
Institute of Physics of Materials, Academy of Science of the Czech Republic, CZ-61662 BRNO, Žižkova 22

Abstract: The present work deals with determination of the sensitivity detached parameters describing fracture behaviour of body with crack with respect to the character change of true stress-strain curve. The typical low carbon cast steel stress-strain curve with dominant region of Lüders deformation was exerted. This paper presents the consideration on the change judgement of J-integral and Q-parameter as the base parameters of two parameters fracture mechanics. The attention is paid on the influence of deformation hardening exponent of the idealised true stress-strain material curve described by the Ramberg-Osgood relation to Q-parameter. All computations are based on the 3D elastic-plastic analyses using FEM and supported by mechanical tests.

Keywords: stress-strain curve, deformation hardening, J-integral, Q-parameter, finite element method

1. Introduction

It seems that for transferring fracture-mechanical data from test specimens to exposed real constructions or to its monitored parts, it is necessary to use two-parameter fracture approach. This requirement involved large investigations, which are considered of the constraint influence near the crack tip to fracture behaviour. Recent extensive investigations on crack tip constraint effects provide a necessity of testing various constraint configurations, such as shallow-cracked SEN(B) specimens. The present paper is contributing to assessment of the influence of constraint effect near the crack tip to fracture parameters.

Determining static fracture toughness of SEN(B) specimens is one of the basic fracture mechanics test. As a result of this test are significant values of static fracture toughness, which depends upon temperature. It must be emphasised that the most important values are critical K-value, in case of using linear-elastic fracture mechanics and critical value of J-integral, in case of using elastic-plastic fracture mechanics. Subsequently we confine our investigation to elastic-plastic material behaviour.

Standard assessments of elastic-plastic fracture behaviour in large engineering structures using laboratory specimen data employ a one-parameter characterization of loading and toughness, most commonly the J-integral or the corresponding value of the Crack Tip Opening Displacement (CTOD). Fracture toughness measured on one specimen size can be directly at some circumstances transformed to another geometry. This approach can be valid only in the case of small scale yielding (SSY).

79

I. Dlouhý (ed.), Transferability of Fracture Mechanical Characteristics, 79–92.

Experimental measurements have shown significant elevations in the elastic-plastic fracture toughness in the transition region for shallow crack specimens of ferritic steels. This apparent increased toughness of ferritic steels has enormous practical implication in defect assessment. Elastic-plastic stress fields along the crack front depend strongly on the specimen geometry, size, loading mode and material flow properties. A more realistic description of crack tip stress and deformations fields has been developed. Approaches are based on two-parameter characterization of crack tip fields, such T- stress and Q-stress. In both approaches, J sets the magnitude of near tip deformation, while the second parameter characterizes the level of stress triaxiality. These J-T and J-Q approaches retain contact with traditional fracture mechanics. Laboratory measurements on the specimens with varying crack length (changing the relation a/W) and with the same ligament showed increasing values of fracture toughness expressed using J_c versus decreasing crack length. Following the idea of Sumpter [1], Kirk and Dodds [2] investigated several possibilities of J-integral and CTOD estimation for SEN(B) specimens with shallow crack. Presented results summing up that critical values of J-integral and CTOD obtained on SEN(B) specimens with different crack length are comparable due to establishing parameter η_p, where $J = J_{el} + J_{pl}$ and $J_{pl} = \eta_p U_p / Bb$. This parameter is depended on ratio a/W and represents the influence of crack length on fracture toughness. For standard crack length is $\eta_p = 2$.

For fracture toughness valuation on the basis of two-parameter fracture mechanics the evaluation of parameters, which expresses this constraint ahead the crack tip in our case Q-parameter, is critical. Several approaches exist:

a) On the base of experimentally determined dependence J_c on a/W the Q calculation comes after from numerically given stress fields received by FEM for every analysed body separately. This approach is time consuming due to experimental work and next modelling and computation.

b) Statistical approach using so called local approach [3]. We limit our focus to a stress controlled, cleavage mechanism for material and adopt the Weibull stress (σ_w) as the local parameter to describe crack-tip conditions. Unstable crack propagation occurs at a critical value of (σ_w) which may be attained prior to or following some amount of stable, ductile crack extension [4].

c) Function $J_c(Q)$ can be found on the basis of so called toughness scaling models. The procedure focuses on an application of the micromechanical model to predict specimen geometry and crack effects on the macroscopic fracture toughness J_c Dodds [5] and Anderson [6]. The procedure requires attainment of equivalent stressed volumes ahead of a crack front for cleavage fracture in different specimens. The D-A model does not reflect such variations, with equal weight attributed to all material volumes having $\sigma_1/\sigma_o > \sigma_c$. This can be done e.g. on the base of Weibull stress, because the Weibull stress incorporates both the effects of stressed volume (as in the D-A model) and the potentially strong changes in the character of the near tip stress fields due to constraint loss and ductile crack extension [7].

The present analyses are generally based on the FEM calculations. To obtain the correct values of stress-strain fields using real materials model seems to be fundamental for effective numerical analysis. As an experimental material manganese cast steel was used (ČSN 42 2707). Two procedures, how to approximate the stress-strain curve, were used. The first model describes how to more precisely express the real curve (experimentally determined) using the incremental theory of plasticity. The curve can be divided into three basic parts. The initial part is linear elastic, next is the part of dominant Lüders deformation and the last part for deformation hardening. The second model is based on the deformation theory of plasticity and constitutes continual hardening material.

The change estimation dependence of J integral and Q parameter on the material stress-strain curve can be very expressive. Therefore this study describes a computational framework to quantify the influence of the stress-strain approximation to the J integral values and Q parameter. The stress is put onto the ability of using the deformation theory of plasticity, which actually describes not correctly the materials with dominant yield stress area, with comparison to incremental theory of plasticity. The reason why to use this approximation is given by the capability of finite elements to transport the high values of plastic strain. The next problem, which is solved, is the selection of proper value of deformation hardening exponent to the stress-strain field ahead the crack tip and to the force-displacement diagram.

2. Q-parameter Theory and its Determination

Two-parameter approaches to elastic-plastic fracture mechanics were introduced to remove some of the conservatism inherent in the one-parameter approach based on the J integral and to account for observed size effect on fracture toughness. This paper presents the consideration on the change judgement of J-integral and Q-parameter as the base parameters of two-parameter elastic-plastic fracture mechanics. Q-parameter is the second parameter in two-parameter approach and it is used for describing constraint effect near the crack tip. In accordance with [8], [9] Q-parameter is defined at point $\theta=0$, $r=2J/\sigma_0$ as the participation of difference stress field $(\sigma_{ij})_{diff}$ and yield stress σ_0:

$$Q = \frac{\left(\sigma_{ij}\right)_{diff}}{\sigma_0} = \frac{\left(\sigma_{ij}\right)-\left(\sigma_{ij}\right)_{ref}}{\sigma_0} \qquad (1)$$

It is possible to say [10], [15] that difference stress field is constant in region $r=<J/\sigma_0$, $5J/\sigma_0>$ for $\theta \leq \pi/2$. This stress value is difference between real stress fields, which are obtained usually from numerical analyzes, and reference stress field. Different methods of the Q-parameter computation are given by using different types of reference stress fields. The most commonly used approaches for obtaining reference stress field values are boundary layer method (BLM) [10], [12] and method that is based on using HRR solution. The second one arises from works of Hutchinson [13], Rice and Rosengren [14], who generalized Williams' solution (for linear elastic material) into nonlinear hardening material. Its behavior is determinated by deformation theory of plasticity.

2.1. HRR SOLUTION

Determination technique of the difference stress field is based on elastic-plastic solution of the specimen with crack. Thus, the real stress (σ_{ij}) near the crack tip is given as an approximate stress in case that sufficiently fine mesh was used. In accordance with [9] the difference stress field is given by the relation:

$$\left(\sigma_{ij}\right)_{diff} = \left(\sigma_{ij}\right) - \left(\sigma_{ij}\right)_{HRR} \tag{2}$$

where $(\sigma_{ij})_{HRR}$ is the reference field stress. Values of the reference stress are tabulated as a function of stress intensity factor K_I, yield stress σ_0 and deformation hardening coefficient n. The value of the Q-parameter is:

$$Q = \frac{\left(\sigma_{\theta\theta}\right) - \left(\sigma_{\theta\theta}\right)_{HRR}}{\sigma_0} \tag{3}$$

Advantage of this method is the fact that $(\sigma_{\theta\theta})_{HRR}$ values are tabulated, but only for a small range of specimens with simple geometry. To the contrary the disadvantage of this method could be the fact that the material behavior is not described well.

2.2. BOUNDARY LAYER METHOD (BLM)

BL method is suitable for any types of geometry configurations. This method is based on analysis of the maximal principal stress ahead of the crack tip using finite element method (FEM). The small local analyzed area with the crack is removed from the whole body (*Figure 1*). The size of the area has to be chosen efficiently because of generating small scale yielding (SSY) conditions on the boundary of this area. The boundary conditions are given by displacements from elastic solution of the whole body [10]:

$$u = \frac{K_I(1+\nu)}{E} \sqrt{\frac{r}{2\pi}} \cos\frac{\theta}{2} (3 - 4\nu - \cos\theta) \tag{4}$$

$$v = \frac{K_I(1+\nu)}{E} \sqrt{\frac{r}{2\pi}} \sin\frac{\theta}{2} (3 - 4\nu - \cos\theta) \tag{5}$$

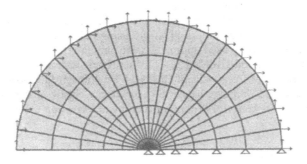

Figure 1. Schema of BL model

where u, v are displacements in directions of X and Y in Cartesian system of coordinates, K_l is stress intensity factor, which was determined from Rice's J integral, E is Young's modulus and v is Poisson's ratio.

As follows the elastic-plastic calculation for the whole body with crack is solved to obtain values of the "real" stress. The difference stress field is given by:

$$(\sigma_{ij})_{diff} = (\sigma_{ij}) - (\sigma_{ij})_{SSY} \qquad (6)$$

where (σ_{ij}) is real stress in the specimen, which is determined by using elastic-plastic FEM solution and $(\sigma_{ij})_{SSY}$ is reference stress field, which is determined from BLM calculation. In accordance with works of [9] and [10] there are two definitions for Q-parameter computation:

$$Q = \frac{(\sigma_{\theta\theta}) - (\sigma_{\theta\theta})_{SSY}}{\sigma_0} , \quad Q_m = \frac{(\sigma_m) - (\sigma_m)_{SSY}}{\sigma_0} \qquad (7, 8)$$

for $\theta=0$, $r=2J/\sigma_0$, where $(\sigma_{\theta\theta})$ is stress in cylindrical system of coordinates and for $\theta =0$ it is the same value as (σ_{yy}) in Cartesian system of coordinates and

$$\sigma_m = \frac{1}{3}(\sigma_{xx} + \sigma_{yy} + \sigma_{zz}) \qquad (9)$$

where σ_{xx}, σ_{yy}, σ_{zz} are stresses in particular axis in Cartesian system of coordinates.

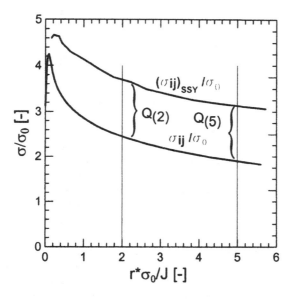

Figure 2. Schema of Q-parameter determination

The theory said that Q-parameter value is constant on defined interval, but in fact the distance between $Q_{(2)}$ and $Q_{(5)}$ have to be less than 10 % (see *Figure 2*).

84

3. Problem and Aim Formulations

For numerical simulation it is necessary to determine model geometry, boundary and load conditions, as well as the way of material model. Three sets of SEN(B) specimens were tested and modelled. The geometry of the specimens is shown in *Figure 3* and its characteristic dimensions are in the TABLE I (all values are in mm).

TABLE I. Dimensions of test specimen

	a/W=0.1	a/W=0.2	a/W=0.5
L	120	140	250
B	25	25	25
W	26	30	50
a	2.5	7	25.25
l	104	120	200

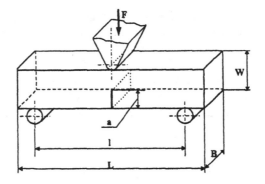

Figure 3. The test specimen

The generalised material behaviour considered in this study is that typically found in low carbon steels. The elastic part is followed by a perfect plateau (this part is commonly called Lüders deformation) and part with work hardening (see *Figure 4*). There are two ways in computational system Abaqus, which can be used for describing material behaviour. The first one is to use déformation theory of plasticity. This material model is trouble-free for numerical calculations. In that case Ramberg-Osgood well-known relation defines material. The second one describes material by using incremental theory of plasticity.

The main question is to show the influence of the stress-strain curve approximation on the fracture parameters. The aim of presented work is confrontation of fracture parameters, which are used for assessment pre-cracked bodies in three-point bending. Attention is paid on the influence of deformation hardening exponent to the history of the idealised true stress-strain material curve.

4. Experiments and Modelling

As an experimental material C-Mn cast steel was used. This material was modelled as homogenous and isotropic with elastic constants $E=2,05.10^5$ MPa and $\nu=0.3$. The average value of yield stress was 360 MPa. The testing temperature was -100 °C.

In the case of using incremental theory of plasticity the curve $\sigma-\varepsilon$ was modelled by 23 points, which were connected to linear parts. These points belong to experimental measured stress-strain curve.

In the case of using deformation theory of plasticity material was described by Ramberg-Osgood relation:

$$\frac{\varepsilon}{\varepsilon_0} = \frac{\sigma}{\sigma_0} + \alpha\left(\frac{\sigma}{\sigma_0}\right)^n \tag{10}$$

where   n   is hardening exponent,
         α is hardening coefficient,
         ε_0 is yield strain,
         σ_0 is yield stress.

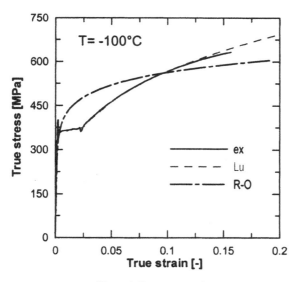

Figure 4. True stress-strain curve

All computations are based on 3D elastic-plastic analysis using FEM, concretely Abaqus version 6.1. 3D model of specimen is shown in the *Figure 5*. It is only one quarter of real body because of using two planes of symmetry. Models were meshed with eight-node hybrid elements provided by the finite element code Abaqus. 15 680 of elements (C3D8H) were used, which means 17884 nodes. The figure below (*Figure 6*) shows the detail with the crack. All solutions have to be evaluated at a distance of cJ/σ_0 ahead of the crack tip (c=1-5). This region of interest is very small and stress and strain gradients are steep. Hence, a very fine mesh is required. Element size is increased as the radial distance is increased from the crack front.

Outer radius of the area (*Figure 6*) is 0.1 mm and the crack tip radius was 0.01mm. Twelve elements were used for dividing this radius. Thus the characteristic element length was $8,3.10^{-4}$ mm. Ten elements in the direction of thickness were used.

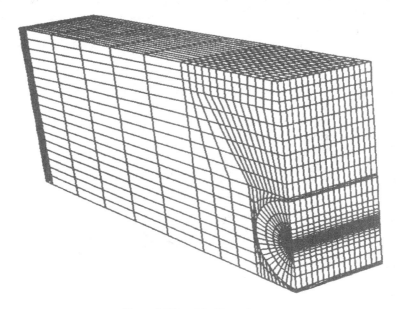

Figure 5. 3D model with mesh

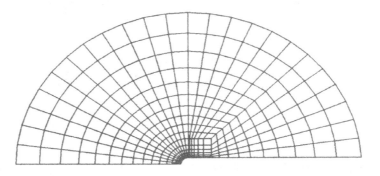

Figure 6. Detail of the crack tip

The real system consists of the specimen, the punch and rollers. These three parts are in contact and that is why the contact between the specimen and punch and between rollers was modelled. The length of elementary face and contact pressure were determined. 3D contact solutions need a long time for computation and that a why the equivalent pressure was applied on the equivalent face (*Figure 7*).

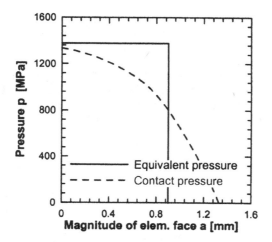

Figure 7. Contact and equivalent pressures (a/W=0.2)

Four calculations were solved for range of value n in case of using deformation theory of plasticity because of choosing the proper value n. In the *Figure 8* there are force-displacement curves for various numerical calculations. The bolt-dashed lines indicate high and low experimental dependencies. It seems that all progressions for linear area (force value to 40kN) are in good agreement with experimental values. To the contrary force-displacement relations are rather different for area with overvalue forces.

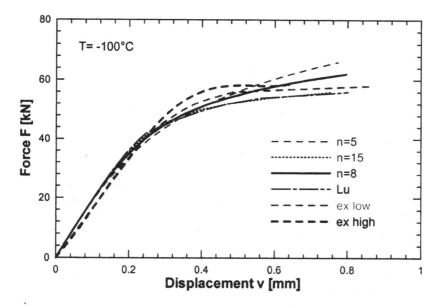

Figure 8. Force-displacement relation

The selection of proper value n was based on comparing the relations between force and displacement. After that value of hardening exponent n=8 was applied as the best fit, as well as the value of hardening coefficient α=1. Hence the model of continual hardening material was determined. Its behaviours are very similar as with real material with Lüders strain region. Fracture forces were in interval 35, 65, 75 kN for specimens with ratio a/W=0.5, 0.2, 0.1. In the case of modelling material behaviour by incremental theory of plasticity fracture forces were too high and solutions were not converged because of destruction of small elements near the crack tip due to achievement high values of plastic strain.

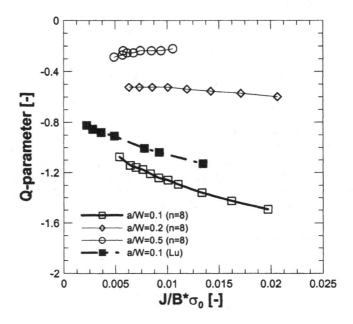

Figure 9. J-Q diagram

Results from numerical calculations are compared in the *Figure 9.* Apparently, there are differences between Q-values due to using two ways of modelling the same material. In this case, the incremental theory of plasticity could not be right because of numerical errors, which appeared near the crack tip. It must be emphasised, that Q-parameter value determined as odds between real stress from elastic-plastic solution and reference stress normalized by yield stress value could be taken as parameter for assessment bodies with crack, but only in case, that the same theory of plasticity and the same material stress-strain curve will be used.

Relation in the *Figure 10* was constructed for check calculation of the influence of the specimen geometry to maximal principal stress near the crack tip distribution. Distribution regions of maximal principal stresses from numerical calculations were constructed for fracture forces interval. Essentially, when the applied force increases, the global stress level increases as well. As to differences in maximal principal stresses between specimens, at first maximal principal stress increases when crack length decreases (a = 25 mm (a/W = 0.5) and a = 7 mm (a/W = 0.2)). After that the maximal principal stress decreases in case of specimen with shallow crack a = 2.5 mm (a/W = 0.1).

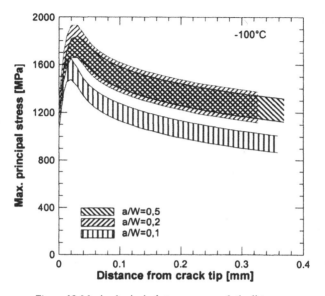

Figure 10. Maximal principal stresses vs. crack tip distance

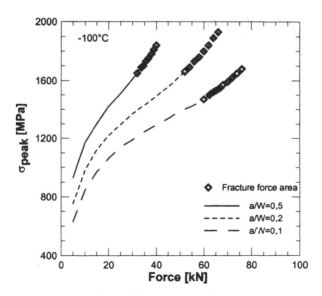

Figure 11. Force-stress relation

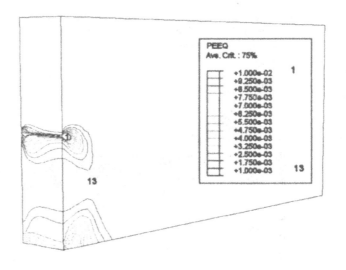

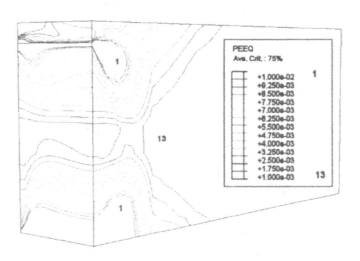

Figure 12. Plastic deformation for standard (a/W=0.5) and small (a/W=0.1) specimens

For better understanding stress trends ahead of the crack tip the relation between maximal stress values and fracture force level was constructed (*Figure 11*). These trends are given by changing boundary conditions near the crack tip. There are no changes in the direction of specimen thickness, because the same thickness of specimen and yield stress value is maintained. The same situation is in the direction of crack propagation, because the same ligament is sustained too. The main serious change is reducing material volume against of crack propagation direction. Plastic zone monitoring is remarkably interesting for systematic and complex assessment in sphere of elastic-plastic fracture mechanics. It stands to reason, that larger plastic zone appeared in case of shallow crack due to higher fracture forces. The magnitude of the plastic zone decreases with increasing crack length. Account of stress decreasing in case of shallow crack is not unambiguous. But the main difference between standard and small specimens is shown in the *Figure 12*. The plastic deformation is still closed in material of standard specimen, but to the contrary plastic deformation flows on the surface of small specimen and stress relaxation process is available.

5. Conclusions

It could be said that on the basis of numerical stress-strain analyses bodies with crack it may be possible to determine suitable substitution of "real" material by continuously hardening material model. This material model is trouble-free for all numerical applications. Remarkably significant is the fact that using of deformation theory reduced computational time more than ten times. Results obtained from incremental method could be remarkably inaccurate arising high plasticity near the crack tip. It must be stressed, that Q-parameter value determined as odds between real stress from elastic-plastic solution and reference stress normalized by value σ_0 could be taken as parameter for assessment bodies with crack, but only in the case, that the same theory of plasticity and the same material stress-strain curve will be used in all solved problems. A new important piece of knowledge is the fact that the maximal principal stress ahead of the crack tip for the short crack specimen ($a/W = 0.1$) is markedly lower to the other crack lengths. Geometric changes of specimens advert to possible change of the critical fracture stress. This hypothesis was sustained by fractography analysis of fracture surfaces, which extend approximately the same distances of initiation areas in all solved geometries.

Acknowledgements

This research was financially supported by grant 101/00/0170 Grant Agency of the Czech Republic and by project Nato SfP 972655 .

References

1. Sumpter, J. D. G. (1987) J_c Determination for Shallow Notch Welded Bend Specimens, *Fatigue and Fract. of Eng. Mater. Struct.* **10**, 479-493.

2. Kirk, M. T. and Dodds, R. H. (1993) J and CTOD Estimation Equations for Shallow Crack in Single Edge Notch Bend Specimens, *Journal of Testing and Evaluation* 21, 228-238.
3. Beremin (1983) A Local Criterion for Cleavage Fracture of a Nuclear Pressure Vessel Steels, *Metallurgical Transaction*, 14A, 2277-2287.
4. Kozák, V., Dlouhý, I. (1998) Local Approach and FEM, *Computer Assisted Mech. and Eng. Sci.* 5, 193-198.
5. Dodds, R. D., Shih, C. F., Anderson, L. (1993) Continuum and Micromechanics Treatment of Constraint in Fracture, *Int. Journal. of Fract.* 64, 101-133.
6. Dodds, R.H., Tang, M., Anderson, L. (1995) *ASTM STP 1299*, 101-133.
7. Ruggieri, C., Dodds, R. D. (1996) A Transferability Model for Brittle Fracture Including Constraint and Ductile Tearing Effects: a probabilistic approach, *International Journal of Fracture* 79, 309-340.
8. Shih, C. F., O'Dowd, N. P. (1994) Two-Parameter Fracture mechanics: Theory and Applications, Fracture Mechanics, *ASTM STP 1207*, 21-47, 379-386.
9. Henry, B. S., Luxmoore, A. R. (1997) The Stress Triaxiality Constraint and the Q-Value as a Ductile Fracture Parameter, *Eng. Fracture Mech.* 57, 375-390.
10. Nevalainem, M., Dodds, R.H. (1995) Numerical Investigation of 3-D Constraint Effects on Brittle Fracture, SE(B) and C(T) Specimens, *Int. J. of Fract.* 74, 131-161
11. Shih, C. F., O'Dowd, N. P., Kirk, M. T. (1993) A Framework for Quantifying Crack Tip Constraint, Constraint Effects in Fracture, *ASTM STP 1171*, 2-20.
12. Landes, J. D. (1997) Application of a J-Q Model for Fracture in the Ductile-Brittle Transition, *Fatigue and Fracture Mechanics* 27, *ASTM STP 1296*, 27-40.
13. Hutchinson, J. W. (1968) Singular Behaviour at the End of Tensile Crack Tip in a Hardening Material, *Journal of the Mechanics and Physics of Solids*, 13-31.
14. Rice, J. R., Rosengren, G. F. (1968) Plane Strain Deformation Near a Crack Tip in a Power-Law Hardening Material, *J. of the Mech. and Physics of Solids*, 1-12.
15. Anderson, T. L. (1995) Fracture Mechanics Fundamentals and Applications, *Second Edition CRC Press*.
16. Hibbit, Karlsson & Sorenson Inc. (2000) *ABAQUS User's Manual, Version 6.1*.
17. O'Dowd, N. P. (1995) Applications of two Parameter Approaches in Elastic-Plastic Fracture Mechanics, *Engineering Fracture Mechanics* 52, 445-465.
18. Boothmann, D. P., Lee, M. M. K., Luxmoore, A. R. (1997) A Shallow Crack Assessment Scheme for Generalised Material Behaviour in Bending, *Engineering Fracture Mechanics* 57, 493-506.
19. Zhang, Z. L., Thaulow, C., Hauge, M. (1997) Effects of Crack Size and Weld Metal Mismatch on the Haz Cleavage Toughness of Wide Plates, *Engeneering Fracture Mechanics* 57, 653-664.
20. Kim, Y. J., Schwalbe, K. H. (2001) On the Sensitivity of J Estimation to Materials' Stress-Strain Curves in Fracture Toughness Testing Using the Finite Element Method, *Journal of Testing and Evaluation* 29, 18-30.
21. Chlup, Z. (2001) *Disertation*, ÚFM AV ČR
22. Pavankumar, T. V., Chattopadhyay, J., Dutta, B. K., Kushwaha, H. S. (2000) Numerical investigation of crack-tip constraint parameters in two-dimensional geometries, *Int. Journal of Pressure Vessels and Piping* 77, 345-355.
23. Betegón C., Hancock J. W. (1991) Two-parameter Characterization of Elastic- Plastic Crack-Tip Fields, *Journal of Applied Mechanics* 58, 104-110.

CONSTRUCTION OF *J-Q* LOCUS FOR MATERIAL OF REACTOR PRESSURE VESSEL WWER 440 IN THE DUCTILE-BRITTLE TRANSITION REGION

D. LAUEROVÁ[1)], Z. FIALA[2)], J. NOVÁK[1)]

[1)] *Nuclear Research Institute Řež, plc., 250 68 Řež, Czech Republic*
[2)] *Institute of Theoretical and Applied Mechanics, Academy of Sciences of the Czech Republic, Prosecká 76, 190 00 Praha, Czech Republic*

Abstract: In the paper effect of shallow crack (in-plane constraint) on fracture toughness of reactor pressure vessel WWER 440 is examined, using two-parameter fracture mechanics approach. For this purpose, experimental tests were performed on 18 specimens of SEN(B) type and 18 specimens of SEN(T) type, for three relative crack depths $a/W = 0.09$, 0.16 (0.17), 0.5, for each type of loading. The tests were performed at the temperature near reference temperature T_0 determined according to Master Curve concept, in the ductile-brittle transition region. Based on 2D and 3D FE calculations, values of two fracture parameters J and Q were determined. Due to different forms of stress fields found for the two types of loading, the Q-parameter was examined for different values of normalized distance from crack front, i.e. $r.\sigma_0/J = 2, 4, 6, 8$ and 10. Based on a reasonable criterion, the J-Q locus for the examined steel was suggested. Besides of this, also the shift of Master Curve reference temperature due to shallow crack effect was examined and compared with similar results valid for steel A533B.

Keywords: reactor pressure vessel, ferritic steel, two parameter fracture mechanics, J-Q locus, constraint, Master Curve, ductile-brittle transition region

1. Introduction

In this paper, the problems associated with fracture of reactor presser vessel WWER 440 steel, i.e. ferritic steel 15Ch2MFA, in the transition region are dealt with. Two-parameter fracture mechanics, in particular the J-Q approach [1,2], represents one way how to predict fracture toughness J_c for a structure, in dependence on constraint level expressed by the Q-parameter. Despite the fact that to date the RPV integrity assessments are based on a one-parameter conservative approach using only fracture toughness obtained from tests with full constraint (specimens with deep cracks, preferably under conditions of plane strain) the attempts are made how to decrease the conservatism contained in this approach by establishing and using the real fracture toughness dependence on constraint level. In this paper we focused on investigating the

I. Dlouhý (ed.), Transferability of Fracture Mechanical Characteristics, 93–104.

effect of in-plane constraint (i.e. effect of shallow cracks) on fracture toughness together with examining the effect of type of loading (bending vs. tension).

2. Experimental

Fracture toughness tests were performed on 18 specimens of SEN(B) type of dimensions (in mm) 25 (width) x 12.5 (thickness) x 120 (length) and on 18 specimens of SEN(T) type with middle part of length 40 mm reduced so as to have dimensions 12.5 (width) x 10 (thickness) in order that loading capacity is sufficient for failure of the specimens (*Figure 1*). For each loading type, three crack depths were tested: deep ($a/W \sim 0.5$), shallow ($a/W \sim 0.16$, resp. 0.17) and very shallow ($a/W \sim 0.09$). Thus, 6 specimens were tested for each type of loading and each type of the crack. All specimens were tested at temperature $T = -98$ °C, being equal to the reference temperature T_0 determined for the steel concerned according to multi-temperature Master Curve concept [3]. During testing some of the specimens, mainly those containing shallow crack, underwent small amount of ductile tearing. Summary of mean ductile tearing amounts for individual specimens are attached in the TABLE I. All specimens failed by cleavage. During experiments, both CMOD and force values were recorded. Experimental records for all specimens tested are plotted in *Figure 2*.

TABLE I. Ductile tearing amounts and fracture toughness values for specimens (B=bending, T=tension)

Loading type (specimens No.)	a/W	ductile tearing [mm]	fracture toughness J_c [kJm^{-2}]
B (18 – 23)	0.09	0, 0.04, 0.27, 0.34, 0.34, 0.41	97.6, 115.5, 369, 366, 420.7, 460.4
B (26 – 31)	0.16	0.21, 0.13, 0.34, 0.21, 0.06, 0.08	381.5, 245.5, 370.3, 343.6, 122.1, 170.9
B (32 – 37)	0.5	0, 0, 0, 0.19, 0, 0	38.9, 23.5, 54.5, 254.1, 51.5, 24.2
T (38 – 43)	0.09	0.4, 0.33, 0, 0.22, 0.31, 0.41	537.8, 320.5, 76.3, 231, 387, 434
T (44 – 49)	0.17	0.11, 0.11, 0.11, 0.08, 0.09, 0.11	223.5, 217, 195, 184.8, 281.4, 192
T (50 – 55)	0.5	0, 0.06, 0, 0, 0.07, 0	101.5, 237.3, 97.9, 158.7, 197.2, 105

3. Material Properties

In FE calculation the following values (relevant to temperature $T = -98$ °C) of material parameters were used: the yield stress $\sigma_0 = 618.2$ MPa, ultimate tensile strength $R_m = 792.9$ MPa, uniform elongation $A_m = 10.5$ %, Young modulus $E = 210.0$ GPa and Poisson ratio $v = 0.3$.

Figure 1. Geometry of the tension and bending specimens

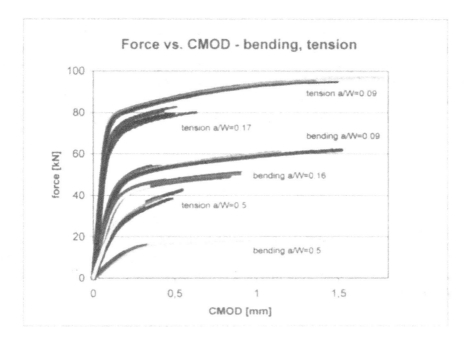

Figure 2. Experimental records - CMOD vs. force values

4. Numerical Analysis of the Experiments

2D and 3D elastic-plastic analyses of the experiments were performed, using FE codes SYSTUS and ANSYS. While for tension specimens the 2D generalized plane strain calculations produced force vs. CMOD curves that were in satisfying accordance with the experimental ones, for bending specimens large discrepancies appeared. The discrepancies disappeared after performing 3D calculations for these specimens.

In 2D quadratic meshes, the crack was modelled as a notch of radius 1 μm, the element size nearest to the crack tip was 0.1 μm. In 3D quadratic meshes the sharp crack was modelled, with radial type of mesh in the first layer of elements adjacent to the crack front and with element size in the vicinity of crack front of 0.01 mm.

Elastic-plastic behaviour of specimens was modelled using flow theory of plasticity with von Mises yield surface and isotropic hardening. Large strains were used.

4.1. METHODS USED IN DETERMINATION OF FRACTURE TOUGHNESS J_C AND Q-PARAMETER

Since the experimental records within one group of six specimens representing certain combination of loading type and relative crack depth did not exhibit large scatter (*Figure 2*), only one FE calculation was performed for each group. Fracture toughness J_c was determined as a critical value of J-integral at the moment of fracture, the CMOD value being used as a criterion. In all cases the accordance between experimental and calculated curve force vs. CMOD was either good or at least acceptable. For determination of J-integral either the 2D Rice contour integral method or the 3D G-θ method was applied. J-values were determined always on the symmetry plane.

The Q-parameter was determined in accordance with J-Q theory using the definition of Q-parameter as follows (σ_{yy} means stress opening the crack, σ_0 denotes the yield stress):

$$Q = \frac{\sigma_{yy} - (\sigma_{yy})_{SSY,T=0}}{\sigma_0} \qquad (1)$$

at $\theta = 0$ and $2 \leq r\sigma_0/J \leq 10$,

where r is distance from the crack front and $\theta = 0$ means that Q is calculated in the symmetry plane.

4.2. FE EVALUATIONS OF EXPERIMENTS

In FE evaluation of experiments small amounts of ductile crack growth were neglected. For tension specimens, practically no path dependence of J-integral was found.

For bending specimens, large path or parameter dependence of J-integral was initially found. This path dependence was later partially removed by considering only paths which do not cross the region of large plasticity arising due to the support of the specimen loaded by 3P-bending (in this region plasticity arises as a consequence of large compressive stresses and is not related to the crack growth).

Fracture toughness values for individual specimens are summarized in Table 1. To determine the Q-parameters, variations of stress opening the crack vs. normalized distance $r\sigma_0/J$ were calculated, together with 2D SSY (with T-stress = 0) reference solution. The resulting variations of Q-parameter vs. $r\sigma_0/J$ are presented in *Figures 3–6*.

From *Figures 3–6* it is seen that while Q-parameters for tension specimens scale well for smaller values of $r\sigma_0/J$ (within some interval enclosing value $r\sigma_0/J = 2$), the Q-parameters for bending specimens scale well for larger values of $r\sigma_0/J$ (within some interval enclosing value $r\sigma_0/J = 8$). In particular, from results for tension specimens with $a/W = 0.09$ (*Figure 5*) it is seen that Q-parameters scale well in interval approximately equal to (1.3, 2.2) where the correct relationship between loss of constraint and loading level (at fracture) takes place: Q is the more negative, the higher is the load that the specimen withstood. Out of this interval this relationship is no more valid.

Also for bending specimens the correct relationship between loss of constraint and loading level (at fracture) is found in some interval of values $r\sigma_0/J$, but the most pronounced loss of constraint occurs in interval approximately equal to (6, 10).

Accepting reasonable criterion that values of Q should be determined in a point (or in an interval, if it is possible) where the scaling effects are most pronounced, the logical conclusion is to calculate the Q-parameters in $r\sigma_0/J = 2$ (or in some point near to 2) for tension specimens and to calculate the Q-parameters in $r\sigma_0/J = 8$ (or in some point near to 8) for bending specimens.

To express the obtained findings in terms of J-Q locus, we attach here two plots of J-Q locus: J-Q locus with Q evaluated in $r\sigma_0/J = 2$ for both tension and bending (*Figure 7*), and J-Q locus with Q evaluated in $r\sigma_0/J = 2$ for tension specimens and in $r\sigma_0/J = 8$ for bending specimens (*Figure 8*).

Similar J-Q loci as in *Figure 8* were constructed also for Q evaluated in $r\sigma_0/J = 2$ for tension specimens and in $r\sigma_0/J = 6$ or 10 for bending specimens, with no significant change in the shape of the locus.

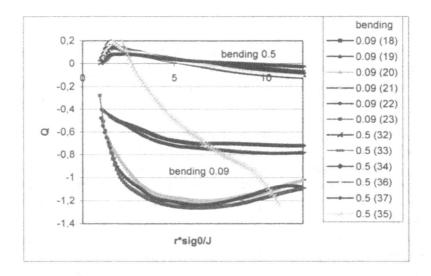

Figure 3. The Q-parameter vs. normalized distance $r\sigma_0/J$ for bending 0.09 and 0.5

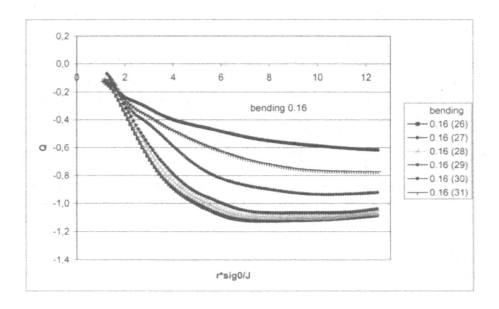

Figure 4. The Q-parameter vs. normalized distance $r\sigma_0/J$ for bending 0.16

5. Relation between Master Curve Reference Temperature and Biaxiality Ratio

Besides constructing the J-Q locus, the obtained experimental data may be used also for determination of relation between Master Curve reference temperature and biaxiality ratio β. Recently some works appeared [4, 5], demonstrating a simple relation between β and reference Master Curve temperature T_0. To perform a similar analysis also for our data, we assume that Master Curve for steel 15Ch2MFA is of the standard shape described in [3]. In what follows the individual series of experiments (tension specimens for 3 values of a/W, bending specimens for 3 values of a/W, each group containing 6 J_c-values) are evaluated. Since the goal of this evaluation is only to determine the shift of T_0 due to shallow crack effect, we neither corrected the data on thickness nor performed censoring of the data, and with respect to this fact we denote the in this way determined reference temperature as T_0^* (instead of T_0) in the following text. Thus, T_0^* was determined using the following formulas:

$$K_i = \sqrt{(J_i E/(1-v^2))}, \ v = 0.3, \ E = 210 \text{ GPa} \tag{2}$$

$$K_0 = 20 + (\Sigma (K_i - 20)^4/6)^{1/4} \tag{3}$$

$$K_{\text{med}} = 20 + (K_0 - 20)(\ln 2)^{1/4} \tag{4}$$

$$K_{\text{med}} = 30 + 70\exp[0{,}019(T-T_0^*)], \ T = -98°C \tag{5}$$

Value of T_0^* was calculated for each of six measurement series. For each series one value of biaxiality ratio β is appropriate, values of β were taken from handbook [6]. The results together with important geometrical characteristics are summarized in Table II.

TABLE II. Values of T_0^* and β for six series of experiments (B=bending, T=tension)

Series	W [mm]	a/W	K_0 [MPa√m]	T_0^* [°C]	β
1-B	25	0.09	280.2	-159.9	-0.385
2-B	25	0.16	257.7	-155.4	-0.295
3-B	25	0.5	164.5	-127	0.118
4-T	12.5	0.09	290.5	-162	-0.494
5-T	12.5	0.17	224.5	-146.6	-0.455
6-T	12.5	0.5	192.4	-137	-0.156

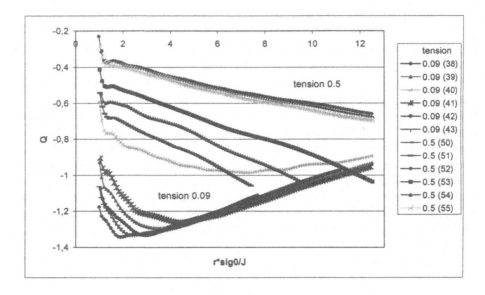

Figure 5. The Q-parameter vs. normalized distance $r\sigma_0/J$ for tension 0.09 and 0.5

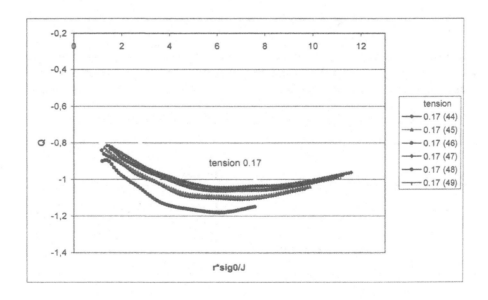

Figure 6. The Q-parameter vs. normalized distance $r\sigma_0/J$ for tension 0.17

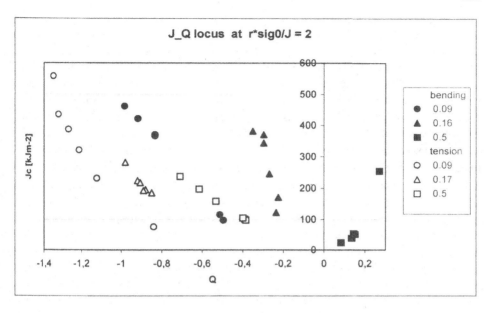

Figure 7. Conventionally determined *J-Q* locus

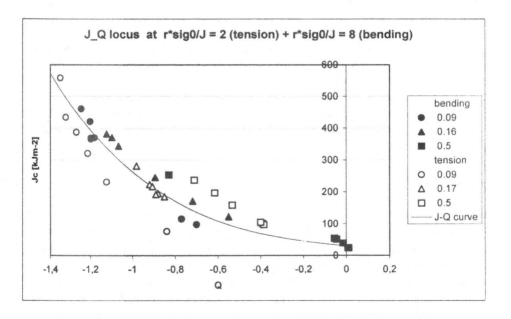

Figure 8. Resulting *J-Q* locus

From *Figure 9* where values of T_0^* vs. β are plotted is seen that a common simple functional dependence cannot be find for both types of loading. This finding is not surprising taking into account different characters of stress and strain fields in the vicinity of crack front for bending specimens on one side and tension specimens on the other side. Confining to bending specimens only (series 1 – 3), qualitatively the same behaviour as that one presented for steel A533B in [4] was found, in particular, a linear relation between β and T_0^*. Quantitative comparison of slopes $\Delta T_0/\Delta\beta$ for steels 15Ch2MFA and A533B is a little more difficult. In *Figure 5* of work [4], values of T_0 vs. β are plotted for specimens CT (1/2T through 1T) and for specimens SEN(B) (0.2T through 1T, both deep and shallow cracks), and it follows from that figure that $\Delta T_0/\Delta\beta = 40°C$. For our data, we obtain $\Delta T_0/\Delta\beta = 65°C$ from comparing series 1 and 3, and $\Delta T_0/\Delta\beta = 69°C$ from comparing series 2 and 3. The difference may be ascribed to the influence of several factors: difference in materials (despite the fact that yield stress and ultimate strength of both materials differ only insignificantly), different sources for determination of β values, and possibly also to the influence of size factor. On the other side, if only selected data of work [4] are considered, in particular, 1/2T SEN(B) data gained from experiments performed at practically the same temperature (-116°C for $a/W=0.52$, and -115°C for $a/W=0.12$), we obtain $\Delta T_0=36°C$, while for our data we obtain $\Delta T_0^*=33°C$ when comparing the series with $a/W=0.50$ and $a/W=0.09$. From this point of view, behaviour of steel 15Ch2MFA seems to be very similar to the behaviour of steel A533B, even quantitatively. The main attention should be focused on answering the question why the same correlation was not found also for tension specimens and, in particular, what type of dependence, if any, of T_0 on geometry of tension specimens exists. The reason for different behaviour of tension specimens from that one of bending specimens may be associated with higher relative level of approaching to limit load in the case of tension specimens in comparison to bending specimens (one tension specimen failed even beyond the limit load, *Figure 2*). This hypothesis requires further examination, including the possibility of using the relative level of approaching to limit load as another parameter in quantifying the effect of constraint on fracture toughness.

6. Conclusions

In this paper two types of J-Q locus for steel of reactor pressure vessel WWER 440 were constructed, in the ductile-brittle transition region. In the first type of J-Q locus (*Figure 7*), the Q-parameters were evaluated in $r\sigma_0/J = 2$ for both tension and bending loading. Within this approach the (J_c, Q) points considered all together do not exhibit common functional dependence. In the second type of locus (*Figure 8*), the Q-parameters are evaluated in $r\sigma_0/J = 2$ for tension loading and in $r\sigma_0/J = 8$ for bending loading. Within this approach all (J_c, Q) points cumulate along a curve with a relatively small scatter. Moreover, intervals of values $r\sigma_0/J$ exist, within which the shape of J-Q

locus does not change significantly when a different value $r\sigma_0/J$ from the interval is selected. Due to these features, the second type of J-Q locus seems to be a more suitable for further possibilities of application in reactor pressure vessel integrity assessment.

As another way of analysis or generalization of results, the shift of Master Curve reference temperature due to both geometry and loading type effects was examined, this way should be also followed further.

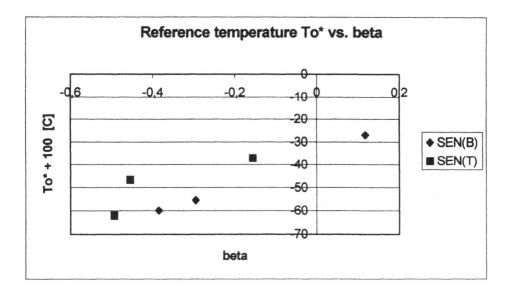

Figure 9. Dependence of reference temperature T_0^* on biaxiality ratio beta

Acknowledgements

The support of the Grant Agency of the Czech Republic through grant No.106/00/1347 is gratefully acknowledged.

References

1. O'Dowd, N. P., Shih, C. F. (1991), Family of Crack-Tip Fields Characterized by a Triaxiality Parameter – I. Structure of Fields, *J. Mech. Phys. Solids* **39**, No.8, 989-1015.
2. O'Dowd, N. P., Shih, C. F. (1992), Family of Crack-Tip Fields Characterized by a Triaxiality Parameter – II. Fracture Application, *J. Mech. Phys. Solids* **40**, No.5, 939-963.

3. ASTM Standard Test Method for Determination of Reference Temperature, T_0, for Ferritic Steels in the Transition Range, ASTM E 1921-97.

4. J.A.Joyce, R.L.Tregoning, (2000), Quantification of specimen geometry effects on the Master Curve and T_0 reference temperature. *13th European Conference on Fracture held in San Sebastián, Spain, 6-9 September 2000.*

5. K.Wallin (2000), The effect of *T*-stress on the Master Curve transition temperature T_0. *13th European Conference on Fracture held in San Sebastián, Spain, 6-9 September 2000*

6. Z.Knésl, K.Bednář, (1997), Two parameter fracture mechanics: Calculation of parameters and their values (in Czech). Edited by Institute of Physics of Materials, Academy of Sciences of the Czech Republic.

THE USE OF THE LOCAL APPROACH FOR THE BRITTLE FRACTURE PREDICTION

V. KOZÁK and A. JANÍK
Institute of Physics of Materials ASCR, Žižkova 22, 616 62 Brno, Czech Republic

Abstract: The aim of the paper is in using the Beremin conception of local approach to fracture resistance assessment. The local approach uses internal microscale variables related to the material damage evolution in order to predict initiation of macroscale cracks in an elastic-plastic regime. This methodology was an invaluable complement to classical fracture mechanics, which is based on a single parameter for characterizing fracture, but not reproduces the transition behaviour satisfactory. The use of Weibull statistics for modelling of defect distribution over characteristic volume of material under a critical state of stress allows predict the probability of failure. Accepting these approaches to the analysis of local criteria for cleavage fracture the location parameter σ_u and shape parameter m were calculated using FEM for notched tensile bars having various type of geometry. The first one was tensile specimen with circumferential notch similar to Charpy V-notch specimen (CVN), the other three types were U shape with radii 1; 0.7; 0.2 mm.

Keywords: Beremin conception, local approach, size effect

1. Introduction

The Weibull stress model was originally proposed by Beremin [1, 2, 3] as a measure of the failure probability of cracked body. It provides a framework to quantify the complex interaction among specimen size, deformation level and material flow properties in fully three-dimensional setting. Weibull stress seems to be a parameter for prediction of cleavage failure of cracked bodies and the study is focused to assess the effects of constraint loss on cleavage fracture toughness (J_c) Pineau [4], Wiesner [5]. The scalar Weibull stress was introduced as a probabilistic fracture parameter, computed by integrating a weighted value of principal tensile stress over the fracture process zone. The Beremin model adopts a two-parameter description for cumulative failure probability (see Eq.1), where m denotes the Weibull modulus and the scale parameter σ_u set the value σ_w at 63.2 percent failure probability.

The Local Approach correlates the fracture probabilities with the stress distribution ahead of the crack or notch tip. Since the cleavage event is predicted by local stresses, the mechanical factors, which influence on cleavage fracture, are included in the calculation of the local criterion. However, the method requires accurate numerical analysis of stress strain state in the structure. The model assumes that the parameters (m, σ_u) describe inherent properties of the material that imply the formation of certain distribution of cracks once plastic deformation occurs as the precursor to cleavage. This

I. Dlouhý (ed.), Transferability of Fracture Mechanical Characteristics, 105–122.

methodology has been an invaluable complement to classical fracture mechanics, which is based on a single parameter for characterizing fracture but not reproduce the transition behaviour satisfactory. Many constitutive models for damage evolution exist for various phenomena of material behaviour. The interest during the last 15 years is restricted to cleavage fracture and ductile tearing of ferritic steels, which occurs either by the formation of microcracks and their extension with little global plastic deformation (brittle or cleavage fracture), or by the nucleation, growth and coalescence of microvoids with significant plastic deformation (ductile rupture). Numerical models exist for both failure phenomena, in particular the Beremin model based on a critical fracture stress concept together with weakest link assumption and Weibull statistics [6], Rice and Tracey void growth [7] and Gurson model [8] or Rousselier model [9] for porous metal plasticity.

The local approach contains application of finite element calculation using very fine mesh to predict fracture. The use of Weibull statistics for modelling of defect distribution over characteristic volume of material under a critical state of stress allows to predict the probability of general unstable fracture. The main steps are determination of the maximum principal stress at the experimentally obtained load level at fracture, calculation of the Weibull stresses at fracture and an iterative maximum likelihood procedure for determination of distribution parameters of the Weibull stress. The effort is concentrated on (i) the use of notched tensile bars and (ii) small test specimen (Charpy V-notch or pre-cracked CVN) for fracture toughness temperature diagram determination including scatter characteristics. An attempt has been made in the present work to compare local parameters generated on specimens with various notch dimensions based on the Beremin conception.

2. Statistical Local Approach

At temperatures in the transition region above the lower shelf the transgranular cleavage fracture develops in the plastic zone from slip-induced cracking of carbides, followed by unstable propagation.

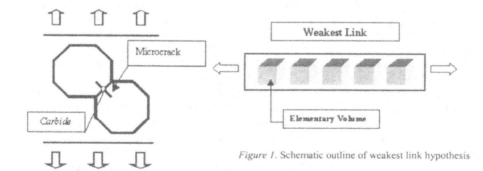

Figure 1. Schematic outline of weakest link hypothesis

The transgranular cleavage fracture is a sequential process involving crack initiation and propagation. In most steels crack nucleation occurs at brittle grain boundary particles (e.g. carbides) due to stress concentration caused by the dislocation pile-ups at these particles. This explains the experimental fact that local plastic deformation always preceded cleavage fracture. The local approach for cleavage is based on the weakest link concept that postulates that failure of the body of a material containing large number of statistically independent volumes is triggered by the failure of one of the reference volume [see *Figure 1*]. Due to microstructural inhomogenity of steel materials, volume-sampling effects play an important role in quantifying the large scatter of the fracture toughness data.

In the local approach to cleavage fracture, the probability distribution (P_f) for the fracture stress of a cracked or notched body is assumed to follow a two-parameter Weibull distribution [1] in the form:

$$P_f(\sigma_W) = 1 - \exp\left[-(\frac{\sigma_W}{\sigma_u})^m\right].$$ (1)

The stress integral over the fracture process zone is denoted as σ_w and is termed the Weibull stress. This stress is defined by:

$$\sigma_w = \left[\frac{1}{V_0}\int \sigma_1^m dV\right]^{1/m},$$ (2)

where m is the so-called Weibull slope, V_0 is the reference volume for the weakest link assumption, the integral is computed over the plastic zone, and σ_1 is the maximum principal stress in elementary volume. The parameters σ_u and m of the Weibull stress σ_w at fracture are material parameters, i.e. independent of the stress state of materials, but may depend on the temperature. Special attention was given to the following problems:

- To find out if the generated local parameters are geometry dependent.
- To test the influence of statistical aspects on the generated local parameters.
- To verify the recommended procedure ESIS P6-98 [11] for the determination of the local parameters.
- Test the scatter in stress-strain behaviour of the material and verify how to model correct material response using finite elements method.

To assess uncertainties in the identification of cleavage local parameters from numerical analysis of notched tensile specimens due to numerical differences and due to different procedures in statistical methods used for parameters evaluation.

3. Material Characteristics and Experiments

In this chapter a short summary of the experimental and theoretical work is given. Manganese cast steel has been used for experiments. Škoda Company has supplied the material as a component part produced for certification procedures of the container of spent nuclear fuel. In order to guarantee the microstructure of specimens used to be the same as that of the inner part of cast body, the computer simulation of cooling of semi-product has been used. As the result of this modelling the optimised plate (size 55 x 90 x 250 mm) for fracture mechanics experiments was produced. From this plate three

108

smaller pieces were reduced (in size 50 x 25 x 240 mm). The dimension of semi-product was the compromise between the cooling rate needed for cooling simulation and the furnace capacity.

True stress-strain curves have been measured using cylindrical smooth tensile specimens with a diameter of 6-mm tested over a temperature range from -196°C to -60°C with crosshead velocity of 2 mm.min^{-1} (*Figure 4*).

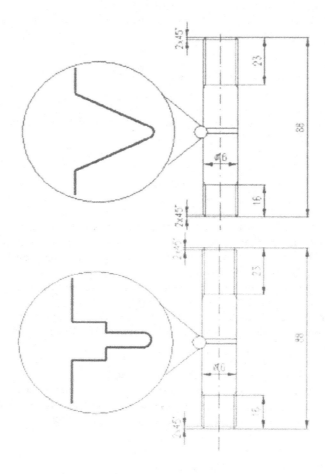

Figure 2. The specimen with the circumferential V-notch – R=0.25 mm and U-notch – R=0.2, 0.7, 1 mm

For one selected temperature in lower shelf region a set of round tensile-notched bars were tested to obtain data for statistical local approach procedures. The diameter of the specimen was 16 mm and the notch depth 4 mm. Three specimens of circumferetial U-notch have been tested having notch tip radius 0.2, 0.7 and 1.0 mm respectively, one circumferetial V–notch (radius 0.25 mm) and Charpy V–notch (*Figure 2 and 3*).

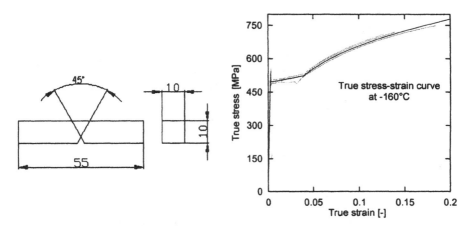

Figure 3. Charpy V-notched specimen   *Figure 4*. True stress-true strain curve at -160°C

As can be seen in the following *Figure 5 and 6*, the selected temperature was -160°C. Increasing scatter for the specimen with sharper notch can be observed. Two types of transducer were used, the first one for the accurate displacement measurement, (MTS 634.12F-51) the second one for the contraction measurement (MTS 632.19F-21). Due to statistical reason, at least 30 experiments were tested for each notch geometry.

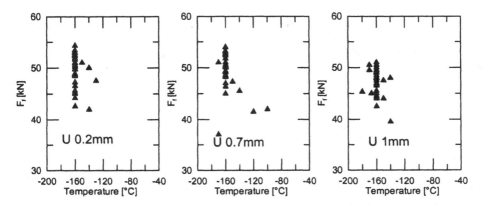

Figure 5. Fracture forces for U-notch tensile specimens R=0.2, 0.7 and 1 mm

The size of the specimens and notch diameters were checked for each experiment before the test. The testing machine Zwick 1382 was used with crosshead velocity of 1 mm.min⁻¹. Only for U-notch R=0.2 mm was not possible to use the contraction measurement due to the shape of the exchangeable contacts. Three points static bending test of standard Charpy specimen was carried out on Zwick testing machine. Before static tests the temperature dependence of fracture forces was determined as can be seen in *Figure 6* in temperature range from -110 to -196°C, with crosshead velocity of

110

2 mm.min⁻¹. According to observation of fracture surfaces the test temperature, for which the cleavage mechanism was dominating, the selected test temperature was -160°C.

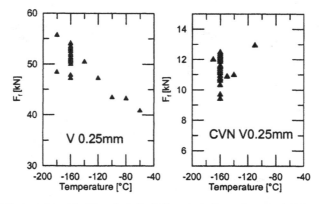

Figure 6. Fracture forces for V-notched (R=0.25 mm) tensile specimens and CVN (0.25 mm)

As was mentioned in the introduction the local approach to cleavage is based on the weakest link concept (be sure about the position in FTDD – see *Figure 7*), which postulates that the failure of the material containing a large number of statistically independent elements is triggered by the failure of one element. The fulfilment of these conditions is necessary to check all fracture surfaces to be absolutely sure, that the weakest link concept is valig preferably using scanning electron microscopy. Example of this work can be seen in *Figure 8 and 9*. In the case of cleavage mechanism the fracture is generated in the area where there is the maximum principal stress, this can be checked by coincidence of actual triggering point and computed stress – strain distribution.

4. Numerical Procedure

The intrinsic model for notched bars is proposed with respect to symmetry as a half of bar. The axisymmetric elements CAX6 from Abaqus FEM package are used. In case of Charpy body, the C3D8I elements were applied and due to symmetry the fourth part of body was solved. Approximately the same element size ahead the notch tip in the region 1 mm is being used because the data from this region are mainly exerted for determination of Weibull stress. The radii of notch were divided at least into 20 parts. Some features of the FEM model can be seen in Table I.

The next step of modelling is to set properly the material characteristics of the cast steel. As can be seen in *Figure 4*, at test temperatures the stress – strain curve has the region where the Lüders deformation is dominating. Therefore the standard relations seem to be not appropriate for the modelling. Incorrectness of standard Ramberg-Osgood or exponential description and then necessity to use piecewise linear description are expressed in case of modelling the body with a notch. The dependence measurement the true stress true strain provides the information below the values of deformation 0.15-0.2. In *Figure 10* one can see the schematic procedure describing the local parameters determination, some details of this procedure were discussed in the following section. Some results of FEM modelling are shown in the *Figure 11-15*. The very different stress and strain distribution for notches with compared in this figures can play an important role in local parameters determination, can be discussed mainly in relation with the elementary volume size used in Beremin equation.

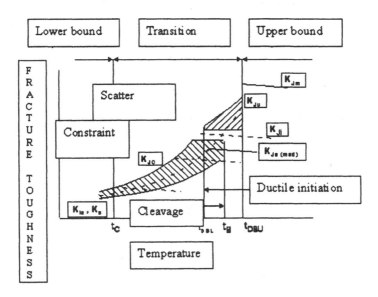

Figure 7. Schematic FTTD (Fracture toughness transition diagram)

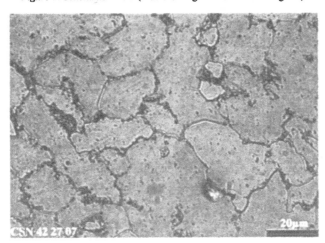

Figure 8. Cast manganese steel and its microstructure (optical microscopy)

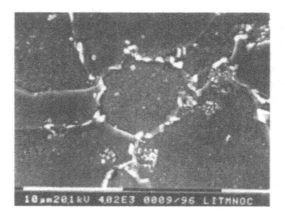

Figure 9. Cast manganese steel and its microstructure (scanning microscopy)

TABLE I. FEM mesh for notched specimens

Geometry	No. of elements	No. of nodes
V-notch	8243	16784
U-1 mm	10279	20784
U-0.7 mm	8755	17836
U-0.2 mm	11303	22966

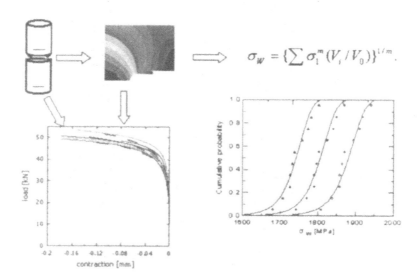

$$\sigma_W = \left\{ \sum \sigma_1^m (V_i / V_0) \right\}^{1/m}.$$

Figure 10. Local parameters determination technique

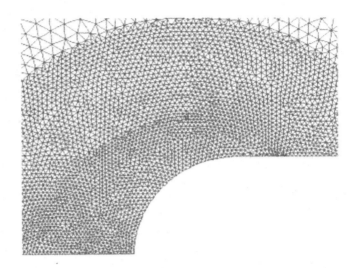

Figure 11. FEM mesh for U-notch R=1 mm and detail of the notch

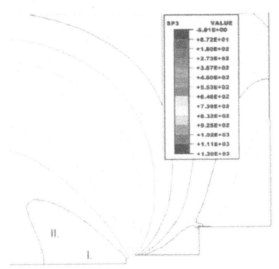

Figure 12a. Maximum principal stress distribution for U-notch R=0.2 mm (for maximum fracture force)

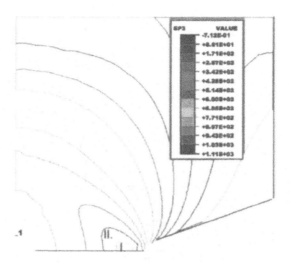

Figure 12b. Maximum principal stress distribution for V-notch R=0.25 mm (for maximum fracture force)

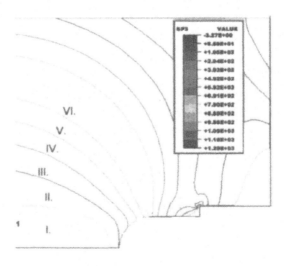

Figure 13a. Maximum principal stress distribution for U-notch R=0.7 mm (for maximum fracture force)

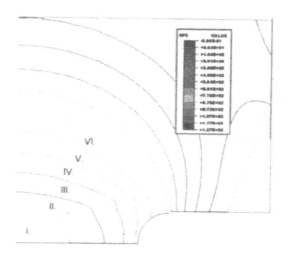

Figure 13b. Maximum principal stress distribution for U-notch R=1 mm (for maximum fracture force)

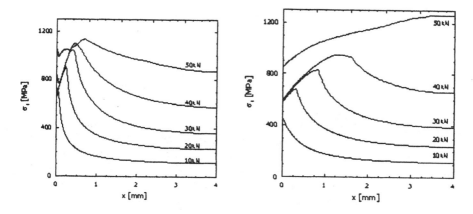

Figure 14. Maximum principal stress distribution ahead notch root for U-notch R=0.2 mm
and U-notch R=1 mm (fracture forces 10-50 kN)

5. Discussion about Local Parameter Determination

The methodology how to generate correctly the local parameters was many times discussed in the scope of the ESIS committees (e.g. [11], [13]). The generation of m and σ_u can be divided in the following steps:

- Test temperature determination.
- At a given temperature at least 15 experiments were performed.
- The FEM computation and valid data selection.

116

- The experimental set verification, where the results from displacement measurement and contraction measurement are compared with the computed values from FEM.
- The application of maximum likelihood method or the least square method. According to [13, 14] the first one method is preferable in order to avoid the influence of outliers.
- The verification of generated local parameters on another notch root geometry.

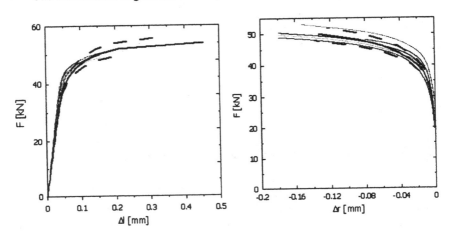

Figure 15. Elongation and contraction for U-notch R=1 mm, the mean value is received from FEM, outer values correspond to the maximum and minimum of fracture forces

As can be seen in [11] the notch root radius for tensile specimens is usually above 1 mm. In the case of using another specimen geometry is important to determine more precisely which data are valid or not. Two assumptions play the main roles in our decision-making. The first one was mentioned in the first lines of this paper, the fulfilling of the weakest link assumption and mechanisms valid for cleavage fracture. The second condition is based on the value of contracting deformation defined as follows:

$$\varepsilon_{pl}[\%] = 200 * \ln[d_s / d_f] \qquad (3)$$

This criterion is described in [15], but there is very little information about its origin, some notices can be found in [16], where ε_{pl} must be in the range <1 - 30%>, see *Figure 16-18*. The base of this criterion, mainly the lower limit, can be found in surface defects, which could not take effect on the damage as well the region of shear bands typical for the plastic damage. Its role can play the stress strain field ahead of the notch radius. It should be noted that this problem could arise in case of using Charpy type specimens for local parameters generation.

Standard iteration procedure based on the maximum likelihood procedure according [17] was applied. The procedure follows from the next equation 4, where m_{est} and σ_{uest} are estimated local parameters in given iteration step.

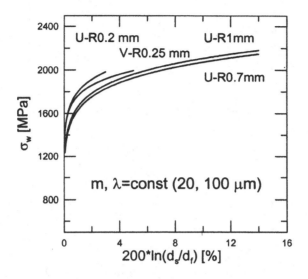

Figure 16. Weibull stress versus ε_{pl} (Eq. 3) for constant value m=20, $V_o = (100 \ \mu m)^3$ for tested notch geometry

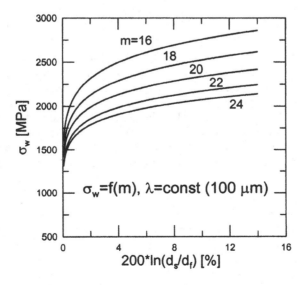

Figure 17. Weibull stress versus ε_{pl} (Eq. 3) for constant value of m, $V_o = (100 \ \mu m)^3$ for tested U-notch R=0.7 mm

118

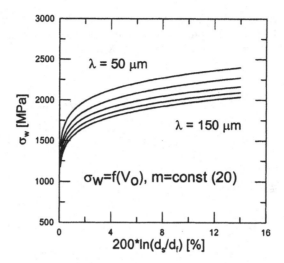

Figure 18. Weibull stress versus ε_{pl} (Eq. 3) for constant value m=20, $V_0 = (\lambda)^3$, $\lambda \in <50,150>$ mm, for tested U-notch R=0.7 mm

$$\frac{\sum_{j=1}^{N} \sigma_{Wj}^{m_{est}} \ln \sigma_{Wj}}{\sum_{j=1}^{N} \sigma_{Wj}^{m_{est}}} - \frac{1}{N} \sum_{j=1}^{N} \ln \sigma_{Wj} - \frac{1}{m_{est}} = 0 \qquad (4)$$

$$\sigma_{u_{est}} = \left[\left(\sum_{j=1}^{N} \sigma_{Wj}^{m_{est}} \right) \frac{1}{N} \right]^{1/m_{est}} \qquad (5)$$

The above described equation can be solved e.g. by bisection method, but in this case the Weibull stress must be computed and therefore this procedure is time consuming. It could be done more effectively by computing for some values of m and then found out where the function crossed over zero. Futher iteration can be used only for precise fixation. By the help of this described procedure the local parameters were determined which can be seen in Table II. However, these steps of this procedure can be seen in the *Figure 19*. As an example the U-notch R=0.7 mm was chosen, where Λ is right hand side of equation 4.

TABLE II. Determined values of m and σ_u

Geometry	m	σ_u [MPa]
U 0.2mm	55	1360
U 0.7mm	20	1850
U 1.0mm	19	2000
V 0.25mm	52	1440

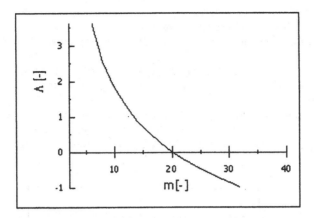

Figure 19. Quality of maximum likelihood procedure for U-notch R=0.7 mm

6. Application

The combination of micromechanical failure models with FEM analysis has resulted in a powerful method for predicting fracture behaviour of structures using single laboratory tests. There are principally two ways how to apply the obtained local parameters:

- Direct application of Beremin model on the real structure using Eq. 1 and 2.
- Extension of local parameters to the fracture mechanics, where one can do (a) direct fracture toughness prediction for SSY (Eq. 6) or (b) using toughness scaling diagram [20] - see *Figure 20* and its application in *Figure 22*.

$$\ln\left[\frac{1}{1 - P_f}\right] = \frac{\sigma_y^{m-4} K_{IC}^4 B C_m}{V_0 \sigma_u^m} \qquad (6)$$

7. Summary

The Beremin model with strain correction (slightly modified [18]) was used for the calculation of σ_w. The reference volume $V_o = (100 \ \mu m)^3$ was used for all computations. The valid local parameters are received on test specimens with notch radius 0.7 and 1 mm. The distribution of maximum principle stress has little influence on the microstructure inhomogenities that can be found in the broken specimens. The Weibull parameters received are close to m=20 (*Figure 21*). As can be seen the local parameters are not independent of geometry, be very careful of using tensile specimens with small notch radius.

120

Recommendation for m and σ_u determinations:
- The stress strain field must be close to homogeneous field
- The material model used in FE computation must be very close to the actual material behaviour
- The quality of the mesh for FE computation is usually important, but our results showed that the difference in the case of using 1000 elements or 10000 elements is only 1 percent
- Validity procedure of experimental data used for local parameters determination is the ground for local approach and correct local parameters determination
- It is important to emphasize, that all is based on the weakest link assumption

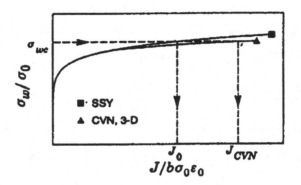

Figure 20. Schematic chart of the toughness-scaling diagram

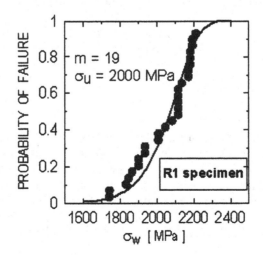

Figure 21. Probability of fracture - fitting for U-notch R=1 mm.

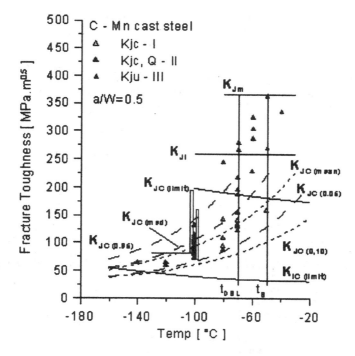

Figure 22. The FTTD based on 1T SENB and transformed values from pre-cracked Charpy

Acknowledgements

This research was financially supported by grant 101/00/0170 Grant Agency of the Czech Republic and NATO Science for Peace program project 972655.

References

1. Beremin, F. M. (1983): A Local criterion for cleavage fracture of a nuclear pressure vessel steels, Metallurgical Transactions Vol.14A, pp.2277-2287.
2. Beremin, F. M. (1981): Cavity formation from inclusions in ductile fracture of A 508 steel, Metallurgical Transactions Vol. 12A, pp. 723-731.
3. Beremin F. M. (1981): Experimental and Numerical Study of the Different Stages in Ductile Rupture: Application to Crack Initiation and Stable Crack Growth, Conf. Three-Dimension Constitutive Relations and Ductile Fracture, Proceedings, pp. 181-205.
4. Pineau, A. (1992): Global and local approaches of fracture-transferability of laboratory test results to components, Topics in Fracture and Fatigue. A. S. Argon. Springer Verley, pp. 197-234.
5. Wiesner, C. S. (1996): The local Approach to Cleavage Fracture (Concepts and application), Abington Publishing, England.
6. Weibull, W. (1951): A statistical distribution function of wide applicability, J. Appl. Mech., pp. 293-297.

7. Rice, J. R. and Tracey, D. M. (1969): On the ductile enlargement of voids in triaxial stress fields, J. Mech. Phys. Solids 17,pp. 201-217.

8. Gurson, A. L. (1977): Continuum theory of ductile rupture by void nucleation and growth: Part I, J. Engng. Materials and Technology 99, pp. 2-15.

9. Rousselier, G. (1986): Les modeles de rupture ductile et leurs possibilitiés actuelles dans le cadre de l'approche locale, EDF, Les Renardieres, Bp 1, F-77250 Moret-Sur-Loing, France.

10. Kozák, V., Dlouhý, I., The transferability of fracture toughness characteristics from point a view of integrity o components with degects, CMEM X, Alicante, 2001, Wit Press, edited by Y. V. Esteve, G. M. Carlomagno and C. A. Brebbia, pp. 757-766.

11. ESIS P6-98 Procedure to Measure and Calculate Local Approach Criteria Using Notched Tensile Specimens, ESIS Document, (1998).

12. Brückner-Foit A., Riesch-Oppermann H.: WEISTRABA, A Code for the Numerical Analysis of Weibull Stress Parameters from ABAQUS Finite Element Stress Analysis, Forschungszentrum Karlsruhe GmbH, 1998.

13. Mudry and C. Sainte Catherine. Draft procedure to measure and calculate local fracture critreria on notched tensile specimens, Karlsruhe, ESIS TC1 meeting. Appendix I, 1995

14. Minami F., Brückner-Foit A., Munz D., Trolldenier B., (1992): Estimation Procedure for the Weibull Parameters Used in the Local Approach, Int. Journal of Fracture, Vol. 54, pp. 197-210.

15. Rosoll A., Tahar M., Berdin C., Piques R., Forget P., Prioul C., Marini B. (1996): Local approach of the Charpy test at low temperature, MECAMAT, Paper 177.

16. Mudry F., Di Fant M. (1994): A Round-Robin on the Measurement of Local Criteria for Fracture, BCR Report EUR 15352EN, ECSC.

17. Riesch-Oppermann H., Brückner-Foit A.: Weistraba (1998): A code for the numerical analysis of Weibull stress parameters from ABAQUS finite element stress analysis.

18. Bernauer G, Brocks W., Schmitt W.: Modifications of the Beremin model for cleavage fracture in the transition region of a ferritic steel, Engineering Fracture Mechanics, Vol. 64, 1999, pp. 305-325.

19. Brückner-Foit A., Munz D., Trolldenier B.: Micromechanical Implications of the Weakest Link Model for the Ductile-Brittle Transition Region, Defect Assessment in Components-Fundamentals and Applications, ESIS/EGF9, 1991, pp. 477-488.

20. Koppenhoefer K. C., Dodds R. H.(1997): Loading Rate Effect on Cleavage Fracture of pre-Cracked CVN Specimens: 3-D studies, Engineering Fracture Mechanics, Vol. 58, pp. 249-270.

ON THE APPLICATION OF THE BEREMIN MODEL FOR PREDICTING THE BRITTLE FRACTURE RESISTANCE

GY. B. LENKEY, ZS. BALOGH, T. THOMÁZY
Bay Zoltán Foundation for Applied Research, Institute for Logistics and Production Systems, Iglói u. 2., H-3519 Miskolctapolca, Hungary

Abstract: The Beremin model of brittle fracture was implemented for a cast ferritic steel materials using the MARC finite element code. The effect of different material parameters (like yield stress, strain hardening exponent) was investigated on the Beremin model's parameters. Also it was analysed how the applied numerical procedure affects these results. As it was found, the strain hardening exponent has small effect on the model parameters, the yield stress has stronger effect. The applied fitting method can cause larger differences. The numerical values of the Weibull-parameters depends strongly on the scatter of the measured fracture probability values. Then the fracture behaviour of Charpy-V specimens was predicted by applying the obtained Beremin-model's parameters. The accuracy of the prediction mostly depends on the variation of the yield stress.

Keywords: local approach, brittle fracture prediction, Beremin-model, cast ferritic steel

1. Introduction

The brittle fracture resistance of the materials can be described by statistical methods. Nowadays the micromechanical modelling of the fracture behaviour is widely applied and studied, when size, geometry and temperature dependent material parameters can be determined. The Beremin-model [1] uses two parameters for describing the fracture process connecting the microscopic defects and the stress state with the fracture probability. The applicability conditions of this model are widely investigated. The aim of our work was to analyse the effect of different material parameters on the applicability of the Beremin-model for predicting the brittle fracture resistance of cast ferritic steel used for spent nuclear fuel containers.

2. Beremin-Model of Brittle Fracture

The Beremin-model was developed by a French research group and was first published in 1983 [1]. The main characteristics of the model are the followings:
- Statistical approach is applied (weakest link model),
- It applies local criteria based on maximum principal stresses,
- It delivers size and geometry independent material parameters, which gives the possibility for transferring small specimen data to real structures.

I. Dlouhý (ed.), Transferability of Fracture Mechanical Characteristics, 123–134.
© 2002 *Kluwer Academic Publishers. Printed in the Netherlands.*

The failure probability within a reference volume V_o – where the stress state can be considered homogenous and can be characterised by the maximum principal stress$\sigma 1$ – can be given with eq. 1:

$$p(\sigma) = \int_{l_0^c}^{\infty} P(l_0)dl_0 = \left(\frac{\sigma_1}{\sigma_u}\right)^m \tag{1}$$

where m and σ_u are material parameters, and l_0^c is the length of a critical microcrack.

If the solid body is divided into n volume element V_i, the probability of brittle fracture P_f can be given as:

$$\ln(1 - P_f) = -\sum_{i=1}^{n} \frac{V_i}{V_0} \left(\frac{\sigma_1^i}{\sigma_u}\right)^m. \tag{2}$$

Introducing the Weibull-stress σ_W:

$$\sigma_W = \sqrt[m]{\sum_i \left(\sigma_1^i\right)^m \frac{V_i}{V_0}}, \tag{3}$$

the eq. 2. can be written as:

$$P_f = 1 - \exp\left[-\left(\frac{\sigma_W}{\sigma_u}\right)^m\right]. \tag{4}$$

σ_u and m are the so called Weibull-parameters of the Beremin-model.

3. Determination of the Beremin-Model's Parameter for Cast Ferritic Steel

The Weibull-parameters have been determined by applying a proposed standard procedure which was elaborated by the European Structural Integrity Society (ESIS) [2]. The algorithm of the calculation is presented in *Figure 1*. The MARC finite element code has been used for the finite element calculations.

Tensile experiments of notched cylindrical specimens were performed at low temperature (T = -160 °C) at the Institute for Physics of Materials, Brno. In this study the results of the specimens with 1 mm notch radius were used. Elastic-plastic constitutive equation determined at −160 °C was first applied in the FEM calculations (Table I). Then the constitutive equations were modified artificially, simulating the possible scatter of the material parameters, i.e. strain hardening exponent (n) and yield stress (σ_y). The modified stress-strain curves and the material parameters are shown in *Figure 2*. and Table II. respectively.

Figure 3. shows the calculated force-strain curves using different constitutive equations comparing with experimental results. As it can be seen the best coincidence could be obtained for n=0.2 and $\sigma_y = \sigma_{y\text{-measured}}$-50 values. At first the Weibull-parameters were determined applying the maximum likelihood method for fitting the Weibull-distribution function and considering the fracture strain as the basis of Weibull-stress calculation (Table III.).

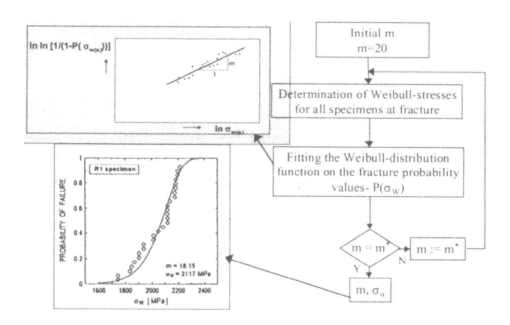

Figure 1. Algorithm for determination of the Beremin-model's parameters

TABLE I. Measured material parameters (T = -160 °C)

ε	σ	Equation
	Measured material parameters (average of three experiments)	
0-0.002434	498±10	$\sigma = 205\ 000 * \varepsilon$
0.002434-0.03981	524	$\sigma = 695.6336 * \varepsilon + 496.3068$
0.03981-0.18		$\sigma = 1151 * \varepsilon^{0.24361}$

The ESIS procedure allows also to use the average fracture stress as fracture parameter, and the fitting of Weibull-distribution function can be done by linear regression as well. So several calculations have been performed applying the combinations of these different possibilities and different constitutive equations, to study the effect of the applied calculation methods and material parameters. The results are summarised in *Figure 4. and 5.*

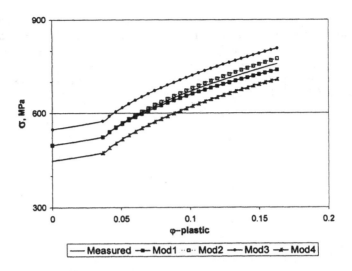

Figure 2. Stress-strain curves used in the finite element calculations

TABLE II. Modified material parameters

ε	σ	equation
	Modified2	
0-0.002434	498	$\sigma = 205\,000*\varepsilon$
0.002434-0.03981	524	$\sigma = 695.6336*\varepsilon+496.3068$
0.03981-0.18		$\sigma = 1151*\varepsilon^{0.2}$
	Modified3	
ε	σ	equation
0-0.002434	498	$\sigma = 205\,000*\varepsilon$
0.002434-0.03981	524	$\sigma = 695.6336*\varepsilon+496.3068$
0.03981-0.18		$\sigma = 1151*\varepsilon^{0.3}$
	Modified4	
ε	σ	equation
0-0.002434	548	$\sigma = 205\,000*\varepsilon$
0.002434-0.03981	470	$\sigma = 695.6336*\varepsilon+546.3068$
0.03981-0.18		$\sigma = 1151*\varepsilon^{0.24361}$
	Modified5	
ε	σ	equation
0-0.002434	448	$\sigma = 205\,000*\varepsilon$
0.002434-0.03981	470	$\sigma = 695.6336*\varepsilon+446.3068$
0.03981-0.18		$\sigma = 1151*\varepsilon^{0.24361}$

TABLE III. Calculated Weibull-parameters with different material equations
(fracture parameter is strain, maximum likelihood method is applied)

Constitutive equation	m	σ_u
Measured	16.53	2394.9
Modified1 (n=0.2)	17.29	2321
Modified2 (n=0.3)	15.72	2486
Modified3 (σ_y+50)	17.1	2553
Modified4 (σ_y−50)	15.5	2279

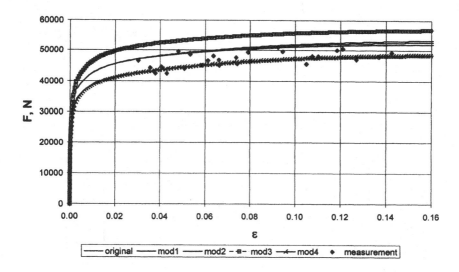

Figure 3. Comparison of calculated force-strain curves with measured values at fracture

On the basis of these results it can be concluded that the strain hardening exponent has a small effect on the model parameters, the yield stress has a stronger effect, and the tendency is opposite when different fracture parameters are considered. The fitting method can cause larger differences. Two examples for the fitted distribution functions together with the measurements based values are shown in *Figure 6*. It can be seen in these figures that the correlation of the fitting function is better for the second case when the fracture stress was used as fracture parameter. When the strain was the fracture parameter due to only one higher fracture strain value the correlation was much worse.

It can be concluded that the numerical values of the Weibull-parameters themselves cannot have much importance. Therefore we examined also their effect on the fracture prediction.

128

Fracture parameter: strain

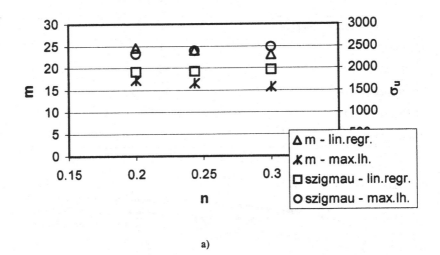

a)

Fracture parameter: strain

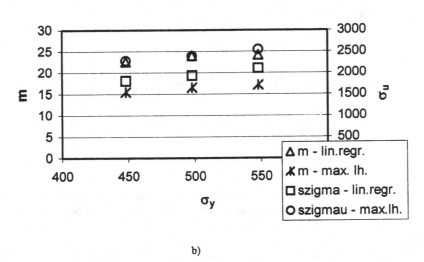

b)

Figure 4. Effect of strain hardening exponent (a) and the yield stress (b)
on the Beremin- model's parameters, applying the strain as fracture parameter

Fracture parameter: stress

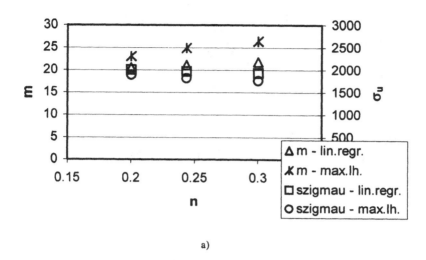

a)

Fracture parameter: stress

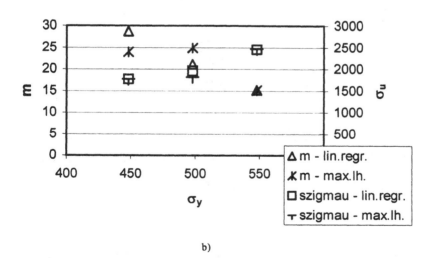

b)

Figure 5. Effect of strain hardening exponent (a) and the yield stress (b)
on the Beremin- model's parameters, applying the stress as fracture parameter

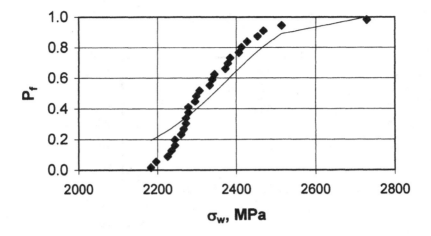

a)

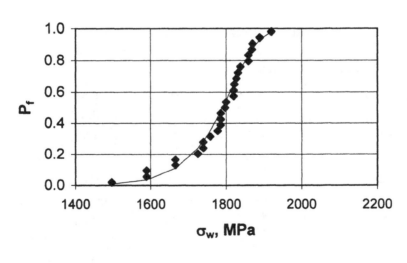

b)

Figure 6. Comparison of the fitted Weibull-distribution function with the failure probabilities based on the measurements using the measured constitutive equations (maximum likelihood method was applied): a) fracture parameter is strain ; b) fracture parameter is stress

4. Prediction of Brittle Fracture Behaviour of V-Notched Charpy Specimens

Since the determined Beremin-model's parameters are size and geometry independent, they can be used for predicting the fracture probability of specimens of other geometry. So we used the Beremin-model for analysing the brittle fracture behaviour of Charpy-V specimens under static loading conditions. The basic principle of this prediction is that fracture probability is the same at a given load level in specimens of different geometry if the Weibull-stresses are the same. So if one determines the change of the Weibull-stress as a function of external loading, then the fracture probability belonging to different Weibull-stresses can be determined using eq. 4. On the basis of this the fracture load can be calculated at different fracture probability level.

Figure 7. shows the Weibull-stress as the function of external load using the measured constitutive equation. The two horizontal lines represent the Weibull-stresses at 5 % and 95 % fracture probability levels. *Figure 8.* and *Figure 9.* show the predicted fracture forces using different constitutive equations comparing with experimental values. It can be stated that there is usually good agreement between the predicted and the measured fracture forces, but the materials parameters, the selection of the fracture parameter and the applied fitting method can have smaller or larger effect: e.g. when the fracture parameter was the strain with the maximum likelihood method the predicted force values shifted toward higher values as compared to the other obtained with linear regression. But when the fracture stress was used as fracture parameter, in both cases the averages of the predicted force values are almost the same. Only their intervals are different. This could be explained by the fact which was explained in the previous section that the correlation of the Weibull-distribution function was much better when the stress was the fracture parameter.

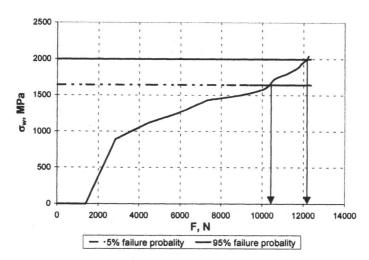

Figure 7. Weibull-stress vs. load for V-notched Charpy specimen – determination of critical forces at 5% and 95 % failure probabilities

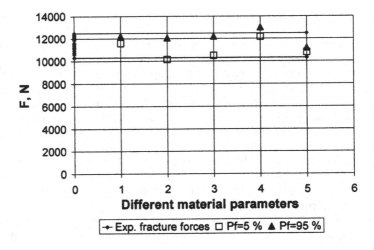

a)

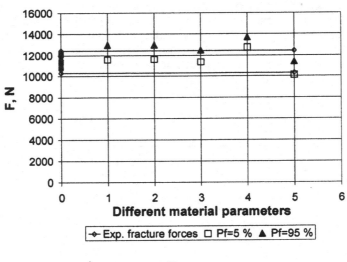

b)

Figure 8. Comparison of the measured and predicted fracture forces
(at 5% and 95 % failure probability) for different constitutive equations,
fracture parameter is strain: a) linear fitting ; b) maximum likelihood method

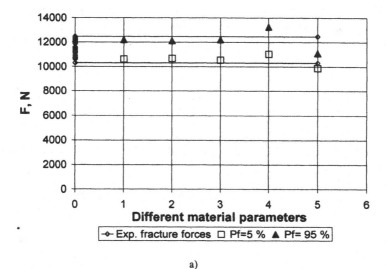

a)

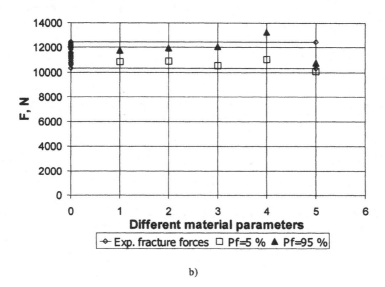

b)

Figure 9. Comparison of the measured and predicted fracture forces
(at 5% and 95 % failure probability) for different constitutive equations,
fracture parameter is stress: a) linear fitting ; b) maximum likelihood method

The strain hardening exponent practically does not have effect on the prediction (cases 1-3 in *Figures 8-9*), but the yield stress modifies the results significantly (cases 4-5 in *Figures 8-9*): with higher yield stress the predicted fracture force can be 5-10 % higher.

5. Summary and conclusions

Analyses on the effect of variation of different material properties, the selected fracture parameter and applied fitting procedure on the Beremin-model's parameters has been performed.

On the basis of the obtained results the following can be concluded:

1. The strain hardening exponent has small effect on the model parameters, the yield stress has stronger effect. The model parameters also depends on which parameter (strain or stress) is selected.
2. The fitting method can cause larger differences. The numerical values of the Weibull-parameters depends strongly on the scatter of the measured fracture probability values.
3. With the application of the Beremin-model it was possible to predict well the fracture force of Charpy-V specimens. The accuracy of the prediction mostly depends on the variation of the yield stress.

Acknowledgement

The financial support of NATO SfP 972655 project is greatly acknowledged.

References

1. Beremin, F. M. (1983) A Local Criterion for Cleavage Fracture of a Nuclear Pressure Vessel Steel, Metallurgical Transactions A, Vol. 14A, pp. 2277-2287.

2. ESIS P6 – 98 (1998) Procedure to measure and calculate material parameters for the local approach to fracture using notched tensile specimens, ESIS procedure – final document, 1998.

PHYSICAL FUNDAMENTALS OF LOCAL APPROACH TO ANALYSIS OF CLEAVAGE FRACTURE

S.KOTRECHKO
G.V.Kurdyumov Institute for Metal Physics of National Academy
of Sciences of the Ukraine,
36 Vernadsky Blvd., Kyiv-142, 03142, Ukraine

Abstract: Physical interpretation of regularities of metal fracture ahead of notch and sharp crack is supplied based on results of statistical modelling of microcrack nucleation and propagation in polycrystal. It is shown that not only local stress but also local plastic strain influences significantly fracture probability ahead of notch. This is due to both the influences of dislocation microstresses on the crack nucleus unstable equilibrium and the crack nucleus density dependence on the plastic strain value. It is demonstrated that plastic strain influences also sensitivity of local fracture stress of iron to the process zone volume. Main factors affecting the process zone size are ascertained. The findings on influence of grain sizes and plastic strain on Weibull parameters are discussed.

Keywords: cleavage fracture, Local Approach, statistical model, microcrack nucleation, scale effect, process zone, microstructure effect, Weibull parameters.

1. Introduction

Recently, "Local Approach to Fracture" is one of the most strongly developing directions in fracture mechanics. It is based on methods of micromechanics and enables a description of the fracture process close to its physical nature [1]. Three scientific areas may be indicated where the use of such approach is effective:
1. Fracture mechanics (prediction of fracture toughness temperature dependence according to the results of tests by small specimens; relation between impact toughness and fracture toughness).
2. Materials science (effect of microstructure on structural integrity).
3. Fracture physics (peculiarities of fracture initiation and propagation micro-mechanism under the conditions of high stress and strain gradients as well as at multiaxial stress state).
The first direction is recently the most intensively developing one. Two others are the perspective tasks of the near future.

A statistical criterion for fracture of separate cell ahead of notch is the key problem of Local Approach as it is obvious from [2-5]. For cracked bodies or for sharply notched specimens exhibiting steep stress-strain gradient, the stressed region can be

I. Dlouhý (ed.), Transferability of Fracture Mechanical Characteristics, 135–150.
© 2002 Kluwer Academic Publishers. Printed in the Netherlands.

divided into cells. Each cell is subjected to a quasi-homogeneous stress-strain state. A failure probability of separate cell F_{V_j} (elementary probability) is calculated as follows:

$$F_{V_j} = F(\sigma_1 > \sigma_c; \ \bar{e} > e_{in}) \tag{1}$$

where σ_1 is the value of principal local (but macroscopic) stress that acts in cell; σ_c is stress of unstable equilibrium of the crack nucleus in cell; $\bar{e}$ is local equivalent plastic strain in cell; e_{in} is the value of crack nucleation strain. The cumulative probability of failure for entire notched specimen that consists of M cells may be written in the following way:

$$1 - \prod_{j=1}^{M} [1 - F_{V_j}] = F_c \tag{2}$$

where F_c is a fracture probability tolerance. In most cases the following values are accepted in calculations: $F_c = 0.01$ (low limit for fracture stress); $F_c = 0.99$ (its upper bound); $F_c = 0.5$ (at symmetric distribution density function this F_c value corresponds to realisation of fracture stress mean value).

Determination of probability function for fracture of cell is the main problem of such approach. Now, Weibull distribution is employed for these purposes:

$$F_{V_j} = 1 - \exp\left[-\frac{V_j}{V_0} \left(\frac{\sigma - \sigma_{th}}{\sigma_u} \right)^m \right] \tag{3}$$

where σ_{th} is threshold stress (low limit of strength), m is parameter that determines the shape of probability function (shape factor), σ_u is scale stress (parameter); V_0 is a reference unit volume.

This approach postulates that Weibull parameters σ_u, σ_{th} and m are metal constants i.e. they are independent on stress-strain state of metal ahead of notch.

In spite of the fact that Weibull distribution is widely adopted for prediction of fracture of notched specimens, great number of experimental data exists that are not in accordance with assumption concerning independence of Weibull parameters on stress-strain state of metal ahead of notch. This appears evidently at change in stress and strain gradients ahead of notch over a wide range i.e. when comparing data on fracture of notched and on unnotched specimens. For example, it was shown [5], that increase in local fracture stress with decreasing notch radius from 0.1mm to 1.0mm at 77K may be described by certain fitting of Weibull distribution parameters. However, in this case the calculated fracture stress value for unnotched specimen is 1.3 times higher than appropriate experimental value. The dependence of Weibull parameters on notch geometry has been determined [6]. Comparison of Weibull parameter values for notched tensile, notched (Charpy-type) four point bend and fracture mechanics specimens have shown that the value of parameter m for notched bend specimens may be two times higher than for notched tensile specimens and four times higher than corresponding value for the pre-cracked specimens. On the contrary, the values of scale stress, σ_u, for notched bend specimens are approximately 1.5 times lower than those for notched tensile specimens and 1.8 times lower than σ_u for fracture mechanics specimens. Besides, Weibull parameters for notched bend specimens and pre-cracked specimens change with test temperature growth. It has been found [7] that substantial change in Weibull parameters occurs when notch radius of tensile specimens decreases from 0.2 to 1.0 mm. It means that stress-strain state of metal ahead of notch affects the values of Weibull parameters. As a result, it is necessary to propose a local fracture criterion that takes into account the effect both of metal structure and of stress-strain state ahead of the crack/notch tip. Such criterion may be developed on physical basis by modelling crack nucleation and growth.

Statistical model of the brittle fracture of polycrystalline metals was suggested in [8, 9]. Possibility of its application for prediction of a fracture based on the "first principles", i.e. on analysis of the crack nucleus formation and propagation was demonstrated. This model allows to develop physical fundamental of local fracture criteria.

The aim of this paper is both to analyse basic factors affecting the unstable equilibrium of the crack nucleus ahead of notch and to ascertain regularities of influence of both microstructure and local plastic deformation on the values of Weibull parameters.

2. Physical Model of Brittle Fracture of Polycrystalline Metal

2.1 CRACK NUCLEATION

The essential parameter for any statistical model of fracture is a microcrack size distribution which type is, usually, postulated. In most cases, Gauss, power-series or beta distribution laws are applied [1, 2].

According to physical nature, the crack nucleus triggering is due to microplastic inhomogeneity in polycrystal. This conception allows to derive the expression for a

138

crack nucleus length a [8] in a single-phase polycrystalline materials or steels with fine carbide particles:

$$a = \frac{p}{\gamma} \left(\frac{\delta \tau_c}{\beta} \right)^2 \frac{d}{\bar{\varepsilon}} \qquad (4)$$

where p, δ and β are the constants; γ is the specific energy of lattice fracture at the nucleus crack tip dependent on the lattice parameters; d is grain size; $\bar{\varepsilon}$ is equivalent microplastic strain value in the grain of size d.

Here:

$$\bar{\varepsilon} \geq \varepsilon_{in} \qquad (5)$$

where ε_{in} is the minimum value of microscopic strain that is required for the crack nucleation.

In polycrystal the grain size d and the plastic strain $\bar{\varepsilon}$ are random quantities. Accordingly, a is also a random quantity. Its distribution is presented on *Figure 1* [8]. Specific feature of this distribution lies in dependence of distribution density function for

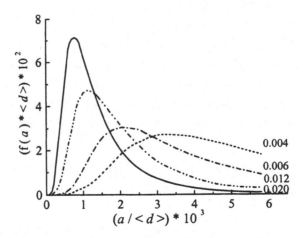

Figure 1. Density distribution function $f(a)$ for the crack nucleus lengths at different values of plastic strain; $\langle d \rangle$ is average grain size.

the crack nucleus not only on the grain size distribution but also on the value of plastic strain. It should be underlined that maximum length of the crack nucleus is independent on the value of plastic strain. However, there is maximum length of the crack nucleus that predetermines low limit of fracture stress.

2.2. CRACK PROPAGATION

The crack nucleus is of submicroscopic size. As a rule, its length doesn't exceed $0.05 - 1\ \mu m$. Therefore, microstresses influence greatly the unstable equilibrium of such cracks and their growth. Unfortunately, this problem is not studied until now. As it has been shown in [9], microstresses ξ_{ij} induced by elastic deformations of grains and microstresses ξ_{ij}^p created by dislocations should be differentiated depending on their effect on the crack nucleus. In the first approximation, microstresses ξ_{ij} are uniform within the grain and change from one grain to another. The value of their variance is predetermined by macroscopic stress state acting on polycrystalline aggregate i.e. it depends on macroscopic principal stresses σ_1, σ_2, σ_3. The variance of these stresses can be estimated by dependencies:

$$D(\xi_{11}) = D_I\sigma_1^2 + D_{II}(\sigma_2 + \sigma_3)^2 + 2[\mu_I(\sigma_1\sigma_2 + \sigma_1\sigma_3) + \mu_{II}\sigma_2\sigma_3] \qquad (6)$$

$$D(\xi_{22}) = D_I\sigma_2^2 + D_{II}(\sigma_1 + \sigma_3)^2 + 2[\mu_I(\sigma_2\sigma_1 + \sigma_2\sigma_3) + \mu_{II}\sigma_1\sigma_3] \qquad (7)$$

$$D(\xi_{33}) = D_I\sigma_3^2 + D_{II}(\sigma_1 + \sigma_2)^2 + 2[\mu_I(\sigma_3\sigma_1 + \sigma_3\sigma_2) + \mu_{II}\sigma_1\sigma_2] \qquad (8)$$

here $D_I, D_{II}, \mu_I, \mu_{II}$ are coefficients dependent on lattice elastic anisotropy and grain orientation distribution. For example, for non-textured polycrystalline iron $D_I = 1.7 \cdot 10^{-2}$, $D_{II} = \mu_{II} = 0.66 \cdot 10^{-2}$, $\mu_I = 0.72 \cdot 10^{-2}$ [9].

As derived from Eq.(6) – Eq.(8), even at *uniaxial* tension of polycrystal $(\sigma_1 > 0, \sigma_2 = \sigma_3 = 0)$, the *multiaxial* stress state exists in grains $(\xi_{11} \neq 0, \xi_{22} \neq 0, \xi_{33} \neq 0)$. For example, at *uniaxial* tension of iron the magnitude of microstress ξ_{11} changes from $0.6\sigma_1$ to $1.4\sigma_1$ but the values of ξ_{22} and ξ_{33} change within the range from $-0.24\sigma_1$ to $+0.24\sigma_1$. This is due to both elastic anisotropy of grains and their random orientations.

Dislocation stresses acting inside the region where the crack nucleus forms only (and at the moment of their nucleation only) affect the microcrack unstable equilibrium. In addition, it may be kept in mind that dislocation structure characteristics in the region where crack nucleates (for example, at grain boundary region) may be essentially different from those for initial one. Pile-ups of the same sign are main peculiarity of dislocation structure in the region where cracks nucleate. Microstresses created by such dis-

140

location structures have non-homogeneous spatial distribution, so their influence on unstable equilibrium of the crack nucleus is determined by the value of effective tensile microstresses $\bar{\xi}$. Its magnitude depends on the crack nucleus length and gradient of microstress field:

$$\bar{\xi} = \frac{2}{\pi a} \int_0^a \xi_{11}^P(x) \sqrt{\frac{x}{a-x}} \, dx \qquad (9)$$

where $\xi_{11}^P(x)$ is distribution function for tensile stresses on plane where the crack nucleus grows.

General regularity of $\bar{\xi}$ change consists in the fact that $\bar{\xi}$ rises with increase in plastic strain ($\bar{\xi} \sim \sqrt{\bar{\varepsilon}/d}$) over the interval of strains that are typical for quasi-brittle fracture, namely, from 0.005 to 0.10. This effect is the cause of existence of decreasing curve branch on the dependence of fracture stress on the value of plastic strain (*Figure 2*) [8]. The values of brittle fracture stress R_{MC} corresponding to individual steels used are shown in TABLE I.

TABLE I. The values of R_{MC} and e_c for different steels shown in *Figure 2*

Material	Brittle fracture stress R_{MC}, MPa	Critical strain, e_c [10]
1 – iron	625	0.016
2 – steel (0.7%C), annealing	700	0.012
3 – steel (0.3%C), annealing	640	0.040
4 – steel (0.45%C), annealing	900	0.045

Respectively, the critical value of tensile microstress ξ_C for the crack nucleus unstable equilibrium is given by the following expression:

$$\xi_C = \frac{K}{\sqrt{a}} \varphi(\theta, \eta) - \bar{\xi} \qquad (10)$$

where $\bar{\xi}$ is the effective microstresses induced by dislocations; η is stress state mode parameter ($\eta = \xi_{22}/\xi_{11}$, ξ_{11} and ξ_{22} are the principal tensile microstresses), θ is angle between the crack plane and direction where principal tensile microstresses act; K is the coefficient that characterises resistance of the crystal to the microcrack propagation; $\varphi(\theta, \eta)$ is the function that describes the influence of microstress state η and orientation of the crack θ on the critical stress. In the first approximation, we may accept $\varphi(\theta,\eta) = 1/\sqrt{\sin^2\theta + \eta\cos^2\theta}$ [8].

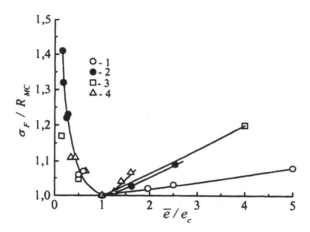

Figure 2. Effect of plastic strain $\bar{e}$ on the value of brittle fracture stress at uniaxial tension σ_F; e_c is critical strain that corresponds to minimum value of brittle fracture stress R_{MC} over the ductile-brittle transition region [8].

Eq.(10) allows to obtain expression for the distribution density $f_1(\xi_c)$ and to evaluate a probability of unstable equilibrium for the crack nucleus at given level of macrostresses σ_F [8]:

$$F_{V_0}(\sigma_F) = 0.5 \int_{\xi_c^{\min}}^{\xi_c^{\max}} f_1(\xi_c)\left[1 - \operatorname{erf}\left(\frac{\xi_c - \sigma_F}{\sqrt{2}\, I(\xi_{11})\sigma_F}\right)\right]d\xi_c \quad , \qquad (11)$$

where

$$f_1(\xi_c) = \frac{dF_1(\xi_c)}{d\xi_c}. \qquad (12)$$

Here $I(\xi_{11})$ is the coefficient of variation of microstresses ξ_{11} created by elastic deformations (i.e. $I(\xi_{11}) = \sqrt{D(\xi_{11})}/\langle\xi_{11}\rangle$, where $D(\xi_{11})$ is the microstress variance, $\langle\xi_{11}\rangle$ is the mean value of microstress, its value is equal to the principal macrostress σ_1); $F_1(\xi_c)$ is probability of the crack nucleus unstable equilibrium at the given value of normal microstresses ξ_c.

142

Eq.(11) characterises probability of unstable equilibrium for *one crack nucleus* at the given level of macrostresses σ_F. For cell where N_a microcracks are nucleated probability of growth of *at least one of them*, $F_{V_j}(\sigma_F)$ follows formula:

$$F_{V_j}(\sigma_F) = 1 - [1 - F_{V_0}(\sigma_F)]^{N_a} \tag{13}$$

The microcracks does not exist in metal initially. They arise from plastic deformation. It means that N_a is not a constant, as it is assumed in conventional models, but it depends also on the plastic strain [9].

Ideas concerning mechanism of brittle fracture of polycrystalline metal considered above enable to point out the most important factors that influence brittle strength of metal:

a) the crack nucleus sizes that are predetermined by grain sizes[1];
b) the distribution of crack nucleus orientation that depends on grain orientation distribution;
c) the crack nucleus density that is specified by the plastic strain value reached at fracture;
d) the magnitude of dislocation microstresses within the region where the microcrack nucleate; it depends on the plastic strain reached at fracture.

Regularities of influence of these factors on the fracture of metal at uniaxial tension are analysed in detail elsewhere [8].

3. Regularities of Brittle Fracture of Metal Ahead of Notch

Steep stress and strain gradients characterise stress-strain state acting on metal ahead of notch. These gradients result in localisation of fracture triggering in extremely small region usually called as "process zone".

3.1. EFFECT OF PROCESS ZONE SIZE

The size of process zone influences significantly the value of local fracture stress σ_F. It is due to scale effect observed at brittle fracture [2]. The magnitude of this effect depends on the crack nucleus density inside the process zone as well as on the size of process zone [9,11]. As the value of plastic strain rises the crack nucleus density increases, and, consequently, the susceptibility of fracture stress to process zone volume

---

[1] In steels containing coarse carbide particles the crack nucleus sizes are controlled by carbide particle sizes.

diminishes. Consequently, the plastic strain influence on the crack nucleus density is a reason for dependence of scale effect in metal on the plastic strain value. This is specific feature of scale effect in metal, unlike ceramics and other brittle materials.

Change in volume V affect significantly the value of fracture stress only for extremely small V values (V<100mm^3) i.e. at the values typical for sizes of process zone ahead of notches and sharp cracks, as it is shown in *Figure 3*.

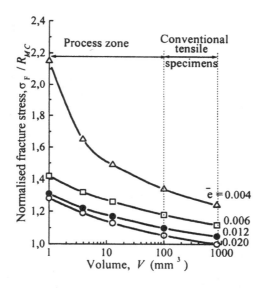

Figure 3. Dependence of normalised brittle fracture stress of iron on the stressed volume at the different values of small plastic strain $\bar{e}$ for fracture [8].

Within the framework of statistical approach the volume of process zone is a region ahead of notch where probability of fracture is higher than zero. *Figure 4* shows influence of the notch radius ρ and test temperature T both on local stress-strain distribution and on length of process zone X_{P-Z} in minimum cross-section of cylindrical specimens of iron [9, 11]. Beremin's model [2] supposes that the plastic zone and process zone sizes are comparable. According to data obtained in [9,11], these zones coincide only at small scale yielding conditions. In general case, process zone size X_{P-Z} is comparably smaller than size X_Y of local plastic area (*Figure 5*). This is, mainly, because of both fracture stress dependence on the plastic strain value (*Figure 2*) and non-homogeneous strain distribution within plastic zone.

144

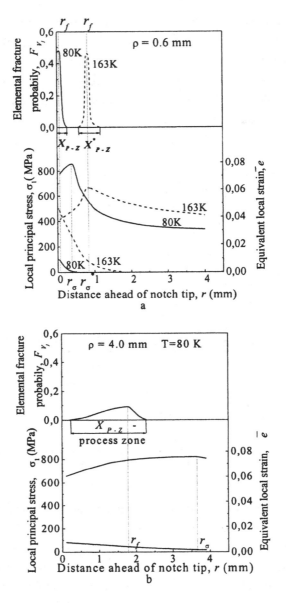

Figure 4. Distribution of principal tensile stresses σ_1, equivalent plastic strains $\bar{e}$ and probability of fracture F_{V_j} ahead of notch root for notched cylindrical specimen of diameter $D = 14\,\mathrm{mm}$: X_{P-Z} is process zone length in minimum cross-section of specimen; r_σ and r_σ^* - are distances from notch tip to maximum tensile stress σ_1 location at temperatures 80K and 163K, respectively; r_f and r_f^* are distances from notch tip to locus of maximum of fracture probability F_{V_j} at temperatures 80K and 163K, respectively (iron, average ferrite grain size is 97 μ m): (a) – for notch radius of $\rho = 0.6\,\mathrm{mm}$, notch depth h = 2.5 mm ; (b) – for notch radius of $\rho = 4.0$ mm , notch depth h = 3.0 mm

The value of plastic strain and its gradient are specified by the notch radius and relative loading of notched specimen σ_{NF}/σ_Y (where σ_{NF} is nominal fracture stress of notched specimen; σ_Y is yield stress). In compliance with data in *Figure 6*, the value of X_{P-Z} may change by nearly 10 times depending on the notch radius. As it is shown in [9], the effect of the plastic strain and its gradient on X_{P-Z} can be described, in the first approximation, by dependence of X_{P-Z} on relative gradient of equivalent local plastic strain G ($G = |grad(\overline{e})|/\overline{e}$; where $\overline{e}$ is the value of local equivalent plastic strain) (see also the *Figure 6*).

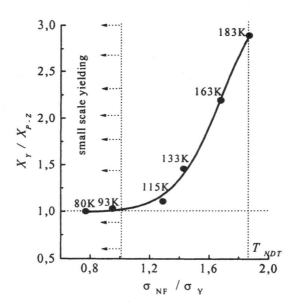

Figure 5. The influence of the relative value of nominal fracture stress σ_{NF}/σ_Y on the interrelation between lengths of local yielding region X_Y and process zone X_{P-z} in the minimum cross-section of notched ($\rho = 0.6$ mm) tensile specimen; T_{NDT} is nil ductility temperature.

3.2. STATISTICAL INTERPRETATION OF LOCAL FRACTURE STRESS

In deterministic fracture criteria the maximum value of principal tensile stress ahead of notch at fracture σ_1^{max} is applied as a measure of local fracture stress σ_F. According to data obtained in [9, 11] such definition of σ_F is correct only at brittle fracture under the conditions of general yielding (*Figure 7*). At lower temperatures location co-ordinates

146

of maximum fracture probability and maximum tensile stress do not coincide. It means that the most probable values of fracture nucleus location co-ordinates may not coincide with co-ordinates of point where tensile stress reaches the maximum value. Such conclusion is in good agreement with data of fractographic investigations of fracture nuclei ahead of notch [12]. As it follows from statistical model proposed, this effect is due to dependence of fracture probability not only on the level of tensile stresses but also on the plastic strain value. The cause of plastic strain effect is that the crack nucleus number within process zone and the level of dislocation microstresses (term $\bar{\xi}$ in Eq.(10)) rise with increase in plastic strain value. That gives rise to interchange in co-ordinates of maximum of fracture probability function nearer to notch root i.e. to the region of higher strains (*Figure 4.a*). Calculations show that difference between maximum value of principal tensile stress σ_1^{max} and the value of tensile stresses at the point of maximum of fracture probability function is not higher than 10-15% at fracture of cylindrical notched specimens with radius $\rho \geq 0.6$mm. However, values of local plastic strains at these points may differ by nearly 10 times. Distances from notch tip to corresponding points r_σ and r_f may differ by several times (*Figure 7*).

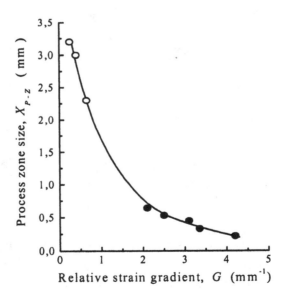

Figure 6. Dependence of process zone length X_{p-z} on the absolute value of relative strain gradient G; Solid circles represent data from specimen with notch radius 0.6 mm; open circles represent data from specimen with notch radius 4 mm

Therefore, in general case, local fracture stress σ_F should be understood as the most probable fracture stress. Its value equals to the principal tensile stress σ_1 at point ahead of notch where probability of fracture initiation reaches maximum ($F_{V_j}(x,y,z) = F_{V_j}^{max}$). The same approach was applied in [13] to determine the value of local fracture stress. However, effect of plastic strain was not accounted in this case. It gives rise to overestimation of distance from notch tip to the point where $F_{V_j}^{max}$.

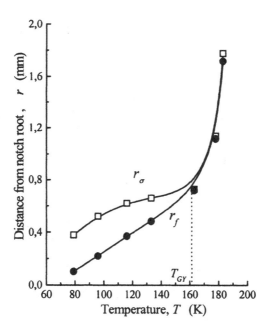

Figure 7. Temperature dependences of distances from the notch root to maximum tensile stress position r_σ and to position of maximum fracture initiation probability r_f for notched tensile specimen with $\rho = 0.6\,\text{mm}$; T_{GY} is general yield temperature.

4. Effect of Metal Microstructure and Plastic Deformation on Weibull Parameers

All these effects may be expressed in terms of Weibull parameters. Three-parameter Weibull distribution is sufficiently "flexible" function. Therefore, a function of cumulative probability of failure for *reference volume* $F_{V_0}(\sigma_F)$ obtained by computer simulation was fitted by three-parameter Weibull distribution [14]. It permits to ascertain relations between Weibull distribution parameters and such microstructural parameters as the most probable grain size and grain size variance (*Figure 8*). It has been exhibited that at uniaxial tension and fixed value of plastic strain the value of shape parameter m

148

is actually independent on the value of the most probable grain size. The value of shape parameter m diminishes with rise in logarithmic grain size variance $D_{\ln d}$. Shape parameter m is linear function of $\sqrt{D_{\ln d}}$ at that case.

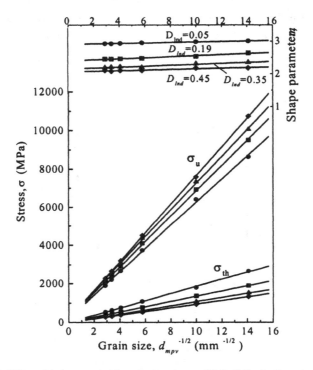

Figure 8. Effect of the most probable grain size d_{mpv} on Weibull distribution parameters; $D_{\ln d}$ is value of variance of grain size logarithm.

The threshold stress σ_{th} is proportional to $d_{mpv}^{-1/2}$ (d_{mpv} be a most probable value of grain size) and depends on grain size variance. This stress is approximately equal to $0.7R_{MC}$ value of shape parameter m is actually independent on the value of the most probable grain size. The value of shape parameter m diminishes with rise in logarithmic grain size variance $D_{\ln d}$. Shape parameter m is linear function of $\sqrt{D_{\ln d}}$ at that case. The threshold stress σ_{th} is proportional to $d_{mpv}^{-1/2}$ (d_{mpv} be a most probable value of grain size) and depends on grain size variance. This stress is approximately equal to $0.7R_{MC}$ (R_{MC} be a minimum brittle fracture stress of unnotched specimens over ductile-brittle temperature region). The normalised scaling stress σ_u/σ_{th} is a linear function of logarithmic grain size variance.

Regularities of plastic strain effect on Weibull distribution parameters are presented in *Figure 9* [14]. By applying this approach, only the shape parameter m is nearly constant over the interval of small plastic strains (i.e. those which do not exceed critical one e_c - *Figure 2*). As strain grows, threshold stress decreases approximately by 1.3 times. The value of σ_u becomes 2.7 times greater.

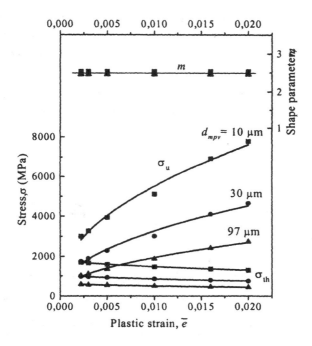

Figure 9. Effect of the plastic strain on Weibull parameters for different grain sizes.

Dependence of σ_{th} and σ_u on the plastic strain value is due to effect of dislocation stresses $\bar{\xi}$ on the crack nucleus unstable equilibrium. As it is follows from Eq.(10), rise of these stresses results in critical stress ξ_c decrease. These results enable to elucidate change in Weibull distribution parameters depending on the test temperature and notch sharpness discovered experimentally [6, 7]

5. Conclusions

Statistical Local approach to brittle fracture takes into account main specific feature of fracture of notched specimens – it is scale effect caused by localisation of fracture initia-

150

tion within extremely small region ahead of notch, process zone. However, Beremin Local Approach does not consider the fact that the extent of scale effect in metal depends on small local plastic strain.

Plastic strain influence on scale effect in metal is due to dependence of the crack nucleus density on the plastic strain value.

Assumption of Beremin's Local Approach version concerning equality of sizes of process zone and local plastic strain region is correct solely when small-scale yielding prevails. At brittle fracture under general yielding process zone size may be several times less than depth of local yielding region. The process zone size is controlled by the value of relative gradient of local plastic strains.

Weibull parameter (threshold stress σ_{th} and scale parameter σ_u) dependence on the plastic strain value is due to dislocation stress influence on the crack nucleus unstable equilibrium.

References

1. Fant, M.Di, et al.: (1996) Development of a simplified approach for using the local approach to fracture, *Journal de physique IV Coloque C6. - supplement au Journal de Physique III* 6, 503-512.
2. Beremin, F.M. (1983) A local criterion for cleavage fracture of a nuclear pressure vessel steel, *Met. Trans.* 14A, 2277-2287.
3. Wiesner, C.S. (1996) *The 'Local Approach' to Cleavage Fracture - Concepts and Applications*, Cambridge, Abington Publishing.
4. Xia L., Fong C. Shih (1996) A fracture model applied to the ductile/brittle regime, *Journal de physique IV Coloque C6. - supplement au Journal de Physique III* 6, 363-372.
5. Kune, K. (1982) *Einfluß des Spannungszustandes und des Gefüges auf die Spaltbruchspannung von Baustahlen*, Aachen, BRD.
6. Wiesner C. S. and Godthorpe M. R. (1996) The effect of temperature and specimen geometry on the parameters of the "Local Approach" to cleavage fracture, *Journal de physique IV Coloque C6. - supplement au Journal de Physique III* 6, 295-304.
7. Kozak V., Dlouhy I., Holzmann M.(2001) The fracture behaviour of cast steel and its prediction based on the local approach, *Nuclear Engineering and Design*, in print.
8. Kotrechko, S.A. (1995) Statistic model of brittle fracture of polycrystalline metals, *Phys. Metals*, 14, 1099-1120.
9. Kotrechko, S.A., Meshkov, Y.Y. (2000) On foundation of physical theory of quasi-brittle fracture of polycrystalline metals in the non-homogeneous fields created by stress concentrations, *Met. Phys. Adv. Tech.* 18, 1393-1412.
10. Kotrechko, S.A., Meshkov, Y.Y., Mettus G.S. (1995) Physical Nature of Strength of Polycrystalline Metal near the Embrittlement Point, *Met. Phys. Adv. Tech.* 14, 1205-1210.
11. Kotrechko, S.A., Meshkov, Y.Y. (1999) Mechanics and physics of quasi-brittle fracture of polycrystalline metals under the conditions of stress concentration, *Strength of Materials* 31, 223-231.
12. Sun J., Boyd, J.D.(1997) Cleavage stress on the delamination plane of a plate steel, *Ninth International Conference on Fracture*, Sydney, Australia, 507.
13. Tsann Lin, Evans A.G. and Ritchie R.O.(1986) Statistical analysis of cleavage fracture ahead of sharp cracks and rounded notched, *Acta metall.* 34, 2205-2216.
14. Kotrechko, S.A., Meshkov, Y.Y, Dlouhy, I. (2001) Computer simulation of effect of grain size distribution on Weibull parameters, *Theoretical and Applied Fracture Mechanics* 35, 2575-260.

PROBABILITY OF BRITTLE FRACTURE IN LOW ALLOY STEELS

B. STRNADEL [1]), I. DLOUHÝ [2])
[1]) *Technical University of Ostrava, Department of Materials Engineering,*
17. listopadu 15, 708 33 Ostrava, Czech Republic
[2]) *Institute of Physics of Materials ASCR, Žižkova 22, 616 62 Brno,*
Czech Republic

Abstract: This paper presents a method for quantifying probability of brittle fracture in low alloy steels. The proposed method shows how the probability of brittle fracture varies with stress intensity factor, temperature, deformation characteristics and microstructural parameters of low-alloyed steel. Application of this method on Ni-Cr steel demonstrated very good agreement of predicted temperature dependence of scatter in brittle fracture toughness values with experimental results. The method enables also to calculate characteristic distance as a radial dimension from the crack tip where microcrack initiation is the most probable. The characteristic distance of investigated Ni-Cr steel was found to decrease with increasing temperature. Microstructural mechanisms of initiation and propagation of brittle fracture were identified from results of fractography analysis. The proposed procedure represents a foundation for systematic control of relationship between stress-strain behaviours, toughness and reliability of steel engineering parts.

Keywords: Failure, micromechanics, crack propagation, probabilistic method, brittle fracture, fracture toughness, microstructural effect

1. Introduction

Mechanical integrity assessment of steel structures frequently requires knowledge of their resistance to catastrophic failure by fast, unstable crack propagation. Then knowledge about material fracture behaviors is obvious benefit to engineers designing steel components, for operation at lower temperatures in particular. However, the determination of steel fracture toughness is accompanied by large scatter, which causes a high risk of catastrophic failure that has to be tolerated. Sources of the scatter in toughness values are firstly experimental and secondly, and more importantly, are caused by microstructural or metallurgical inhomogeneities of steel. Under such circumstances any integrity assessment based on experimentally determined fracture toughness data should include some allowance for this scatter by applying not only statistical analyses of toughness data, but also micromechanical modelling of fracture process formation. To predict the susceptibility of steel structures to catastrophic failure

I. Dlouhý (ed.), Transferability of Fracture Mechanical Characteristics, 151–166.

by fast, unstable crack propagation methods for quantifying probability of fracture respecting metallurgical state of steel microstructure are needed.

Experimental investigations of low energy cleavage fracture in carbon and/or low-alloyed steels showed that carbides act as microstructural inhomogeneities affecting fracture toughness [1-4]. In particular, the slip-induced cracking of carbides, caused by heterogeneous plastic deformation, may result in the initiation of microcracks and their propagation into the surrounding matrix whenever the local component of applied stress exceeds the cleavage strength [4,5]. The subsequent coalescence of initiated microcracks creates a main crack, and ultimately induces fracture instability [6]. Experimental investigations of low temperature brittle fracture in steels have been complemented, over the past years, by attempts to model the fracture process by methods of mathematical statistics [6-14], using local criteria for the initiation of microcracks. These approaches can reveal and quantify the relationship between the microstructural parameters, plastic behaviours of steel and its fracture toughness, and can also describe the temperature dependence of scatter in fracture toughness values computing an integral probability of cleavage failure.

Wallin [15] proved that the theoretical scatter in sharp crack toughness at low temperatures is constant and equals four when three parameters Weibull distribution of fracture toughness was considered. The third datum parameter in statistical distribution of fracture toughness corroborated the existence of a lower limiting toughness value. In this work Wallin [15] also showed that a difference between theoretical and experimental scatter in fracture toughness is caused mainly due to inadequate number of experimental data. Neville and Knott [16] proposed a model to explain the variation of toughness in inhomogeneous materials. They acknowledged the existence of the lower limiting fracture toughness value, which is a function of sampled volume ahead of crack tip or notch. Similarly, Chen et al. [17,18] proposed in their model that the scatter of measured toughness is caused not only by size distribution of carbides but also by distribution of their locations ahead of crack tip.

In all mentioned models parameters influencing the scatter of fracture toughness are discussed separately, and not in their mutual connections.

Then, the present work attempts to compute effects in spatial and size distributions of parameters influencing fracture toughness in their complexity and demonstrates that it is possible to predict the observed fracture toughness variation as a function of temperature. Also, purpose of this work is to find a way, how to estimate the lowest value of fracture toughness at a given temperature. The work thus offers a method for control of microstructural parameters of carbon and low alloy steels during heat treatment so as to ensure an optimum level of their fracture resistance and the least possible scatter of fracture toughness.

2. Experimental Material and Procedures

The material of commercial quality used for the present study was commercial Ni-Cr steel, of composition shown in Table I.

TABLE I. Chemical composition of Ni-Cr steel (in wt. %)

C	Mn	Si	P	S	Ni	Cr	Fe
0.14	0.53	0.28	0.011	0.008	3.52	0.82	balance

The steel was heat treated to give a structure of tempered martensite and roughly spherical carbides. To vary carbide size, the steel was austenitised at 820°C for 4 hours, water cooled, and then spheroidised at 630°C or 550°C for either 100 or 500 hours followed by cooling in still air. Transmission electron microscopy at a magnification of 13.500× was employed to investigate the statistical distribution of carbide sizes $\phi(d_p)$ and area density of carbide particles N_A. The relative frequencies of incidence of various carbide sizes were subjected to statistical processing and the statistical distribution of carbide sizes in the following analytical form was found

$$\phi(d_p) = 1 - \exp[-(d_p / \alpha)^\beta]. \qquad (1)$$

A microstructure with coarser carbides (termed T), with a size parameter $\alpha_T = 0.263$ μm and a shape parameter $\beta_T = 2.28$, was obtained by spheroidising at 630°C for 100 hours, and a microstructure with finer carbides (termed I), with parameters $\alpha_I = 0.212$ μm and $\beta_I = 2.08$, by spheroidising at 550°C for 500 hours. The area density of carbide particles in T and I structures were $N_{AT} = 1.45 \ 10^{12} \ m^{-2}$ and $N_{AT} = 2.29 \ 10^{12} \ m^{-2}$, respectively.

The studied steel was mechanically tested. True stress-strain curves and the yield stress variatiᵣns over the low temperature range from 93 K to 153 K and at room temperature were assessed from uniaxial tensile tests at a strain rate of $\dot{\varepsilon} = 3 \times 10^{-4} \ s^{-1}$. At room temperature 293 K, yield stress values were $\sigma_{0T} = 406$ MPa and $\sigma_{0I} = 541$ MPa for T and I microstructures. Shapes of true stress-strain curves detected in the low temperature range led to the evaluation of strain hardening exponents, $n_T = 5.2$ and $n_I = 5.5$ for T and I structures of the investigated steel. The Young's modulus and the Poisson ratio were taken as $E = 207$ GPa and $v = 0.3$, respectively.

Plane strain fracture toughness, K_{Ic} was evaluated at temperature range of lower shelf and transition region on pre-cracked single-edge-notched specimens 25 mm thick

154

tested in three point bending. The tests were carried out in accordance with standard ASTM E 399-90 at a stress intensity rate of $\dot{K}_I = 2\,\text{MPam}^{1/2}\text{s}^{-1}$. Experimental results of fracture toughness measurements of studied steel with both T and I microstructures are plotted in *Figures 1* and *2*.

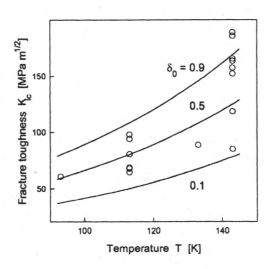

Figure 1. Experimentally determined fracture toughness for Ni-Cr steel with T microstructure, and predicted temperature dependence of fracture toughness scatter band.

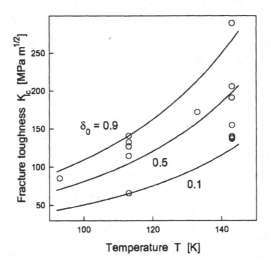

Figure 2. Experimentally determined fracture toughness for Ni-Cr steel with I microstructure, and predicted temperature dependence of fracture toughness scatter band.

Fracture surfaces of fracture toughness specimens tested in the temperature range from 93 K to 153 K were examined by scanning electron microscopy. Fractography investigation indicated that the fracture path of all tested specimens was transgranular. Fracture surfaces of specimens with both T and I structures broken at temperature of - 100 °C (173 K) were found to consist of a short ductile failure zone formed by dimples and directly adjacent to a narrow stretched zone. Brittle fracture zone beyond the ductile zone consisted of cleavage facets. Below 143 K no ductile zone was observed on fracture surfaces and stretched zone emanated from fatigue crack tip was directly followed by brittle fracture morphology.

At lower magnification, the fracture surfaces of bend test specimens exhibited macroscopic river patterns bellow the stretched zone and ductile region. For the microstructure with finer carbides (I) the river patterns have been traced to one discrete initiation region (*Figure 3a*). For the microstructure with coarser carbides (T) this pattern originates at one small spot in the fracture surface (see *Figure 3b*) that contains one or more often a few cleavage facets. In both cases the centre of the river pattern was assumed to be the cleavage initiation origin.

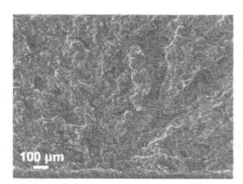

Figure 3a. Fracture appearance below crack tip stretched zone (at bottom) for microstructure I, tested at −100 °C; NiCr steel.

Figure 3b. Fracture appearance below crack tip stretched zone for microstructure T, tested at −100 °C; NiCr steel.

At higher magnification (SEM), the following types of initiation origins were identified:

(i) One cleavage facet without distinct triggering point inside, typical for I state. Average size of the individual facets was about 20 to 30 μm and corresponds about to typical grain size. Dislocation micromechanisms of cleavage triggering was supposed in these cases.

(ii) Cleavage facet with carbide particles (increased metal content in particles in comparison to matrix chemical composition has been proved by Edx microanalysis). Carbide micromechanism of cleavage triggering was supposed in these cases that were observed essentially in microstructures I. In *Figure 4* star the shaped cleavage facet is evidently initiated by local stress concentration in its centre. Here, carbide triggering the facet was presumably dislodged during the rapid fracture event and it is not visible.

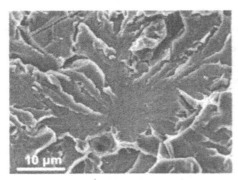 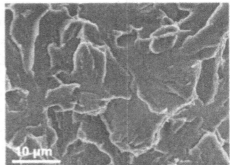

Figure 4. Star shaped cleavage facet nucleated by local stress concentration in its centre, microstructure I; tested at –100 °C.

Figure 5. Cleavage facet morphology, cleavage being triggered by decohesion of carbide particle, microstructure T; tested at –100 °C.

(iii) Two or three cleavage facets arranged in initiation site, all facets without clear discrete triggering point. Cleavage nucleation by dislocation micromechanisms starting from grain boundaries or triple points was supposed in these cases; this behaviour was typical for microstructure T.

(iv) Facets with carbide particles as triggering origin have been observed also in this microstructural type (the carbide particles have been evidenced by increased content of carbide forming elements determined by microanalysis).

Figure 5 documents the predominant failure morphology observed in microstructure T - a smooth microcrack initiated by decohesion of carbide and matrix interface and further propagation of this microcrack forming the cleavage facet. Decohesion of the carbide and matrix, the microcrack extension and the cleavage facet formation are evidently three stages of cleavage in the studied steel. The critical step in this process is microcrack extension appears to be controlled by carbide size.

(v) In both microstructures, I and T, some cleavage facets contained one or a few smaller inclusions located in the middle of the facet.

3. Micromechanical Model

As it follows from fractographic analysis, the dynamic microcrack extension into matrix and smooth facet formation is the controlling micromechanism of brittle fracture in tested Ni-Cr steel. The penny-shaped microcrack initiated by carbide de-cohesion in steel matrix extends rapidly, provided that the local stress σ exceeds the local strength S_p given in the following form of Griffith-Orowan condition:

$$\sigma \geq S_p = \sqrt{\frac{\pi E \gamma_{eff}}{(1-v^2)d_p}} \tag{2}$$

where d_p is carbide particle sized microcrack at carbide-matrix interface caused by de-cohesion and γ_{eff} is effective surface energy.

For small scale yielding conditions of a power hardening material that satisfies the constitutive law of $\varepsilon / \varepsilon_0 = \alpha_0 (\sigma / \sigma_0)^n$, the effective stress at distance r ahead of a macrocrack (*Figure 6*) is given by the HRR singular solution [19]:

$$\sigma_e(r, \theta) = \sigma_0 \left[\left(\frac{1-v^2}{I_n} \right) \left(\frac{K_I}{\sigma_0 \sqrt{r}} \right)^2 \right]^{1/(n+1)} \overline{\sigma_e}(n, \theta) \tag{3}$$

where σ_0 and ε_0 are the yield stress and yield strain, α_0 is a material constant of order unity, n is the work hardening exponent, I_n is dimensionless parameter weakly dependent upon the work hardening exponent n, $\overline{\sigma_e}(n, \theta)$ is an angular function of n and K_I is Mode I stress intensity factor.

Since carbides are in structure of steel precipitated randomly obeying distribution function given by Eq.(1), the probability of microcrack initiation can be expressed as:

$$p_f(\sigma) = \Pr(d_p \geq d_{pf}(\sigma)) = 1 - \phi(d_{pf}(\sigma)) \tag{4}$$

where $d_{pf}(\sigma)$ is the critical size of initiated microcrack in a carbide given for any stress by Eq.(2). To find the probability of brittle fracture instability in pre-cracked body, firstly the elementary probability of a microcrack initiation in volume elements ahead of the macrocrack tip was found. As it is obvious from *Figure 6*, two kinds of volume elements were chosen: isostressed volume element with the constant value of

158

the effective stress and cylindrical volume element with the constant radial co-ordinate. The elementary probability that a microcrack will initiate in at least one carbide within isostressed volume element $\delta V(\sigma)$ ahead of a macrocrack tip (*Figure 6*) can be, in the first approximation, investigated as follows:

$$\delta F(\sigma) = 1 - \exp[-N_V \, p_f(\sigma)\delta V(\sigma)]. \tag{5}$$

Isostressed volume element is easily an integral of Eq.(3) in the form of

$$\delta V(\sigma) = \frac{2(n+1)(1-v^2)^2 bK_I^4 \sigma_0^{2(n-1)}\delta\sigma}{I_n^2 \sigma^{(2n+3)}} \int_0^{\varphi} \overline{\sigma_e}(n,\theta)^{2(n+1)} \, d\theta \tag{6}$$

where φ is an angle of the wedge active region ahead of the macrocrack tip (*Figure 6*), N_V is volume density of carbides, b is the characteristic width of the crack front. [8].

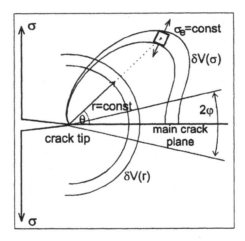

Figure 6. Schematic illustration of isostressed and cylindrical volume elements and wedge active zone ahead of macrocrack tip.

This characteristic width is the maximum distance between two points, along of a crack front, where a cleavage microcrack must be initiated if unstable propagation of the macrocrack is to follow. The distance between these two points recognizes that unstable propagation of the macrocrack requires the contemporary activation of several microcracks initiated along the crack front. The characteristic width could be regarded as some multiply of the grain diameter or might be dictated by the specimen width.

Similarly, the elementary probability of a microcrack initiation within a cylindrical volume element $\delta V(r)$ ahead of the macrocrack tip (*Figure 6*) is:

$$\delta F(r) = 1 - \exp[-2bN_V r\delta r \int_0^{\varphi} p_f(\sigma_e(r,\theta))d\theta].$$ (7)

The integral probability of brittle fracture initiation can now be established by integrating eqn.(5) within the limits of the lowest $S_{p\min}$ and the highest $S_{p\max}$ local strength

$$F = 1 - \exp[-N_V \int_{Sp\min}^{Sp\max} p_f(\sigma)\frac{\delta V(\sigma)}{\delta\sigma}d\sigma].$$ (8)

The lowest $S_{p\min}$ and the highest $S_{p\max}$ local strengths can be calculated using Eq.(2) from the largest $d_{p\max}$ and smallest $d_{p\min}$ carbides given by statistical distribution of them. Another way how to calculate the integral probability of brittle fracture is integration of Eq. (7) in limits of the minimum, $r_{\min}$ and the maximum, $r_{\max}$ distance from the original crack tip given by Eq. (3) for maximum $S_{p\max}$ and minimum $S_{p\min}$ strengths. Using this way the integral probability of brittle fracture initiation is as follows:

$$F = 1 - \exp[-2bN_V \int_{r\min}^{r\max} r\int_0^{\varphi} p_f(\sigma_e(r,\theta))d\theta dr].$$ (9)

Equations (8) and (9) show that the integral probability of brittle fracture initiation is always dependent on stress intensity factor K_I. For any selected probability of cleavage F, these equations allow us to ascertain the critical magnitude of this factor, the fracture toughness, K_{Ic} as a function of the stress distribution ahead of pre-crack tip, of the deformation characteristics of the steel and the statistical distribution of carbides. Since the integral probability of brittle fracture initiation F equals to the complement of distribution function of K_{Ic} values to unity, survival probability $\delta_0 = 1 - F$ represents a selected level of fracture toughness in its distribution.

4. Results and Discussion

The foregoing micromechanical model enabling to predict the brittle fracture toughness was applied on the experimental data of Ni-Cr steel. Firstly, the probability of a

160

microcrack initiation $p_f(\sigma)$ using Eq. (4) was calculated. Variations in $p_f(\sigma)$ with local stress σ for both T and I structures of tested steel at temperature of 133 K are given in *Figure 7*. These calculations were performed supposing that effective surface energy depends on temperature according to equation $\gamma_{eff} = 9.17 + 0.19\exp(0.0183T)$

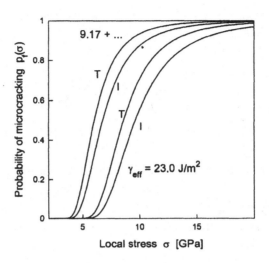

Figure 7. The dependence of microcracking probability on the local stress for two values of the effective surface energy; Ni-Cr steel at 133 K.

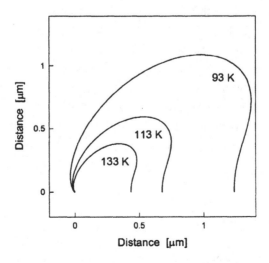

Figure 8. Curves ahead of macrocrack tip where microcracking probability equals to 0.5, for the tested Ni-Cr steel with T microstructure.

[10] or when it is independent on temperature then $\gamma_{eff} = 23$ Jm^{-2} [20]. As it is apparent from *Figure 7* the increase in effective surface energy and refinement of carbides very rapidly lower the probability of microcrack initiation at a given stress. Inserting effective stress σ_e ahead of the crack tip (Eq. (3)) into right side of Eq. (4), a curve

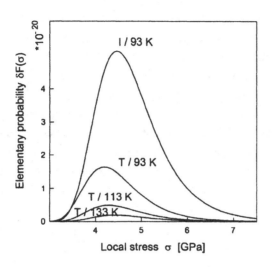

Figure 9. Variation of the elementary probability of a microcrack initiation with local stress, temperature and microstructure of investigated Ni-Cr steel.

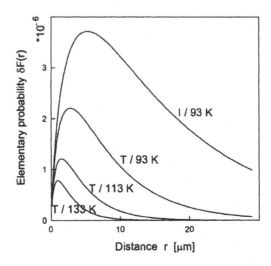

Figure 10. Variation of the elementary probability of a microcrack initiation with radial distance from the macrocrack tip, temperature and microstructure of investigated Ni-Cr steel.

162

encompassed macrocrack tip for any selected probability $p_f(\sigma)$ can be calculated. Graphically, for three different temperatures, for temperature dependent effective surface energy and for $p_f(\sigma) = 0.5$ such curves are given in *Figure 8*. As it is obvious from this figure increasing temperature rapidly reduces volume around crack tip where probability of brittle fracture initiation is larger than 0.5. Above 143 K, as it was proved by fractographic observations, brittle fracture is preceded by ductile one, and also the zone around crack tip in which the probability of brittle fracture initiation is larger than 0.5 becomes extinct. The elementary probability that a microcrack will be initiated by at least one carbide within isostressed volume element $\delta V(\sigma)$ (Eq. (5)) or within cylindrical volume element $\delta V(r)$ (Eq. (7)) was found not to be monotonic function of stress or distance from the macrocrack tip. At some characteristic distance r_f^* from the crack tip, the maximum elementary probability of a microcrack initiation occurs. How the elementary probabilities $\delta F(\sigma)$ or $\delta F(r)$ alter with local stress and distance from macrocrack tip is illustrated for both T and I structures of tested steel and for three different temperatures in Figs. 9 and 10. The characteristic distance controlled by following relationships:

$$\frac{d\delta F(\sigma_e)}{d\sigma_e}\frac{d\sigma_e}{dr} = 0 \quad \text{or} \quad \frac{d\delta F(r)}{dr} = 0 \qquad (10)$$

with increasing temperature diminishes as it is documented in *Figure 11*.

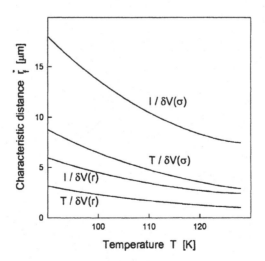

Figure 11. The temperature dependence of characteristic distance calculated from elementary probability of microcrack initiation in isostressed and circular volume elements of Ni-Cr steel in both structural conditions.

Decline of the characteristic distance with increasing temperature indicates that the critical cracking event of a microcrack initiation is closer to the macrocrack tip at higher temperatures. Since the characteristic distance is a result of competition between the size of elementary volume ahead of macrocrack tip and the probability of a microcrack initiation (see Eqs. (5) and (7)) the coarser carbides distribution of T structure exhibits smaller characteristic distance than I structure.

Similarly, characteristic distance computed from elementary probability of microcrack initiation in isostressed volume elements (Eq. (5)) is larger than that computed from probability in cylindrical volume elements (Eq. (7)). This is due to cylindrical volume element that is for the same average actual stress larger than isostressed one.

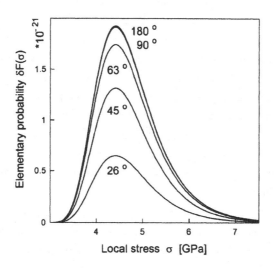

Figure 12. How the dependence of the elementary probability of a microcrack nucleation on the local stress varies with an angle of the wedge active zone ahead of macrocrack tip in Ni-Cr steel with T microstructure at 133 K.

It is evident from the experimental results that direction of macrocrack propagation cannot deflect from the main cross sectional plane more than $\varphi = 5°$. Then for this angle of the wedge active zone elementary probability of microcracking (Eqs. (5) and (7)) was calculated. The effect of the active zone widening is obvious from *Figure 12*. The increase of angle φ influences the elementary probability of microcracking more significantly at lower angles. Then elementary probabilities of microcracking $\delta F(\sigma)$ for angles of $\varphi = 90°$ and $\varphi = 180°$ are almost the same, and the microcrack initiation in the zone bounded by these two angles is practically excluded.

Since initiation of one microcrack along the whole crack front is usually insufficient to cause fracture instability the characteristic width b was introduced to calculate elementary probability of microcracking. Even if in previous papers [8,11,12] the characteristic width was estimated as a multiply of grain size, it is very difficult to find it experimentally. However for a unity characteristic width volume density of carbides N_V could be replaced by the ratio of N_A/b. This simplification is more appropriate for technical applications of the model.

The predicted temperature dependencies of fracture toughness of tested steel in both T and I structural conditions are compared to the experimental results in *Figures 1* and *2*. These figures clearly demonstrate that the predicted statistical variation in fracture toughness (Eqs. (8) and (9)) at levels $\delta_0 = 0.10$, 0.90 of its distribution function confidently embrace the experimental data. *Figures 1* and *2* are completed by a temperature dependence of a median level of K_{Ic}, i.e. for survival probability of $\delta_0 = 0.5$. At higher temperatures above about 150 K some experimental points are outside of the predicted statistical scatter band of fracture toughness. This is in accordance with experimental results of this study proving that around this temperature the controlling mechanism of fracture changes and fracture is started by ductile failure mechanism followed by brittle fracture. Then further work may possibly turn this method into a foundation for the statistical interpretation of ductile crack growth before brittle failure, and for prediction of fracture toughness of steels in transition temperature domain.

Even though the present model concerns with brittle fracture in low-alloyed steels possessing a spheroidised microstructure where the crack propagation is initiated by extension of carbide sized microcracks into a surrounding matrix, the model can also be applied to steels where fracture is controlled by cracking of other microstructural elements. Alternation of the microcrack initiation criterion shown by Eq. (2) may be required.

5. Conclusions

A statistical model for prediction of scatter in low temperature brittle fracture toughness of low-alloyed steels has been presented.

Considering the critical cracking event is dynamic carbide sized microcrack extension into matrix, the elementary probability of brittle fracture initiation has been quantified as a function of microstructural parameters of steels and localization ahead of macrocrack tip.

Elementary probability that a microcrack will initiate within volume element ahead of macrocrack tip has been found to be not monotonic function of stress or distance from the crack tip. The characteristic distance as the radial distance from the macrocrack tip where the microcrack initiation is most probable has been found to diminish with increasing temperature.

Knowing statistical distribution of carbides and stress-strain characteristics of Ni-Cr steel the predicted temperature dependence of fracture toughness and its scatter were found in very good agreement with the experimental results.

The developed procedure represents a foundation for systematic control of microstructural parameters so as to ensure the optimum relation between strength characteristics and fracture toughness of steel.

Acknowledgements

The financial supports of Grant Agency of the Czech Republic under projects Nr. 101/00/0170 and 106/01/0342 to this research are gratefully acknowledged.

References

1. McMahon, C.J. and Cohen, M. (1965) Initiation of cleavage in polycrystalline iron, *Acta Metallurgica* 13, 591-604.
2. Gurland, J. (1972) Observations on the fracture of cementite particles in a spheroidised 1.05 %C steel deformed at room temperature, *Acta Metallurgica* 20, 735-741.
3. Rawal, S.P. and Gurland, J. (1977) Observations on the effect of cementite particles on the fracture toughness of spheroidised carbon steels, *Metallurgical Transactions* 8A, 691-698.
4. Curry, D.A. and Knott, J.F. (1978) Effect of microstructure on cleavage fracture stress in steel, *Metal Science* 12, 511-514.
5. Hahn, G.T. (1984) On influence of microstructure on brittle fracture toughness, *Metallurgical Transactions* 15A, 947-959.
6. Lin, T., Evans, A.G. and Ritchie, R.O. (1986) A statistical model of brittle fracture by transgranular cleavage, *J. Mechanics Physics Solids* 34, 477-497.
7. Curry, D.A. and Knott, J.F. (1979) Effect of microstructure on cleavage fracture toughness of quenched and tempered steels, *Metal Science* 13, 341-345.
8. Evans, A.G. Statistical aspects of cleavage fracture in steel, *Metallurgical Transactions* 14A, (1983), p. 1349.
9. Beremin, F.M. (1983) A local criterion for cleavage fracture of a nuclear pressure vessel steel, *Metallurgical Transactions* 14A, 2277-2287.
10. Wallin, K., Saario, T. and Törönen, K. (1984) A statistical model for carbide induced brittle fracture in steel, *Metal Science* 18, 13-16.

166

11. Lin, T., Evans, A.G. and Ritchie, R.O. (1986) Statistical analysis of cleavage fracture ahead of sharp cracks and rounded notches, *Acta Metallurgica* 34, 2205-2216.
12. Lin, T., Evans, A.G. and Ritchie, R.O. (1987) Stochastic modelling of the independent roles of particle size and grain size in transgranular cleavage fracture, *Metallurgical Transactions* 18A, 641-651.
13. Strnadel, B. and Mazanec, K. (1991) Low temperature transcrystalline failure, *Acta Metallurgica et Materialia* 39, 2461-2468.
14. Meizoso, A.M., Arizorreta, I.O., Sevillano, J.G. and Perez, M.F. (1994) Modelling cleavage fracture of bainitic steels, *Acta Metallurgica et Materialia* 42, 2057-2068.
15. Wallin, K. (1984) The scatter in K_{Ic} – results, *Engineering Fracture Mechanics* 19, 1085-1093.
16. Neville, D.J. and Knott, J.F. (1986) Statistical distributions of toughness and fracture stress for homogenous and inhomogeneous materials, *J. Mechanics Physics Solids* 34, 243-291.
17. Chen, J.H., Yan, C. and Sun, J. (1994) Further study on the mechanism of cleavage fracture at low temperatures, *Acta Metallurgica et Materialia* 42, 251-261.
18. Chen, J.H., Wang, G.Z. and Wang, H.J. (1996) A statistical model for cleavage fracture of low alloy steel, *Acta Metallurgica et Materialia* 44, 3979-3989.
19. Hutchinson, J.W. (1968) Singular behaviour at the end of the tensile crack in a hardening material, *J. Mechanics Physics Solids* 16, 13-31.
20. Gerberich, W.W. and Kurman, E. (1985) New contributions to the effective surface energy of cleavage, *Scripta Metallurgica* 19, 295-298.

DAMAGE MECHANISMS AND LOCAL APPROACH TO FRACTURE

Part I: Ductile fracture

C. BERDIN[1], P. HAUŠILD[1,2]
[1]*Ecole Centrale Paris, LMSS-Mat, Grande Voie des Vignes,*
92 295 Châtenay-Malabry, France
[2]*Czech Technical University, Faculty of Nucl. Sci. & Phys. Eng.,*
Dept. of Materials, Trojanova 13, 120 00 Praha 2, Czech Republic

Abstract: This work aims at the application of local approach to ductile fracture. GTN model is used in order to model damage for different materials. The problem of parameter identification is pointed out. Mesh size dependency of the results is also tackled. GTN model is found to be able to predict ductile fracture for different materials with different damage mechanisms, provided that damage mechanisms are correctly determined: cast iron with high initial void volume fraction, duplex stainless steel with void nucleation as a major part of damage, low alloy steel, with a combination of small initial void volume fraction and void nucleation.

Keywords: local approach, ductile fracture, damage modeling, ductile cast iron, duplex stainless steel, low alloy steel

1. Introduction

Most of the structural components are designed in order to remain in the elastic, reversible part of their mechanical behaviour. For structures submitted to high cycle fatigue, the design is performed in order to keep the structure in such a stress state that the endurance limit is not reached. However, fail safe design needs to predict the fracture of components. This is important for components such as automotive engine supports, which act as breaking pieces in accidental case, but also when the damage tolerance concept has to be applied. This paper concerns the prediction of ductile fracture of materials used for such components.

For cracked structures, the global approach is currently applied. Linear or non-linear fracture mechanics theory (Irwin, Rice) is used in order to predict the propagation of a crack under the "in-service" loading. In this approach, a global parameter (G or J) which is related to the local stress field, is compared to a critical value which depends only on the material properties (fracture toughness). Material mechanical behaviour (characterised by e.g., Young's modulus, Poisson's ratio, yield stress and plastic hardening exponent) and the specimen geometry (with the crack) are sufficient to obtain

167

I. Dlouhý (ed.), Transferability of Fracture Mechanical Characteristics, 167–180.
© 2002 *Kluwer Academic Publishers. Printed in the Netherlands.*

the global parameter which represents the damage level in the structure. Nevertheless, various difficulties arise from this approach:

(a)   the global parameter is only well-known for simple geometry,
(b)   the critical value has to be measured with specific experiment,
(c)   the critical value is dependent on specimen geometry,
(d)   the complex mechanical behaviour of the material (kinematic hardening, loading rate dependence) is not really taken into account.

For these reasons, the transferability of the global approach from laboratory specimens to complex structural components is not ensured.

An alternative approach is the local approach to fracture [1,2] which has been more recently developed. It is based on the following steps:

a)   identification of the physical damage mechanisms and damage variables,
b)   establishment of the relations between the macroscopic local stress-strain fields and the damage variables,
c)   computation of the relevant field variables.

The last step can be achieved using the widespread finite element method. The local approach needs a very good knowledge of the local mechanical behaviour and the physical damage mechanisms of the material. However, it leads to a geometry independent approach, and allows damage assessment for any structural component.

This paper is dedicated to various applications of the local approach to ductile fracture. Comparisons are made with experimental results for different materials presenting very different damage mechanisms. The application of local approach to brittle fracture is proposed in the second part of this paper.

Ductile damage is classically considered to initiate by void nucleation followed by void growth and void coalescence leading to a macroscopic crack. Hence, the damage variable can be simply represented by the void volume fraction. The first model for void growth was proposed by Mac Clintock [3] who computed the exact solution for cylindrical void growth into an infinite rigid plastic matrix under generalised plane strain and axisymmetric loading. An approximate solution was found for spherical void growth into an infinite rigid plastic matrix by Rice and Tracey [4]. It was improved by Budiansky et al. [5] and later by Huang [6]. These models pointed out that stress triaxiality and (plastic) strain are the relevant mechanical variables controlling damage. Ductile fracture is then assumed to occur at a critical void volume fraction, which is supposed to be only material dependent. The applications of this model to experimental results [7,8] showed a correct prediction of the ductile fracture strain evolution versus stress triaxiality.

However, these models are not relevant neither for intermediate and high void volume fraction nor to model crack propagation. In these cases the damage controlling mechanical variables are dependent on the damage level; this needs the so-called "coupled local approach", in which damage variables directly appear in the constitutive equations modelling the mechanical behaviour of the material. The most famous model for ductile fracture modelling is the Gurson's model [9]. Following the same procedure as Rice and Tracey did, this model is based on a micro-mechanical approach for porous material. It was subsequently modified in a more or less heuristic manner by Tvergaard and Needleman [10,11] in order to correctly predict void growth and fracture for hardening material.

This constitutive model requires to identify quite a few parameters: this is one of the main difficulties for the applications of the model [12]. Different ways can be followed for this identification: fully phenomenological method, *i.e.* determination of the damage parameters from macroscopic mechanical response [13], unit cell numerical computation [14], or direct damage kinetics assessment [15]. Any combination of these three methods can be used.

In the following, we propose to review different applications of this model, in order to enlighten the relation between damage mechanisms and model parameter's identification. The size of the characteristic volume for the physical damage mechanisms is an additional parameter for any local approach to fracture [16,17]. This parameter becomes very important when large mechanical gradients are present in the structure to be designed as in the case of pre-cracked component. In a finite element analysis, it is directly related to the element size ahead of the crack tip. The choice of the mesh size at the crack tip will be discussed.

After a short review of the GTN model, three different cases are presented: prediction of fracture toughness of cast iron, strength prediction of pre-cracked bend specimens made of cast duplex stainless steel, and finally prediction of the ductile fracture preceding cleavage in the ductile to brittle transition temperature (DBTT) range for a low alloy steel.

2. Gurson Tvergaard Needleman Model (GTN Model)

The yield function of the GTN model is [10,11] :

$$\phi = \frac{\Sigma_{eq}^2}{\sigma_y^2} + 2q_1 f^* \cosh\left(q_2 \frac{3\Sigma_m}{2\sigma_y}\right) - \left(1 + q_1^2 f^{*2}\right) \tag{1}$$

q_1 and q_2 are parameters, Σ_{eq} the macroscopic von Mises equivalent stress, Σ_m the macroscopic hydrostatic stress, σ_y the flow stress of the matrix, which depends at least on the equivalent plastic strain of the matrix, ε_y, and f^* is the parameter related to the void volume fraction f, by :

$$f = f_0 \qquad \text{before any loading} \tag{2.a}$$

$$f^* = f \qquad f \le f_c \tag{2.b}$$

$$f^* = f + \delta(f - f_c) \qquad f \ge f_c \tag{2.c}$$

$$\delta = \frac{1/q_1 - f_c}{f_f - f_c} \tag{2.d}$$

f_0, f_c, f_f are respectively the initial void volume fraction, the critical void volume fraction at void coalescence and the void volume fraction at fracture. Fracture occurs when $f = f_f$ or $f^* = 1/q_1$. The stress carrying capacity of the material is then lost.

Both nucleation of new voids and growth of existing voids contribute to the increase of the void volume fraction, so its evolution can be expressed as:

$$\dot{f} = \dot{f}_{growth} + \dot{f}_{nucleation} \tag{3}$$

$$\dot{f}_{growth} = (1 - f) \; \mathrm{tr}\left(\underline{\underline{\dot{E}}}^{\,p} \right) \tag{4}$$

$$\dot{f}_{nucleation} = A_n(\varepsilon_y) \, \dot{\varepsilon}_y \tag{5}$$

$\underline{\underline{\dot{E}}}^{\,p}$ is the macroscopic plastic strain rate, and $A_n(\varepsilon_y)$ is a function which depends on the mode of void nucleation.

The parameters q_1 and q_2 are related to the hardening of the matrix and they are difficult to identify because of their strong interaction with the stress-strain behaviour of the matrix which is generally unknown. In this study the classical values (q_1=1.5 and q_2=1)[11] where chosen.

The Abaqus/standard code [18] was used for numerical modelling. The GTN model, not readily available in the software package, was implemented via a 'user subroutine' provided by GKSS [19]. The stress carrying capacity loss of the material at fracture was modelled by a drastic drop of the elastic rigidity of the finite element.

3. Applications

3.1. FRACTURE TOUGHNESS OF NODULAR CAST IRON

In nodular cast iron, since specimen thickness exerts a large influence on toughness [20,21], global approach is difficult to apply. So, deformation and damage mechanisms were investigated, and the local approach was applied in order to predict fracture toughness.

In this material, damage occurs mainly by debonding of the interface at the pole cap of the nodules [22]. Debonding propagates along the interface between matrix and graphite nodules thus inducing void nucleation (*Figure 1*). This step is followed by void growth and void coalescence. In fact, depending on the material, damage kinetics can be different [23]. Two cast irons were studied; GS52 with 10% of graphite nodules and GGG40 cast iron with 7.7% of graphite nodules. The chemical composition of ferritic matrix in the both materials differed mainly by the Si and Ni contents, which play a major role in plastic hardening.

The damage evolution during tension was observed thanks to in-situ tensile tests *i.e.* tensile tests performed into a scanning electron microscope [22]. For the GGG40 cast iron, most of the graphite nodules were debonded from the matrix in the very early stages of macroscopic yielding. For the GS52 one the decohesion of the graphite/matrix interface seemed to start later, at about 2% of macroscopic strain with a more heterogeneous damage spatial distribution.

As the presence of voids modifies the stiffness of the material, the evolution of the Young's modulus was measured as a function of plastic strain (*Figure 2*). This has been achieved by tensile loading followed by unloading after various strain increments according to the procedure described in reference [22].

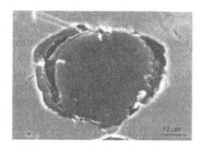

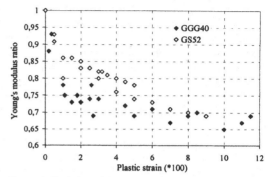

Figure 1. Damage in nodular cast iron: debonding of graphite nodule during tensile test (the direction of tension is horizontal).

Figure 2. Evolution of the Young's modulus ratio versus plastic strain during a tensile test.

Damage kinetics can consequently be precisely characterised by Young's modulus measurements associated with in situ tensile test observations. In GGG40 from the results presented in *Figure 2*, one can consider that all the graphite nodules are debonded when the macroscopic yield strength of the material is reached; so the initial void volume fraction can be the graphite nodule volume fraction, and only void growth can be considered for damage mechanism.

Once damage kinetic is established, it is then possible to identify the hardening law of the matrix $\sigma_y(\varepsilon_y)$ using an inverse method: the parameters of a Ludwik's law were adjusted in order to give a good fit of the hardening of the damaged material, *i.e.* the mechanical behaviour in tension.

The void volume fraction at coalescence, f_c, was identified in order to predict the fracture diameter of notched specimen (f_c=0.12). Fracture elongation measured by tensile test cannot be used to determine the fracture parameters [24] because it is strongly dependent on the viscoplasticity of the material, so that the strain rate dependence of the mechanical behaviour has to be precisely established. The diameter at fracture for notched specimen appears to be less sensitive to strain rate. The measurement of the area ratio occupied by dimples over total fracture surface observed with a scanning electron microscope allowed the determination of void volume fraction at fracture (f_f = 0.2 or δ=6.8)

The mesh size at the crack tip of the CT specimen used for fracture toughness assessment was chosen in order to account for the experimental crack opening displacement at crack initiation measured by the multiple specimen method [25]. It was found to be 100*100 (microns)^2. This is about 3 times the mean nearest neighbour distance between graphite nodules, whereas Steglich et al. [26] found about 6 times of this distance for very similar materials. In fact, it should be noted that in this work, the distance between integration points of finite elements is about the mean nearest neighbour distance between graphite nodules: the volume associated to the integration point is therefore corresponding to the elementary representative volume with respect to the damage process.

For the GS52 cast iron, direct (in-situ tensile test) and non-direct (Young's modulus measurements) damage observations showed a more progressive damage kinetics than in GGG40 (see for example *Figure 2*): a void nucleation process with constant void nucleation rate can be assumed: $A_n(\varepsilon_y)=A_n$. the initial void volume fraction was obtained from Young's modulus measurements via a relation between Young's modulus ratio and void volume fraction [27] ($f_0=0.04$). The void nucleation rate and the matrix hardening law were numerically adjusted in order to correctly predict the tensile and CT tests. Since the mean nearest neighbour distance was not so different between both cast iron, the same values of f_c, f_f and mesh size at the crack tip as for GGG40 cast iron were used.

Prediction of P_Q [28] is correct as it can be seen in TABLE I; for 2D computation, experimental value lies between the results from the two extreme assumptions, that is plane stress assumption for the lower bound, and plane strain assumption for the upper bound [23].

The difference in the fracture toughness of these two cast irons seems to be due to the existence of different damaged zone sizes, related to different damage kinetics : the spatial distribution of the graphite nodules is pointed out as the main origin of the difference in damage kinetics for the two materials as it was modelled. In GS52, the spatial distribution of voids is less homogeneous than in GGG40. This is emphasised by the higher graphite content in GS52. Aggregates of graphite nodules lead to a more localised damage. In the mechanical modelling, this is taken into account in a phenomenological way, thanks to the void nucleation law.

TABLE I. Comparison between experimental values of P_Q [28] and computed values

P_Q (kN)	GGG40	GS52
Experiment	12	17.5
Computation	11.3-13*	17.8**

* 2D-plane stress and 2D-plane strain computation
** 3-D computation

3.2. FRACTURE TOUGHNESS OF DUPLEX STAINLESS STEEL

CF8M duplex stainless steels are widely used for temperature applications because of their good mechanical properties and corrosion resistance. The fracture toughness of this material was investigated in order to compare the influence of shrinkage cavities aggregate and an equivalent crack on the mechanical behaviour of a 3-point bend specimen [29].

The microstructure of CF8M is composed of interconnected ferritic lathes and austenitic islands (*Figure 3*). The material under study contained 25% volume fraction of ferrite. During ageing below 400°C, spinodal decomposition of ferrite occurs and increases ferrite brittleness. Fracture of the thermally aged material under monotonic loading is induced by crack nucleation in embrittled ferrite (*Figure 4*), followed by ductile tearing of the austenite ligaments between micro-cracks (e.g. [30]).

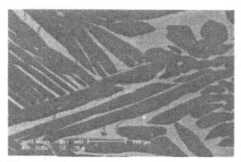 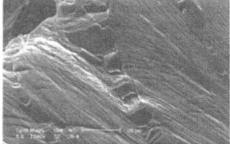

Figure 3. Microstructure of cast duplex stainless steel (dark grey: ferrite – white grey: austenite)   *Figure 4*. Damage mechanisms: cracking into a ferrite lath under the main fracture surface

The physical mechanisms of deformation and damage being correctly identified, a local approach can be applied. Recently Besson et al. [30] following [15] have proposed a model of duplex stainless steel mechanical behaviour based on the local approach to ductile fracture, which was used in [29]: void nucleation (i.e. cleavage microcracking) was considered to be driven by plastic strain, as suggested by experimental observations [15] with a constant rate (as for cast iron GS52). No initial void volume fraction was considered.

For that material, matrix flow stress was first determined thanks to compression loading experiments assuming that no damage occurs during this test. The damage parameters, specially A_n, were then adjusted in order to predict tensile test results. No void coalescence law (Eq. 2) was needed, so $f_c = f_f = 1/q_1$.

Modelling a pre-cracked three points bend specimen [29], a relationship was found between the void nucleation rate and the mesh size in front of the crack (*Figure 5*). For the damage mechanisms under consideration, the elementary representative volume should contain a sufficient number of ferrite lathes with favourable crystallographic orientations for cleavage. Since the nucleation rate is an average coefficient over this representative volume, there is a strong correlation between this parameter and the mesh size. Therefore, delayed crack initiation due to the use of a coarse mesh can be compensated by an increase of the nucleation rate taking into account the higher actual plastic deformation at the crack tip: for a given nucleation rate, the predicted strength of the pre-cracked bend specimen increases with the mesh size at the crack tip (*Figure 5.a*), whereas for a given mesh size, increasing the nucleation rate induces an earlier damage and a lower predicted strength of the modelled specimen (*Figure 5.b*).

174

For 3D modelling, the element size at the crack tip was imposed as 0.5x0.5x1.25 mm³ in order to reduce the computation time. A nucleation rate was identified in order to account for the strength of the pre-cracked specimen. The optimised value is $A_n=1$, compared to 0.6 which was the value determined in [30] for tensile test prediction; the associated mesh size was 0.25x0.25mm² in section, which is approximately two or three times the mean distance between the ferrite laths. Since it has been identified for a given mesh size, the nucleation rate becomes a fully phenomenological parameter and is no more available for tensile tests prediction. However, this procedure can be interesting for the modelling of large structural cracked component, even though the initial physical basis of the void nucleation rate is partly lost.

The results of 3D numerical modelling of the pre-cracked specimen is shown in *Figure 6* and compared with experimental results (load drop corresponds to an unloading imposed by the operator as soon as crack initiation occurred). This comparison shows that the load versus deflection is correctly predicted.

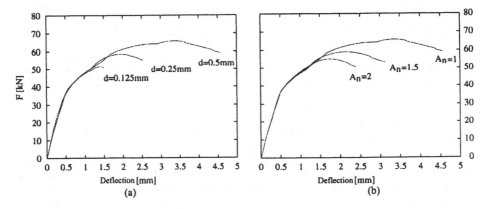

Figure 5. Influence of mesh size and void nucleation rate on the prediction of the mechanical behaviour of precracked specimens (2D plane strains). (a) constant void nucleation rate: An=1; (b) constant mesh size constant: d=500x500 µm².

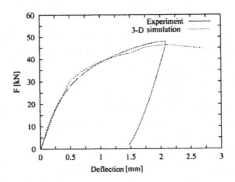

Figure 6. Comparison between experiment and 3-D modelling for a precracked specimen (An=1, d=500x500 µm² in section).

3.3. DUCTILE FRACTURE PREDICTION IN THE DBTT RANGE

The examples presented above concerned only crack initiation and fracture toughness prediction. The following work focuses on ductile crack propagation modelling in the DBTT range in a low alloy bainitic steel (16MND5). In this temperature range, ductile fracture precedes cleavage. The second part of this paper deals with the cleavage fracture prediction. Ductile crack advance is shown to modify significantly the mechanical field thus inducing a significant increase of the maximum principal stress [31,32,33] in Charpy V-notch (CVN) specimen, as well as an increase of the loaded volume susceptible to contain potential cleavage initiation sites. So, for cleavage fracture prediction, ductile fracture has to be well represented.

Three types of inclusions playing a role in ductile fracture were distinguished in this material:

a) large elongated Manganese sulfide (MnS) inclusions (10μm in diameter of short section, 100μm in length) usually formed clusters,

b) smaller spheroïdal MnS inclusions and oxydes (1μm in diameter),

c) carbides (0.1μm)

According to the chemical composition (see part II) the volume fraction of carbides was about 6.10^{-3}, and the volume fraction of MnS together with part of oxydes was estimated at 5.10^{-4} [34]. For the mechanical test considered, the elongated inclusions were mainly loaded perpendicular to their length which induced early debonding. This can be observed in the *Figure 7*, which is a cross section of a CT specimen loaded at 0°C: large cavities coming from MnS initiating void growth are linked by smaller voids nucleated on finer particles.

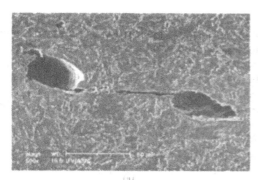 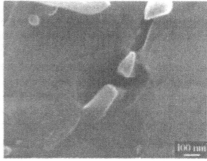

(a)   (b)

Figure 7. Ductile damage in the low alloy bainitic steel. (a) Void growth on large MnS inclusions linked by void nucleation on smaller particles. Cross section of a CT specimen loaded at 0°C (nital etching); (b) nucleation and void growth from carbides near the notch root in a CVN specimen tested at –30°C.

Following schematically these observations the initial void volume fraction was chosen as the volume fraction of MnS with part of oxydes. A gaussian void nucleation law (Eq. 6) was considered [35], in order to account for the second population of cavities. Three parameters are included in this function: the volume fraction of nucleated void f_n, which is related to the particle volume fraction inducing void, the mean plastic strain ε_n of the matrix for which the nucleation rate is maximum and the standard deviation s_n, indicating the range of plastic strain inducing void nucleation; the total volume fraction of nucleated voids, f_n, was taken as 25% of the carbides content [36]. For the other nucleation parameters, classical values were used : ε_n=0.3 and s_n=0.1 [35].

$$A_n\left(\varepsilon_y\right)= \frac{f_n}{s_n\sqrt{2\pi}}\exp\left[-\frac{1}{2}\left(\frac{\varepsilon_y-\varepsilon_n}{s_n}\right)^2\right] \tag{6}$$

The mechanical behaviour of the matrix was modelled by a Cowper-Symonds law to account for the high strain rates applied on Charpy impact specimens and identified on a large experimental data base [31,32]. A Linear temperature dependence was also included. So σ_y was defined by the following expression:

$$\sigma_y\left(\dot{\varepsilon}_y,\varepsilon_y,T\right)= \sigma_y^{(1)}\left(\varepsilon_y\right)\left(1+\left(\frac{\dot{\varepsilon}_y}{D}\right)^{1/p}\right)\left(\alpha-\beta T\right) \tag{7}$$

where D, p, α, β were parameters, and $\sigma_y^{(1)}$ the static hardening law. Parameters values can be found in [31,32].

The parameters for void coalescence and fracture prediction (f_c and δ) were identified in a phenomenological way on smooth and notched tensile test results. They were adjusted, in order to correctly predict the diameter at fracture of specimens tested at 0°C (f_c=0.04, δ=4 which is still a common value for obtaining a high void growth acceleration with a good numerical stability). The parameters associated with ductile damage were supposed to be independent on temperature and strain rate for the ranges concerned here (10^{-4}-1000 s^{-1} and –90°C-0°C). It should be noted that matrix hardening was found to be independent on these variables, so that q_1, q_2 which depend on matrix hardening can be considered as constant too.

A mesh size of 100*100µm^2 in section provides a correct ductile crack growth rate for CT specimens tested at 0°C (*Figure 8.a*). In this study the mesh size identification for ductile crack extension modelling in CT specimen is validated by the ductile crack prediction for Charpy specimens (*Figure 8.b*). Furthermore, the ductile crack growth is correctly predicted all along the loading. Quantitative microstructure analysis [37] revealed a 50µm mean nearest neighbour distance between large MnS inclusions, which is about half of the mesh size.

Figure 9 shows the comparison between macroscopic fractography of a Charpy specimen and iso-contour of the maximum principal stress which is the mechanical variable associated with brittle damage (part. II). It can be seen that the ductile crack advance is very well predicted (region where maximum principal stress value is about zero) with crack tunnelling propagation. However, shear lips are not predicted, which is a classical result with that model, since no void growth can occur under pure shearing. Consequently, lateral necking is slightly overestimated. The same figure shows that the cleavage initiation point lies in the region of higher maximum stress value; this is in agreement with the classical theory of brittle fracture (Part. II).

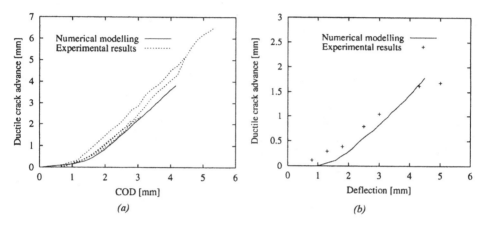

(a)   (b)

Figure 8. Ductile crack advance versus loading (deflection for KCV specimen and COD, crack opening displacement for CT specimen). Comparison between experiment and modelling. (a) side-grooved CT specimen tested at 0°C (2-D modelling); (b) KCV specimen tested at –30°C (3-D modelling).

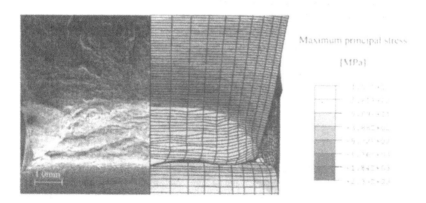

Figure 9. Comparison between fractography of a Charpy specimen tested at -30°C, and iso-contour of the maximum principal stress.

4. Conclusions

Different materials with different damage kinetics were modelled with the GTN model:
a) cast iron with high initial void volume fraction,
b) duplex stainless steel with void nucleation as a major part of damage,
c) low alloy steel, with a combination of small initial void volume fraction and void nucleation.

The identification of the damage parameters and hardening laws of matrix were performed thanks to a combination of direct damage measurements (in situ tensile tests, fractography analysis), non-direct damage measurements (Young's modulus measurements, comparison between compressive and tensile tests results), and in a phenomenological way, comparing macroscopic mechanical test results and the model prediction. The last procedure was essentially used for the determination of the void volume fraction at void coalescence, f_c, the void growth acceleration slope δ, as well as for fitting the mesh size. This last parameter can be defined only thanks to cracked specimen mechanical results. Mesh size is closely related to the elementary representative volume associated with the damage mechanisms. It shall be larger at least than this characteristic volume, what is effectively found in these studies.

However, the relation between mesh size and microstructure is not so clear, and such a mesh size dependence of the modelling is a real problem for fracture prediction of cracked components, since mesh size shall be chosen in order to resolve the mechanical fields, but may be still dependent on the problem size and on a reasonable computation time. These criterion can be in contradiction with a mesh size allowing a correct prediction of ductile crack initiation and growth by the GTN model. So, the best way to resolve that sort of conflict is certainly to use non-local approach of ductile fracture [38] which includes a length scale directly into the constitutive equations of the model, associated with adaptive meshing. However, waiting for these developments, a relation can be found between damage parameters and mesh size as it was demonstrated in the case of duplex stainless steel.

Finally, The GTN model allows good fracture prediction in very different cases if the parameter's identification is performed on a sufficiently large database. Damage parameters were found to be strain rate and temperature independent (at least in the ranges of identical plasticity mechanisms, i.e. outside the adiabatic shear and creep domains). Nevertheless, this result should be confirmed on other materials.

References

1. Lemaitre, J. (1986) Local approach of fracture, *Engineering Fracture Mechanics* 25, 523-537.
2. Mudry, F. (1987) A local approach to cleavage fracture, *Nuclear Engineering and Design* 105, 65-76
3. Mc Clintock (1968) On the ductile enlargement of voids in triaxial stress fields, *J. of Mechanics Physics and Solids* 17, 201-217.
4. Rice, J.R. and Tracey, D.M. (1969) On the ductile enlargement of voids in triaxial stress fields, *J. of Mechanics and Physics of Solids* 17, 201-217.
5. Budiansky, B., Hutchinson, J.W. and Slutsky, S. (1982) Void growth and collapse in viscous solids, in Mechanics of Solids, eds Hopkins H.G., Sewell M.J., 13-45.
6. Huang, Y. (1991) Accurate dilatation rates for spherical voids in triaxial stress fields, *J. of Applied*

Mechanics 58, 1084-1086.

7. Hancock, J.W. and Mackenzie, A.C. (1976) On the mechanisms of ductile failure in high strength steels subjected to multi-axial stress-states, *J. of Mechanics and Physics of Solids* 24, 147-169.

8. Marini, B., Mudry, F. and Pineau, A. (1985) Experimental study of cavity growth in ductile rupture, *Engineering Fracture Mechanics* 22, n°6, 989-996.

9. Gurson, L.A. (1977) Continuum theory of ductile rupture by void nucleation and growth, *J. of Engineering Materials and Technology*, Jan., 2-15.

10. Tvergaard, V. (1981) Influence of voids on shear band instabilities under plane strain conditions, *International J. of Fracture* 17, 389-407.

11. Tvergaard, V. (1982) On localization in ductile material containing spherical voids, *International J. of Fracture* 18, n°4, 337-252.

12. Zhang, Z.L. and Hauge, M. (1998) On the micro-mechanical parameters, ASTM-STP 1321.

13. He, R., Steglich, D., Heerens, J., Wang, G.W., Brocks, W. and Dahms, M. (1998) Influence of particle size and volume fraction on damage and fracture in Al-Al3Ti composites and micromechanical modelling using the GTN model, *Fatigue and Fracture of Engineering Materials* 21, 1189-1201.

14. Brocks, W., Sun, D.Z. and Hönig, A. (1996) Verification of micromechanical models for ductile fracture by cell model calculations, *Computational Material Science* 7, 235-241.

15. Joly, P., Pineau, A. and Meyzaud, Y. (1993) Fracture micromechanisms of an aged duplex stainless steel; application to the simulation of fracture of notched tensile and compact specimen, in Proceedings Mecamat 93: International seminar on micromechanisms of materials, 210-221.

16. Lemaitre, J. and Dufailly, J. (1987) Damage measurements, *Engineering Fracture Mechanics* 28, n°5-6, 643-661.

17. Rousselier, G. (1987) Ductile fracture models and their potential in local approach to fracture, *Nuclear Engineering and Design* 105, 97-111.

18. HKS ABAQUS/Standard, Version 5.8 (1998), Theory Manual, Hibbit, Karlsson and Sorensen Inc., USA.

19. Siegmund, T. and Brocks, W. (1997) A user-material subroutine incorporating the modified Gurson-Tvergaard-Needleman model of porous metal plasticity into ABAQUS finite element program, Technical report GKSS/WMG/97/2, GKSS, Geesthacht, Germany.

20. Bradley, W.L. and Srinivasan, M.N. (1990) Fracture and fracture toughness of ductile iron and cast steel, *International Materials Reviews* 35, n°3, 129-161.

21. Kobayashi, T. and Yamada, S. (1994) Evaluation of static and dynamic fracture toughness in ductile cast iron, *Metallurgical and. and Materiala Transactions* 25A, 2427-2437.

22. Dong, M.J., Prioul, C. and François, D. (1997) Damage effect on the fracture toughness of nodular cast iron: Part I and Part.II, *Metallurgica and Materiala Transactions* 28A, 2245-2262.

23. Berdin, C., Dong, M.J. and Prioul, C. (2001) Local approach of damage and fracture toughness for nodular cast iron, *Engineering Fracture Mechanics* 68, 1107-1117.

24. Bernauer, G. and Brocks, W. (2000) Micromechanical modelling of ductile damage and tearing – Results of a european numerical round robin, Report GKSS 2000/15.

25. Dong, M.J., Berdin, C., Béranger, A.S. and Prioul, C. (1996) Damage effect in the fracture toughness of nodular cast iron, *J. de physique* IV, 65-74.

26. Steglich, D. and Brocks, W. (1998) Micromechanical modelling of damage and fracture of ductile materials, *Fatigue and Fracture of Engineering Materials* 21, 1175-1188.

27. Bompard, P. and François, D. (1984) Damaging effects of porosity on fracture of sintered Nickel Proceedings of ICF6, New Dehli, 1279.

28. ASTM, Standard method for measurement of fracture toughness, E1820-99.

29. Haušild, P., Berdin, C., Bompard, P. and Verdière, N. (2001) Ductile fracture of duplex stainless steel with casting defects, *International J. of Pressure Vessels and Piping* 78, 607-616.

30. Besson, J., Devillers-Guerville, L. and Pineau, A. (2000) Modeling of scatter and size effect in ductile fracture: application to thermal embrittlement of duplex stainless steels, *Engineering Fracture Mechanics* 67, 169-190.

31. Rossoll, A. (1998) Détermination de la ténacité d'un acier faiblement allié à partir de l'essai Charpy instrumenté, Ph.D thesis, Ecole Centrale Paris, France.

32. Rossoll, A., Berdin, C. and Prioul, C. (2002) Determination of the fracture toughness of a low alloy steel by the instrumented Charpy impact test, *International J of Fracture*, in press.

33. Haušild, P., Nedbal, I., Berdin, C., and Prioul, C. (2002) The influence of ductile tearing on fracture energy in the ductile-to-brittle transition temperature range, *Material Science and Engineering A*, in press.

34. Mäntylä, M., Rossoll, A., Nedbal, I., Prioul, C. and Marini, B. (1999) Fractographic observation of cleavage fracture initiation in a bainitic A508 steel, *Journal of Nuclear Materials* 264, 257-262.
35. Chu, C.C. and Needleman, A. (1980) Void nucleation effects in biaxially stretched sheets, *Journal of Engineering Materials and Technology* 102, 249-256.
36. Fisher, J.R. and Gurland, J. (1981) Void nucleation in spheroidized carbon steels- part1. Experiment, *Metals Science* 15, 185-192.
37. Haušild, P. (2001) Transition ductile-fragile d'un acier faiblement allié, Internal Report, Ecole Centrale Paris, L.MSS-Mat, France.
38. Tvergaard, V. and Needleman, A. (1995) Effects of nonlocal damage in porous plastic solids, *International J. of Solids and Structures* 32, n°8-9, 1063-1077.

DAMAGE MECHANISMS AND LOCAL APPROACH TO FRACTURE

Part II: Brittle fracture prediction in the ductile to brittle transition

P. HAUŠILD[1,2], C. BERDIN[1]
[1]*Ecole Centrale Paris, LMSS-Mat, Grande Voie des Vignes,*
92 295 Châtenay-Malabry, France
[2]*Czech Technical University, Faculty of Nucl. Sci. & Phys. Eng.,*
Dept. of Materials, Trojanova 13, 120 00 Praha 2, Czech Republic

Abstract: In this work, the use of statistical local approach to fracture in the ductile-to-brittle transition region is examined. The Beremin model is applied to predict the cleavage fracture probability. The Weibull parameters are identified from instrumented Charpy test results and with the finite element analysis. These parameters are then used for the prediction of probability distribution of the fracture toughness in the DBTT range. The Weibull parameters were found to vary with temperature. However, taking into account the evolution of cleavage micromechanisms in a phenomenological way, i.e. evolution of Weibull parameters with temperature is not sufficient to predict correctly the fracture toughness from the Charpy impact tests. The influence of other variables as e.g. strain rate on the fracture criterion is discussed.

Keywords: local approach, ductile to brittle transition, brittle fracture prediction, Weibul parameters, Charpy test, strain rate effects

1. Introduction

The first part of the present paper was dedicated to the challenge of ductile fracture prediction; the second part contributes to brittle fracture, especially cleavage fracture modelling. The fracture prediction of low alloy steels in the ductile-to-brittle transition temperature (DBTT) range is complicated, because two fracture mechanisms are in competition here – ductile fracture characteristic for high temperatures, and transgranular cleavage characteristic for low temperatures.

Cleavage fracture can be characterised as a sequential stochastic process of crack nucleation and crack propagation by a local re-initiation. The growth stage can further be divided into initial growth of microcracks and consequent growth of a macrocrack as the first obstacles are encountered. It is generally recognized that the development of cleavage can be described by the Griffith's criterion, in which the critical stress, σ_c, is inversely proportional to the square root of the length of the microcrack [1]:

I. Dlouhý (ed.), Transferability of Fracture Mechanical Characteristics, 181–194.
© 2002 *Kluwer Academic Publishers. Printed in the Netherlands.*

$$\sigma_c = \sqrt{\frac{const}{l_c}} \qquad (1)$$

where *const* depends on Young modulus, E, effective surface energy, γ_{eff}, and the shape of the crack.

In low alloy mild steels, the critical fracture event is commonly assumed to be a propagation of carbide microcracks into the surrounding ferrite matrix [2,3]. The role of other second phase particles as MnS was mentioned in e.g. [4-6]. A model relating the fracture toughness of mild and/or low alloy steels to its yield and fracture stresses through a microstructurally determined characteristic distance has been proposed by *Ritchie, Knott and Rice* [7]. The fracture criterion is given by the condition that the stress must exceed the critical stress over some critical distance, X_o, ahead of the crack tip.

In general, the characteristic distance has to be determined empirically since no simple relation exists between it and microstructural parameters, e.g., grain size [3].

The probability of sampling the particles of critical size to fulfill the Griffith criterion of crack propagation was taken into account by *Curry and Knott* [8] and later by *Tsann Lin et al.* [9]. The orientation of microcracks nucleated in carbide particles prior to cleavage planes in the matrix was considered by *Strnadel et al.* [10]. A model based on the presence of two independent barriers for cleavage propagation (the particle/matrix interface, and the grain boundary) was proposed by *Martin-Meizoso et al.* [11]. With these approaches, the increasing scatter of K_{Ic} data can be described. The probability of failure induced by cleavage can also be predicted using the statistical local approach to fracture proposed by *Beremin* [12] and the fracture toughness in the lower shelf was successfully predicted from instrumented Charpy tests by *Rossoll* [13].

Nevertheless, none from the above mentioned models could explain the sharp upturn in the ductile-to-brittle transition. To overcome this discrepancy, a variation of parameters with temperature was introduced in some models as for example increasing microcrack arrest capacity at the carbide/ferrite interface $K_{Ia}^{c/f}$ [11,14].

In this paper, the probability of cleavage fracture in the DBTT range is modelled by Weibull statistics. For that purpose, local mechanical fields are computed by finite element method. Weibull parameters are identified on Charpy impact tests and used for fracture toughness prediction. The numerical results are compared to experimental values. The influence of temperature on the fracture criterion is studied. The influence of other variables as e.g. strain rate is discussed.

2. Local Approach to Brittle Fracture

The classical fracture mechanics using K_{Ic} or J_c concepts can hardly describe the unstable brittle fracture in the DBTT range, where ductile and brittle fracture are in competition and stable ductile crack growth may precede final brittle fracture. Hence, the local approach, having better physical basis, becomes an interesting tool as discussed in part I for ductile fracture.

The local approach aims to predict the fracture of any structural component using local criteria, providing that the mechanical fields in the structure are known [15,16].

Cleavage fracture will be predicted by the model developed by *Beremin* [12]. In this model based on the weakest link assumption, the number of microcracks (e.g. *Figure 1*) in the reference volume, V_o, is assumed to be a power law function of the microcrack length (e.g. [16,17]). The probability of finding a microcrack of length between l and $l + dl$ is therefore:

$$P(l)\, dl = \frac{\alpha}{l^\beta} dl \qquad (2)$$

where $\alpha > 0$, and $\beta > 1$.

Assuming the Griffith formula (Eq. 1), the failure probability in a reference volume, V_o, subjected to a stress σ is:

$$p(\sigma) = \int_{l_c}^{\infty} P(l)\, dl = \left(\frac{\sigma}{\sigma_u} \right)^m \qquad (3)$$

where $m = 2\beta - 2$, and $\sigma_u = \sqrt{const} \sqrt[m]{\frac{m}{2\alpha}}$

The cumulative fracture probability of the entire specimen is given by the product:

$$P_F = 1 - \prod_{i=1}^{n} \left[1 - p(\sigma_i) \right] \qquad (4)$$

If all the probabilities $p(\sigma_i)$ are small, Eq. (4) can be rewritten:

$$ln(1 - P_F) = \sum_{i=1}^{n} ln\left[1 - p(\sigma_i) \right] \approx \sum_{i=1}^{n} -p(\sigma_i) \qquad (5)$$

The Fracture probability is then given by a two-parameter Weibull's distribution [18] as:

$$P_F = 1 - exp\left[-\left(\frac{\sigma_w}{\sigma_u} \right)^m \right] \qquad (6)$$

where m and $\sigma_u V_o^m$ are material dependent parameters, and the so called Weibull stress, σ_W, is defined as an integral of the positive maximum principal stress, σ_1, over the plastic volume, V_{pl}:

$$\sigma_w = \sqrt[m]{\int_{V_{pl}} \sigma_1^m \frac{dV}{V_o}} \qquad (7)$$

Some authors (e.g. [18,19]) also use a three-parameter statistic introducing in Eq. (7) a threshold stress, σ_{th}, below which the cleavage fracture probability is zero.

$$\sigma_w = \sqrt[m]{\int_{V_{pl}} (\sigma_1 - \sigma_{th})^m \frac{dV}{V_o}} \qquad (8)$$

This may have some advantages for description of low fracture probabilities. However, if the threshold value is unknown, it should be identified together with other parameters having as a consequence considerable complication of the identification procedure. The two-parameter statistic will therefore be used.

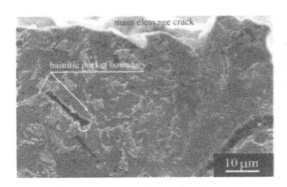

Figure 1. Cleavage microcrack located under the main cleavage crack. (A 508 Cl.3 steel, cross section of CT 25 specimen tested at 0 °C, 2% Nital etching).

3. Experimental Basis

The material chosen for this study is the French 16MND5 pressure vessel steel considered as an equivalent to the American standard A508 Cl.3. The chemical composition is given in TABLE I. The material had undergone a thermal treatment consisting of two austenitisations at 880 °C followed by water-quenching, recovery annealing at 640 °C and a final stress relief treatment at 610 °C.

TABLE I: Chemical composition of A508 Cl.3 steel (wt.%)

C	S	P	Mn	Si	Ni	Cr	Mo	Cu	Al
0.159	0.008	0.005	1.37	0.24	0.70	0.17	0.50	0.06	0.023

The standard compact tension (CT) and Charpy V-notch (CVN) specimens were taken from a nozzle cut-out of a pressure vessel (at ¾ thickness from the inner wall). The specimens were sampled in the T-S (long transverse-short transverse) orientation.

The tests were carried out at various temperatures ranging from -196 °C to room temperature [13]. The upper shelf impact energy reaches 160 J. The ductile-to-brittle transition temperature (defined as the temperature for which the mean fracture energy is half the sum of the upper and lower shelf energy) is –20 °C. Three temperatures (-90 °C, -60 °C and –30 °C) were more detailed studied. About thirty CVN specimens were tested at each temperature for the statistical treatment of the cleavage fracture probability. Twenty-four CT specimens were also tested at -90 °C in order to compare the Weibull's parameters with those identified on CVN specimens.

The fracture probability is assigned to the ith specimen from N tested, ranged in an increasing order of displacements to fracture as:

$$P_F(i) = \frac{i - 0.5}{N} \qquad (9)$$

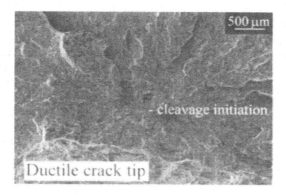

Figure 2. Cleavage crack initiation in front of the ductile crack tip
(A 508 Cl.3 steel, CT 25 specimen tested at 0 C).

4. Numerical Procedure

4.1. DUCTILE FRACTURE MODELLING

The ductile fracture areas situated next to the crack tip and/or notch root are correlated to fracture toughness and/or Charpy impact energy (e.g. [21]). In the DBTT range, fracture toughness and/or Charpy impact energy and corresponding ductile areas have a large scatter. Therefore, a change of stress-strain field caused by ductile crack growth has to be taken into account in the numerical modelling.

The ductile crack growth preceding the unstable fracture (*Figure 2*) will be treated in a deterministic way using the model of *Gurson* [22] modified by *Tvergaard and Needleman* [23], which prescribes material softening linked to the growth and coalescence of voids. Within the finite element analysis the loss of stress carrying capacity corresponds to an increment of ductile crack growth. A more detailed description of this model is presented in the previous part.

A user-material subroutine incorporating the modified GTN model of porous metal plasticity [24] was introduced into the ABAQUS^{TM} software package [25]. The initial void volume fraction is taken as the volume fraction of MnS inclusions given by Franklin formula [26]. The nucleation of new voids is taken into account by *Chu and Needleman*'s description [27]. The parameters of the critical void volume fraction inducing void coalescence, f_c=0.04, and the acceleration of the void volume fraction, δ=4, are chosen in order to fulfil the load drop in the load vs. reduction of diameter diagram of notched tensile specimens NT2 and NT4 tested at 0 °C.

For the simulations of CT specimens the matrix is represented by an elastic-plastic material obeying the Hollomon stress-strain law. The flow stress is assumed to vary linearly with temperature. For the simulations of Charpy impact tests, the Cowper-Symonds formula is used in order to take into account viscous effect due to high strain rates. The parameters of the stress-strain laws were identified in previous research [13] from compressive tests performed on a Hopkinson bar device (strain rate 1000 s^{-}1) and on an INSTRON hydraulic testing machine (strain rate 1 s^{-1} and 4x10^{-3} s^{-1}). The temperature change caused by adiabatic heating during Charpy impact tests was taken into account. Other details on the mechanical behaviour can be found elsewhere [13,21].

Linear elements with selective integration are employed in the finite element analysis. The mesh size on the crack tip and/or notch root is (100 x 100) μm^2 in section. The computations are performed in the framework of finite strains, with an updated-Lagrangian formulation. Previous results of *Rossoll et al.* [28] showed that the amount of kinetic energy is negligible for the considered Charpy impact tests. Consequently, the influence of inertial effects on the total energy (or stress distribution in the plastic zone) can be neglected in the modelling of Charpy impact test in the studied temperature domain. Quasi-static calculations can therefore be used for the ductile crack propagation analysis. On the other hand, the 3D aspect of the ductile crack front cannot be neglected (see the previous part). The ductile crack growth during fracture toughness and Charpy tests will therefore be modelled in a 3D quasi-static formulation.

4.2. CLEAVAGE FRACTURE PROBABILITY

Before the parameters of damage models can be transferred from specimens to structural components, a considerable number of tests with specimens of various geometries must be undertaken. Different geometries of tested specimens results in different mechanical gradients (notch, crack). For this reason, the choice of mesh size, D, and the related damage parameters is delicate. D must be representative in a way that supports arguments coupling the physical and computational model, i.e. to account for e.g. the grain size or the inclusion spacing. Predicted values scale almost proportionally with D for fixed other damage parameters (thicker layer requires more total work to reach the critical conditions). The mapping of one finite element per cell must provide adequate resolution of the stress-strain field in the structure as well as the material's behaviour must be representative for the chosen volume. Finally, the type of finite elements used (linear vs. quadratic) can play an important role [29].

In Eq. (6), the reference volume, V_o, is chosen to be $(100 \ \mu m)^3$, so that it is directly related to the mesh size on the crack tip and/or notch root. The size of the reference volume corresponds to the volume containing several primary austenitic grains.

The finite element analysis is needed to compute the fracture probability at the onset of fracture for each specimen. Since fracture probability depends on m, an iterative procedure must be used to identify the Weibull parameters [30]. For that, the maximum likelihood method (e.g. [31]) is used. The parameter m has to be bias corrected in dependence on the number of tests, N, and the desired confidence level (procedure described e.g. in [32]).

5. Results and Discussion

5.1. STRESS DISTRIBUTION

The distribution of the maximum principal stress (corresponding to the longitudinal tensile stress) along the ligament in the centre plane of the Charpy and CT specimens are shown in *Figures 3* and *4*. As can be seen in these figures, the situation changes once ductile crack has occurred. The stress level is significantly increased at the notch root (*Figure 3*). The location of the stress peak is shifted from the notch root due to the ductile crack advance. Ductile crack initiation and propagation in CT specimens change the stress distribution ahead of the crack tip differently from that observed in Charpy specimens. At first the fatigue pre-crack is blunted by plastic deformation and then the ductile crack initiation and propagation cause a shift of the stress peak at a larger distance from the fatigue pre-crack front (*Figure 4*). The maximum value of the stress (peak value) and the distance of the maximum from the actual ductile crack tip are practically unchanged during ductile crack growth. The width of the stress peak, i.e. the distance at which the stress is higher than a given value (e.g. the yield stress), is larger than the peak width found for Charpy specimens. The temperature rise due to the energy dissipation by the plastic deformation lowers the computed stress at the notch root in Charpy specimens, but exerts no significant influence on the damage process. Ductile crack initiation and growth are practically identical for isothermal or adiabatic analysis.

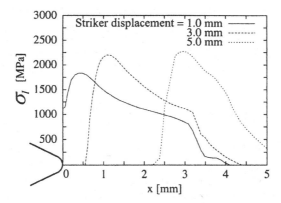

Figure 3. Distribution of the maximum principal stress in the ligament ahead of the notch root during the ductile crack growth (CVN specimen tested at -30 °C). The notch root is represented at the origin of the graph.

Comparing CT and Charpy specimens, the main difference appears to be the value of stress triaxiality which is strongly dependent on specimen geometry. In CT specimens, the maximum value of stress triaxiality ahead of the crack tip is about 3 and decreases slightly to about 2.5 during ductile crack growth. The initial stress triaxiality is only about 1.5 in Charpy specimen and increases progressively up to 2.5 during ductile crack propagation. Higher initial stress triaxiality in CT specimen delays the development of plasticity, and allows an increase of stress.

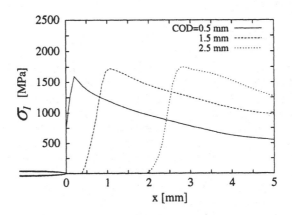

Figure 4. Distribution of the maximum principal stress in the ligament ahead of the crack tip (in the origin of the graph) during the ductile crack growth (CT specimen tested at 0 °C).

On the one hand, the stresses ahead of the crack tip are lowered at temperatures near the DBTT, and on the other hand, the ductile crack occurs. The effect of the size of the volume sampled by the ductile crack propagation has to be introduced. The growing ductile crack is consequently sampling the volume ahead of the crack tip and searching for the pre-existing weak points. Cleavage fracture is then controlled by the occurrence of a critical defect in the process zone. Since stresses are lower, the size of critical defect must be larger according to the Griffith criterion and the probability of finding of the critical defect is commonly assumed to be decreasing with increasing size of defect – Eq. (1).

5.2. FRACTURE TOUGHNESS PREDICTION

The main assumption of the Beremin model is that the Weibull's parameters are not temperature dependent (since cleavage is supposed to be independent of temperature) and geometry. The Weibull's parameters can therefore be identified on one specimen geometry and transferred to another. In our case, the experimental data, on which the identification is carried out, are the onset of fracture of CVN specimens. So that identified parameters are then used for the prediction of fracture probability as a function of fracture toughness. The fracture toughness is computed by the finite element method using the ESIS J_{Ic} method [33].

Figure 5 shows the fracture probability evolution as a function of Charpy impact energy, given by the model and compared to the experimental values obtained by the Charpy impact tests at –90 °C. The predictions of fracture toughness with these parameters are presented in *Figure 6*. As it can be seen in this figure, the prediction of fracture toughness gives correct values at low temperatures but fails in the ductile to brittle transition temperature range.

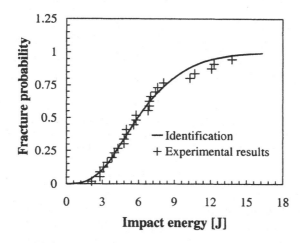

Figure 5. Identification of Weibull's parameters on the experimental values of the fracture probability as a function of Charpy impact energy. Parameters $m = 21$ and $\sigma_u = 2500$ MPa, Charpy impact tests at -90 °C.

This discrepancy has already been mentioned in previous works using the same model and the same type of steel [34,35]. Therefore, considering only the temperature decrease of the yield stress is not sufficient to predict the increase of the mean value and the standard deviation of fracture toughness in the transition domain, even if the ductile fracture preceding cleavage is taken into account. In particular *Rossoll et al.* [34] pointed out that the hypothesis of temperature independence of the Weibull parameters is probably the most questionable assumption of the Beremin model.

Using the Weibull's parameters identified on CVN specimen at low temperature leads to an underestimation of the fracture toughness at temperatures near DBTT. Computations with a threshold stress, σ_{th}, introduced in Eq. (8) were also carried out. The threshold stress level $\sigma_{th} = 1000$ MPa was chosen as the stress level at the crack tip at temperatures where the fracture mode is ductile e.g. 100 °C. However, using of three-parameter Weibull statistic did not lead to improvement of the fracture toughness prediction.

190

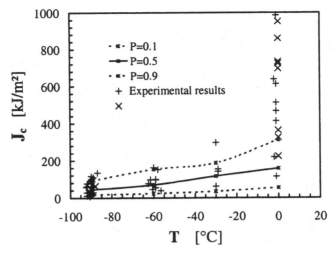

Figure 6. Comparison between the experimental values of the fracture
toughness [13,36] and the predictions given by the Beremin model in the DBTT range.
Parameters m and σ_u identified from Charpy impact tests at -90 °C.

Hence the first attempt is to develop the σ_u parameter as an increasing function of temperature, m being kept constant. The plasticity-based mechanisms inducing cleavage are probably associated with some thermally activated process so that it can justify an evolution of the cleavage stress (γ_{eff} and therefore σ_u) with temperature. Thus the increase of the mean value of fracture toughness can be predicted. On the other hand, this approach fails to predict the increase of the standard deviation in the transition range. In fact, this approach supposes only an increase of the critical cleavage stress with temperature, but the fracture mechanism, i.e. the defect population remains unchanged, which is in contradiction with fractographic observations. At low temperatures, cleavage crack initiates mainly from second phase particles whereas, at higher temperatures (near DBTT), the size of cracked particles is not sufficient to provoke cleavage and another micromechanism induced by plastic deformation takes place [37].

For that purpose, the Weibull's parameters were identified at different temperatures (on CVN specimens) in order to account for the temperature evolution of cleavage fracture mechanism. These identified Weibull's parameters are reviewed in TABLE II. The parameters were extrapolated by an exponential function of temperature up to 0 °C.

$$f(T) = A exp(\frac{B}{T}) \qquad (10)$$

where A and B are constants.

TABLE II: Identified and extrapolated Weibull's parameters used in the numerical modelling. N is the number of experimental tests.

Specimen	T [°C]	N	σ_u [MPa]	m	90% confidence interval for m	90% confidence interval for σ_u	$\sigma_u V_o^{1/m}$
CT	-90	24	2400	24	17.2 - 29.8	2360 - 2440	1800
CVN	-90	28	2500	21	15.5 - 25.8	2460 - 2540	1800
CVN	-60	27	3250	15	11 - 18.5	3170 - 3330	2050
CVN	-30	27	5250	9,5	7 - 11.7	5060 - 5450	2550
Extrapolated values	0	-	6500	8	-	-	2750

Applying this approach, the increasing scatter of fracture toughness can be more or less satisfactorily modelled up to temperatures near the DBTT (*Figure 7*). However, some problems are still remaining. The extrapolated Weibull's parameters yield to inadequate predictions of fracture toughness at the DBTT. The 90% confidence bound is underestimated at 0 °C and seems to be overestimated at –60 °C and –30 °C. However, it should be noticed that the number of fracture toughness tests at –60 °C and –30 °C is small for a good statistical evaluation.

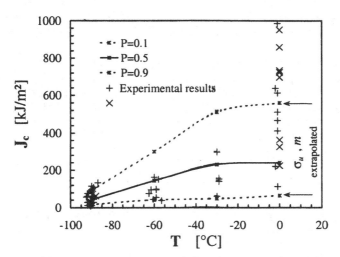

Figure 7. Fracture toughness predictions given by Beremin model. Parameters identified from Charpy impact tests at different temperatures.

The exponential extrapolation can lead to low values of scatter parameter *m*, even though the real value becomes certainly nearly constant in the athermal domain, similarly as the yield stress.

In fact, the influence of strain rate on cleavage fracture mechanisms is not taken into account in this approach, even though the influence of strain rate on mobility of dislocations at low temperatures is well known. In other words, at the same temperature, the fracture mechanisms are supposed to be the same in dynamically loaded CVN specimens as well as in quasi-statically loaded CT specimens. This could explain the overestimation of the 90% confidence bound at –60 °C and –30 °C.

Finally, it has to be noted that the cleavage fracture mechanisms, which are apparently different for low and high values of fracture toughness, i.e. at lower and higher temperatures, can coexist in the transition domain. In this case, a multi-modal fracture model should be used.

6. Conclusions

The stress distributions in CT and CVN specimens computed by the finite element method show that ductile crack initiation preceding cleavage results in a significant increase of the stress level in CVN specimens, which is not the case for CT specimens. The propagating ductile crack causes an expansion of the plastic volume where weak points triggering cleavage can be found (in both types of specimens).

The prediction of fracture toughness from instrumented Charpy test was attempted using the statistical local approach to brittle fracture. The Weibull parameters were found to vary with temperature. Using the Weibull's parameters identified on CVN specimen at low temperature, the fracture toughness can be predicted at the lower shelf but it leads to a significant underestimation of the fracture toughness at temperatures near the DBTT.

Varying the Weibull parameters with temperature allows to account for the increasing scatter of fracture toughness in the transition domain. However, the fracture toughness cannot be correctly predicted from instrumented Charpy test with parameters depending only on temperature, since temperature is not the only variable influencing the change of fracture mechanisms. At least, the influence of strain rate on cleavage fracture mechanisms has to be taken into account.

Acknowledgements

The authors wish to thank Électricité de France (DER EMA Les Renardières) for financial support and supplying the material.

References

1. Cottrell, A.H. (1959) Theoretical aspects of fracture, in B.L. Averbach, D.K. Felbeck, G.T. Hahn, and D.A. Thomas (eds.), *Fracture*, M.I.T. press, Cambridge, Massachusetts, USA, 20-45
2. Smith, E. (1966) The nucleation and growth of cleavage microcracks in mild steel, in *The Physical Basis of Yield and Fracture*, Inst. of Physics and Phys. Soc., Oxford, 36-46.
3. Curry, D.A. (1980) Cleavage micromechanisms of crack extensions in steels, *Metal Science* 14, 319-326.

4. Rosenfield, A.R., Shetty, D.K., and Skidmore, A.J. (1983) Fractographic observations of cleavage initiation in the ductile-brittle transition region of a reactor-pressure-vessel steel, *Metallurgical Transactions* A 14, 1934-1937.

5. Baker, T.J., Kavishe, F.P.L., and Wilson, J. (1986) Effect of non-metallic inclusions on cleavage fracture, *Materials Science and Technology* 2, 576-582.

6. Mäntylä, M., Rossoll, A., Nedbal, I., Prioul, C., and Marini, B. (1999) Fractographic observation of cleavage fracture initiation in a bainitic A508 steel, *Journal of Nuclear Materials* 264, 257-262.

7. Ritchie, R.O., Knott, J.F., and Rice, J.R. (1973) On the relationship between critical tensile stress and fracture toughness in mild steel, *Journal of Mechanics and Physics of Solids* 21, 395-410.

8. Curry, D.A. and Knott, J.F. (1979) Effect of microstructure on cleavage fracture toughness of quenched and tempered steels, *Metal Science* 13, 341-345.

9. Tsann Lin, Evans, A.G., and Ritchie, R.O. (1986) Statistical analysis of cleavage fracture ahead of sharp cracks and rounded notches, *Acta Metallurgica* 34, 2205-2216.

10. Strnadel, B., Mazancová, B., and Mazanec, K. (1990) Effect of cleavage planes orientation on local strength of spheroidized steel, in D. Firrao (ed.), *Fracture behaviour and design of materials and structures, ECF 8*, EMAS, 37-43.

11. Martín-Meizoso, A., Ocaña-Arizcorreta, I., Gil-Sevillano, J., and Fuentes-Pérez, M. (1994) Modelling cleavage fracture of bainitic steel, *Acta Metallurgica and Materialia* 42, 2057-2068.

12. Beremin, F.M. (1983) A local criterion for cleavage fracture of a nuclear pressure vessel steel, *Metallurgical Transaction A* 14, 2277-2287.

13. Rossoll, A. (1998) Fracture Toughness Determination of a Nuclear Pressure Vessel Steel by Instrumented Charpy Impact Test, PhD. Thesis, Ecole Centrale Paris, France.

14. Wallin, K., Saario, T., and Törrönen, K. (1984) Statistical model for carbide induced brittle fracture in steel, *Metal Science* 18, 13-16.

15. Lemaitre, J. (1986) Local approach of fracture. *Engineering Fracture Mechanics* 25, 523-537.

16. Mudry, F. (1987) A local approach to cleavage fracture, *Nuclear Engineering and Design* 105, 65-76.

17. Jayatilaka, A. De S. and Trustrum, K. (1977) Statistical approach to brittle fracture, *Journal of Materials Science* 12, 1426-1430.

18. Weibull, W. (1951) A statistical distribution function of wide applicability, *Journal of Applied Mechanics*, 293-297.

19. O'Dowd, N.P., Lei, Y., and Busso, E.P. (2000) Prediction of cleavage failure probabilities using the Weibull stress, *Engineering Fracture Mechanics* 67, 87-100.

20. Ruggieri, C. (2001) Influence of threshold parameters on cleavage fracture predictions using the Weibull stress model, *International Journal of Fracture* 110, 281-304.

21. Haušild, P., Nedbal, I., Berdin, C., and Prioul, C. (2002) The influence of ductile tearing on fracture energy in the ductile-to-brittle transition temperature range. *Material Science and Engineering A*, in press.

22. Gurson, L.A. (1977) Continuum theory of ductile rupture by void nucleation and growth, *Journal of Engineering Materials and Technology*, Jan., 2-15.

23. Tvergaard, V. and Needleman, A. (1984) Analysis of cup-cone fracture in a round bar, *Acta Metallurgica* 32, 157-169.

24. Siegmund, T. and Brocks, W. (1997) A user-material subroutine incorporating the modified Gurson-Tvergaard-Needleman model of porous metal plasticity into ABAQUS finite element program, Technical report GKSS/WMG/97/2, GKSS, Geesthacht, Germany.

25. HKS ABAQUS/Standard, Version 5.8 (1998), Theory Manual, Hibbit, Karlsson and Sorensen Inc., USA.

26. Franklin, A.G. (1969) Comparison between a quantitative microscope and chemical methods for assessment of non-metallic inclusions, *Journal of The Iron and Steel Institute*, Feb., 181-186.

27. Chu, C.C. and Needleman, A. (1980) Void nucleation effects in biaxially stretched sheets, *Journal of Engineering Materials and Technology* 102, 249-256.

28. Rossoll, A., Berdin, C., Forget, P. Prioul, C., and Marini, B. (1999), Mechanical aspects of the Charpy impact test, *Nuclear Engineering and Design* 188, 217-229

29. Ruggieri, C., Pantonin, T.L., and Dodds R.H. (1996) Numerical modeling of ductile crack growth in 3-D using computational cell elements, *International Journal of Fracture* 82, 67-95.

30. Minami, F., Brückner-Foit, A. and Trolldenier, B. (1990) Numerical procedure for determining Weibull parameters based on the local approach, in D. Firrao (ed.), *Fracture behaviour and design of materials and structures, ECF 8*, EMAS, 76-81.

31. Khalili, A. and Kromp, K. (1991) Statistical properties of Weibull estimators, *Journal of Materials*

Science 26, 6741-6752.

32. ESIS P 6 98 (1998) Procedure to measure and calculate material parameters for the local approach to fracture using notched tensile specimens, K.-H. Schwalbe (ed.), GKSS, Geesthacht, Germany.

33. Landes, J.D. (2001) Evaluation of the ASTM and ESIS multiple specimen J initiation procedures using the EURO fracture toughness dataset, Technical report GKSS 2001/9, GKSS, Geesthacht, Germany.

34. Rossoll, A., Berdin, C., Forget, P. Prioul, C., and Marini, B. (1998) A local approach to cleavage fracture of A508 steel, in M.W. Brown, E.R. De Los Rios, K.J. Miller (Eds.), *Fracture from defects, ECF 12*, EMAS, 637-642

35. Bernauer, G., Brocks, W., and Schmitt, W. (1999) Modifications of the Beremin model for cleavage fracture in the transition region of a ferritic steel, *Engineering Fracture Mechanics* 64, 305-325.

36. Renevey, S. (1997) Approches globale et locale de la rupture dans le domaine de transition fragile-ductile d'un acier faiblement allié, PhD. Thesis, Université Paris XI Orsay, France.

37. Haušild, P., Berdin, C., Bompard, P., Prioul, C., and Parrot, A. (2002) Fracture toughness prediction for a low-alloy steel in the DBBT range, in *ECF 14*, EMAS, in press.

TOUGHNESS SCALING MODEL APPLICATIONS

I. DLOUHÝ, V. KOZÁK, M. HOLZMANN
Institute of Physics of Materials ASCR, Žižkova 22, 616 62 Brno, Czech Republic

Abstract: The applications of toughness scaling diagrams for correction of fracture toughness data in regime of constraint loss have been followed. The applications included adjustment of invalid data from standard specimens and correction of fracture toughness values generated by pre-cracked Charpy type specimens (P-CVN). Standard three point bend specimens have been tested to get the reference values. The Dodds-Anderson toughness scaling model was applied for toughness data correction and transfer from P-CVN specimens being combined with master curve methodology for fracture toughness - temperature diagram quantification. Local parameters, Weibull stress and parameter of scatter have been applied for another toughness scaling diagram calculation. Also in this case good correlation of predicted data with the real fracture behaviour has been found.

Keywords: Fracture toughness transferability, pre-cracked Charpy specimen, toughness scaling models, master curve, cast ferritic steel

1. Introduction

The fracture toughness of material represents an important input parameter for any flaw assessment method. A number of investigations have shown [e.g. 1-4] that present material characterisation practice in the ductile to brittle transition regime provides specimen size dependent fracture toughness values which show a terrific scatter on a temperature dependence. This makes it difficult to assign a unique (unambiguous) meaningful material characterisation without a statistical evaluation of the fracture toughness data.

The development of constraint based fracture mechanics for brittle fracture toughness data treatment is still a topic of research.

The use of different crack length and crack length to specimens width ratios (a/W) in laboratory fracture mechanical tests to simulate structural components with cracks has led to large interests in the role of both. The constraint effects on brittle (cleavage) fracture thus have received considerable attention over the past decade [1,2,7,8 et al.]. Elastic plastic stress field below the crack tip depends strongly on the specimen geometry, size, loading mode and material flow properties. When plastic regions ahead of the crack tip interact with free surfaces, the single parameter characterisation of the crack tip stress, usually scaled in terms of the J-integral, breaks down. Characterisation of the effect of constraint on cleavage fracture toughness using a correlative approach involves a two-parameter description of crack tip stress field [1,2,4,5,9,10]. As shown by e.g. Dodds [11] J sets the size scale of the zone of high stresses and large

I. Dlouhý (ed.), Transferability of Fracture Mechanical Characteristics, 195–212.
© 2002 *Kluwer Academic Publishers. Printed in the Netherlands.*

deformations while Q scales the near tip stress level relative to a high triaxiality reference stress state.

For many applications, the Charpy type specimen with notch or crack supposes test geometry extensively used for fracture toughness evaluation. Due to the wide usage and simple test geometry the potential for the more exact approaches to fracture behaviour of this specimen type is not exhausted and is still kept in focus of interests. In case of V notched specimens the sharp stress gradient below the notch root supposes serious obstacle for local parameters (in sense of Beremin's local approach) determination however [12]. In particular, the scatter parameter m determined from the set of notched specimens has been found not to be simply transferable to pre-cracked ones without any additional adjustments [9]. The P-CVN data usually do not meet the size-deformation limit. The P-CVN specimens lose constraint well before the onset of unstable fracture [4,13]. The predictions for full-scale specimens/components from the P-CVN specimens thus provide unconservative and invalid estimate of conditions required for the onset of unstable fracture. Based on 3D calculations [13,14] the data may be adjusted and used for fracture toughness prediction of structural components.

Recent effort in this area has focused on developing transferability (scale) model for brittle fracture toughness.

According to calculations of Anderson and Dodds [13,14] the shallow cracked specimens loss constraint at very low J values, but the measured fracture toughness values can be corrected for this constraint loss. This correction can also be applied to data from deeply notched specimens that do not meet the recommended size limit. The Dodds and Anderson [4,13] have proposed to quantify the relative effects of constraint variation on the fracture toughness in the form of scaling model diagrams (toughness-scaling models TSM). They solved the problem by postulating the material volume ahead of the crack front over which the principal stress exceeds a critical value as a local fracture criterion without respect for the J integral value. They calculated diagrams relating the parameters J/J_{SSY} vs. $b\sigma_0/J$ for different n values and a/W ratio.

The main principle of toughness scaling diagrams [4] is arising from diagram $J_0/b\sigma_0$ versus $J_{FE}/b\sigma_0$, where b is specimen ligament and σ_0 is yield (flow) stress. Its principle is to transfer the loading parameter from tested (real/small) geometry, where the elasto-plastic fracture toughness in constraint dependent regime is measured, to small scale yielding state (SSY) corresponding to full thickness state (Q = 0, standard specimen geometry).

Nevalainen and Dodds [14] performed extensive 3D analyses of SE(B) specimens to derive J-Q fields (trajectories) across the crack front. They illustrated strong interaction of in plane and trough thickness effects on crack front fields not taken into account in plane strain analyses.

The models of microscale fracture process, e.g. the Dodss and Anderson stressed volume approach [4,5] lead to prediction of relative constrain variations on critical J-values at similar, but unspecified fracture probabilities.

In contrast to J-Q approach, the Weibull stress (σ_W) based local approach provides a mean for prediction of the effects of constraint variations and the distribution of measure values of fracture toughness for different geometries including quantification of the failure probability of components. The treatment of statistical effects on cleavage fracture toughness can be carried out through the process model adopted for microscale fracture. Beremin's local approach based on Weibull stress [15,16] can be accepted as the process model that could be capable to assign the numerical values for fracture probability over the spectrum of J values [17-19]. Since evaluation of the local parameter occurs over a relevant (volume) material at the crack tip (process zone) this approach could reflect 3D model from the outset. The Weibull stress (σ_W) can be adopted as the local parameter describing the crack tip conditions and even as a macroscopic crack driving force. When implemented in finite element code, this model can predict the evolution of Weibull stress with crack tip stress triaxiality.

The scaling diagrams arising from statistical approach has found their justification as shown by Bakker and Koers, Minami et al. Ruggieri et al. and Koppenhoefer et al. [17-21]. As example the method of Koppenhofer et al [21] demonstrates the dependence of Weibull stress σ_W on the crack-tip stress triaxility and the transformation diagram σ_W versus computed value of J_{FE} ($J_{FE}/b\sigma_0$) is being constructed. The idea has been to read the corrected value $J_0/b\sigma$ (corresponding to SSY) from the real one for the same value of probability of failure.

The use of probabilistic approaches in fracture assessments is not without limitation. In particular the Weibull modulus m has to be sufficiently representative of the actual fracture process. As the key point for these methods the scatter parameter m is necessary to be adjusted, in particular when the local parameters are generated from different geometry (notched specimens) comparing to specimens/component (having a crack like defect). Adequate calibration of m parameter is not a simple task but according to works of Gao and newly also Sherry [22,29] corresponding procedures have been suggested.

The aim of the contribution is to verify the methodological approaches described above for transferability of fracture mechanical data from the pre-cracked Charpy type specimens to standard 1T ones and to characterise the fracture resistance of cast ferritic steel predetermined for radwaste casks by this way. The special attention should be paid to the application of toughness scaling models (based on equivalency of critically stressed volume and/or Weibull stress based model) that enable fracture data corrections without loosing information about inherent scatter of material properties.

2. Material and Experimental Procedures

2.1. MATERIAL CHARACTERISATION

Cast manganese - carbon steel has been used for investigations having chemical composition (in wt %): 0.09C, 1.18Mn, 0.37Si, 0.01P, 0.025S, 0.12Cr, 0.29Ni, 0.29Cu, 0.03Mo, and 0.028Al. This material was available as a 270 mm thick plate (Melt I according to nomenclature introduced in paper [24]), the specimens were cut in through thickness direction. The steel followed is typical by fine - grained ferritic microstructure with small islands of bainite and pearlite mixture obtained by intercritical heat treatment.

Tensile properties and true stress-strain curves for FEM calculations have been measured using cylindrical specimens with diameter of 6 mm being loaded over temperature range -196°C to -60°C at a cross-head speed of 2 mm.min^{-1} using clip on gage for elongation measurement.

The cast steel examined exhibits relatively low values of lower and upper yield stress and with decreasing temperature these characteristics increase very slowly (e.g. at −100 °C the yield stress is equal to only 380 MPa).

2.2. MECHANICAL TESTING

For fracture toughness determination two test specimen configurations were used: The standard one (1T SE(B), i.e. specimens having thickness 25 mm) and the "small" pre-cracked Charpy V notch specimens (P-CVN).

For the standard fracture toughness determination the static tests with three-point bend (1T SE(B)) specimens of dimensions 25x50x220 mm^3 have been carried out. These specimens were tested at 1mm/min crosshead speed in the temperature range from −160 to −20 °C. Two sets of specimens (consisting of 25 pcs) were tested at selected temperatures in transition region close to lower shelf (at −100 °C) in order to follow the statistical aspects and the other one close to upper shelf (−70 °C). The fracture toughness values have been determined according to ASTM E 1820-99a and/or similar standards. The experimental details are the same as in case of work [8].

The pre-cracked Charpy type specimens (P-CVN) have been tested statically in three point bending. The evaluation procedures were similar to that above mentioned for standard specimens.

The notched tensile specimens were tested in order to generate data local parameters determination; the test procedures applied and calculations of local parameters have been described in detail in the paper [25].

2.3. CALCULATIONS

The standard FEM code ABAQUS 5.7 was used to model elastic-plastic behaviour (almost in 3D) for the test specimen geometries investigated. The local parameters were determined according to procedure prescribed [26].

3. Results Analyses

3.1. FRACTURE TOUGHNESS TEMPERATURE DIAGRAM

For the static loading and the standard test specimen geometry (SE(B)) the fracture toughness temperature diagram is shown in *Figure 1*. There are all typical areas in the transition and lower shelf region. All the empty triangles represent the K_{Jc} fracture toughness values obtained in elastic-plastic regime without any ductile tearing. The filled rhombi represent K_{Ju} - fracture toughness data with ductile growth larger than 0.2 mm. The K_{Ji} values represent here the value for specimen just with 0.2 mm ductile crack extension and temperature t_{DBL} the lowest temperature of this initiation mechanism occurrence.

For further considerations fulfilling the validity condition for the K_{Jc} values is important. The following equation was used (it is represented by the full thin curve in *Figure 1*):

$$K_{Jc(limit)} = [(Eb\sigma_0)/50]^{1/2} \qquad (1)$$

Data $K_{Jc,Q}$ lying above this limit curve (open circles) represent failures in the regime of constraint loss ($Q \neq 0$), are invalid and without any adjustment cannot be used for further evaluation. Corrections for loss of constraint may be introduced for these data applying the toughness scaling diagram as it will be shown later. The corrected values are shown by filled triangles in *Figure 1*.

Exponential fit for all the fracture toughness values lying under the validity limit including the corrected data $K_{Jc,\,corr}$ was calculated and the curve is shown in figure by dashed line designated as $K_{Jc(mean)}$. The full curve $K_{Jc\,(med)}$ with 90 % probability scatter band (dashed lines) represents the fitted dependences determined applying the master curve methodology [27]. The agreement of both curves, the $K_{Jc(mean)}$ and $K_{Jc\,(med)}$ ones, has been found to be better for the fracture toughness values including the corrected $K_{Jc,corr}$ [27].

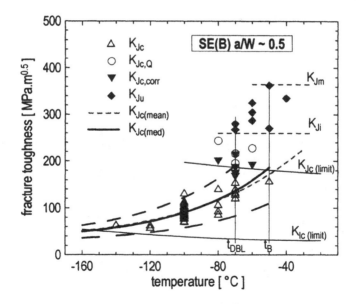

Figure 1. Typical fracture toughness temperature diagram for the steel investigated obtained with 1T SE(B) specimens (cast ferritic steel, Melt I)

For P-CVN specimens the fracture toughness temperature dependence is given in *Figure 2*. The line representing the validity condition, Eq. 1, is shown there. Relatively small number of K_{Jc} data (filled rhombi) fall below this line. The data at –100 °C were intended for investigation of statistical aspects (open rhombi and circles). Data not meeting the validity condition, $K_{Jc(limit)}$, (open rhombi) have to be rejected from further analyses of fracture resistance or could be adjusted for constraint effects as it will be discussed.

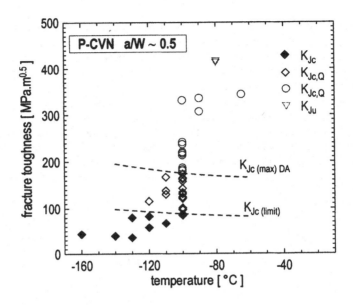

Figure 2. Temperature dependence of fracture toughness obtained with P-CVN specimens tested statically in three point bending

For the constraint loss correction toughness scaling model (TSM) based on equivalency of critically stressed area was applied, as it will be shown later. The diagrams calculated by Nevalainen and Dodds for centre plane have been used for these purposes. But only for the data lying below the line labelled $K_{Jc(max)DA}$ in *Figure 2* this concept may be used, as above this line the level-off of toughness scaling curve has occurred during further loading and, therefore, constraint loss correction may no longer be performed.

3.2. TSM BASED ON EQUIVALENCY OF CRITICALLY STRESSED VOLUME

The concept based on equivalency of critically stressed volume arises from the criterion for cleavage failure introduced by idea [4,5]. The model has assumed that cleavage fracture is controlled by the steel volume ahead the crack tip that is subjected to a high stress level. The larger the stressed volume ahead the crack tip the more likely cleavage will be triggered from a critical defect. The model merely assumes that cleavage fracture is controlled by the principal stress distribution near the crack tip. Comparing to other micromechanics models the Dodds-Anderson methodology does not attempt to make absolute prediction of fracture toughness. The results are scaled to a reference solution. The crack tip stress fields in tested specimen are compared to the small scale yielding conditions, where the plastic zone is small compared to crack length and dimension in direction of crack propagation.

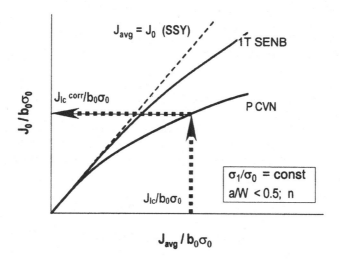

Figure 3. Scheme of toughness scaling diagram enabling correction of data in CL regime

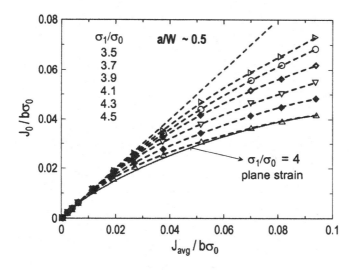

Figure 4. Example of scaling diagram applied for the correction of fracture toughness data obtained with P-CVN specimens; according to [14]

202

One of results of calculation based on this model may be scaling diagram like that shown schematically in *Figure 3*. In the diagram the horizontal axis represents the loading parameter of the tested geometry while the vertical axis correspond the loading parameter of reference stress field (small scale yielding conditions). For the given values of σ_1/σ_0 ratio, strain hardening n and particular specimen geometry (a/W ratio) the curve represents the line for selected value of stress contour σ_1/σ_0. Then applying the measured value on horizontal axis the corrected one can be read from the vertical axis as shown by arrows introduced.

Figure 4 shows example of the curves accepted for direct graphic corrections of P-CVN data.

The correction procedure (for in plane constraint) has been applied on data set obtained at -100 C with P-CVN specimens (in *Figure 2* data represented by open rhombi).

In order to receive the standard specimen behaviour from P-CVN geometry for purposes of comparisons with reference values (really measured on specimens having the standard dimensions) correction for statistical size effects is necessary to introduce. For this purposes the equation like

$$K_{Jc(1T)} = 20 + (K_{Jc(10)} - 20)(\frac{B_{10}}{B_{1T}})^{1/4} \qquad (2)$$

can be used. The same eq. can be also used for opposite way, i.e. for size corrections from 1T to P-CVN specimen size for purposes of comparisons of P-CVN data adjusted for constraint effects with reference values, as shown in *Figure 5*.

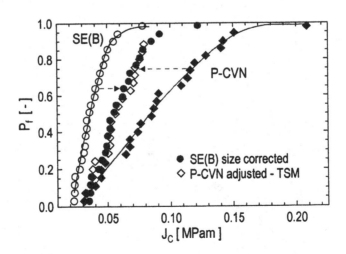

Figure 5. Rank probability diagram for J_c fracture toughness of both specimen types followed including the corrected values

This figure thus shows the rank probability diagram for initial data of both the standard SE(B) specimens and the P-CVN specimens as well as the corrected P-CVN data obtained applying the toughness scaling model described above and SE(B) data corrected for statistical size effects. Very good agreement of both data sets is evident from the figure.

The P-CVN data obtained by both corrections (for constrain loss and statistical size effects) are compared to reference values in fracture toughness temperature diagram shown in *Figure 6*. In addition to both data sets 90% probability scatter band has been introduced for the corrected P-CVN data applying the master curve approach (but any other approach for prediction of full scale specimen transition behaviour with the corrected P-CVN data is applicable).

From data meeting the validity limit according to Eq. (1) and data corrected for constraint loss the reference temperature T_0 and master curve have been obtained. For statically loaded P-CVN specimens the reference temperature T_0 was determined to be - 78 °C, that is in good agreement with the value of T_0 = - 82 °C obtained by means of 1T specimens. The master curve $K_{Jc(med)}$, together with tolerance bounds for 5 and 95% fracture probability are plotted in figure and it is clear that all the data of the 1T specimens (open circles) fall inside the scatter band of P-CVN specimen prediction made by the above-mentioned procedure. For the steel investigated it verifies the potential of utilising small pre-cracked specimens for the fracture toughness evaluation in the transition region.

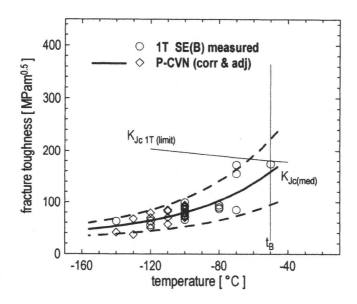

Figure 6. Prediction of 1T specimen behaviour using the corrected P-CVN data

Data from the pre-cracked Charpy type specimens corrected by using the toughness scaling model and combined with the master curve methodology can thus be used for prediction of fracture toughness temperature diagram for 1T specimen in the lower part of the transition region. In such a case the master curve methodology introduces the quantification of fracture probability in lower shelf and transition region.

3.3. INCORPORATING THE SPECIMENS THICKNESS EFFECT INTO TOUGHNESS SCALLING MODEL (IN-PLANE AND OUT-OF PLANE CONSTRAINT CORRECTION)

Initially the in plane constraint correction procedure using the critically stressed area equivalency calculated for centre plane and for given σ_1/σ_0 ratio has been used as described in previous chapter. Even more realistic result involving the stress distribution ahead the crack tip throughout the specimen thickness can be obtained if this effect is accounted for in correction procedure. For purposes of such out-of- plane constraint loss corrections a model based on effective thickness was suggested by Nevalainen and Dodds [14] and will be applied for evaluation of this phenomenon using our data set.

The procedure incorporating both the in-plane and the out-of-plane constraint corrections has been as follows:

Firstly the J_0 value for SSY deformation state (i.e. the J value corrected for in plane constraint loss) is necessary to determine for given loading level $J_{avg}/b\sigma_0$. For these purposes the same toughness scaling diagrams as used in previous chapter (see the *Figure 4*) are to be applied.

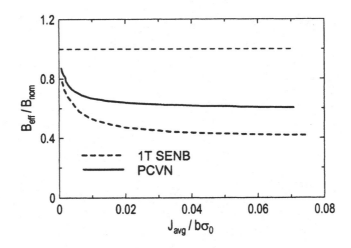

Figure 7. Diagram enabling to determine the ratio of effective thickness against nominal thickness B_{eff}/B_{nom}; according to [14]

Then the ratio B_{eff}/B_{nom} is necessary to get where B_{eff} is the equivalent fraction of the specimen thickness which is subjected to the same high stress condition as a centre plane (as introduced by Nevalainen and Dodd model [14]). Toughness scaling diagram enabling to obtain this characteristic is shown in *Figure 7*.

For evaluation of $\bar{J}_{0(B_{nom})}$ incorporating the in-plane and out-of-plane component of constraint loss the following equation is then used

$$\bar{J}_{0(B_{nom})} = J_0 \left\{ J_{avg} \right\} \left[\frac{B_{eff}}{B_{nom}} \right]^{1/4} \qquad (3)$$

The obtained values of $\bar{J}_{0(B_{nom})}$ have been then converted to its equivalent stress intensity factor values $K_{\bar{J}_{c(Bnom)}}$ and converted to 1T specimen thickness applying the corresponding equation

$$K_{\bar{J}_{c(1T)}} = 20 + \left[K_{\bar{J}_{c(Bnom)}} - 20 \right] \left(\frac{B_{nom}}{1T} \right)^{1/4} \qquad (4)$$

Applying this procedure for P-CVN specimen ($B_{nom} = 0.01$ m) the $K_{\bar{J}0(1T)}$ values have been obtained and plotted as closed triangles in rank probability diagram, *Figure 8*. As seen the distribution of $K_{\bar{J}0(1T)}$ exhibits lower values than $K_{Jc(1T)}$ ones. This results shows that even standard 1T specimens experience out-of-plane constraint loss giving rise higher $K_{Jc(1T)}$ values. Hence, selected single higher value of K_{Jc} (from the standard SE(B) specimen set) has been adjusted for out-of-plane constraint loss. As example, one filled circle shows the result of correction in *Figure 8*. The agreement with the P-CVN data adjusted for constraint loss $K_{\bar{J}0(1T)}$ thus appears to be quite perfect.

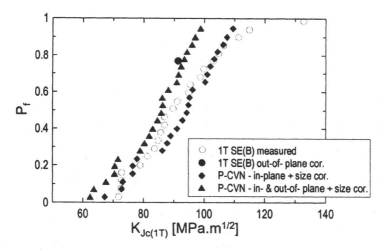

Figure 8. Rank probability diagram for data corrected for both in-plane and out-of- plane constraint loss

From the above application of toughness scaling models it follows that
- the steels with higher cleavage fracture stress σ_{CF} ($\sigma_{CF} = \sigma_1$) and higher hardening coefficient n exhibit significantly lower in-plane constraint loss during loading of cracked body;
- the out-of-plane constraint loss for the given specimen geometry is primarily function only of hardening coefficient of steel n and as almost independent of σ_{CF} / σ_0 ratio.

3.4. WEIBULL STRESS BASED TSM

As mentioned in introduction, to quantify the synergistic effects of specimen size, constrain loss and material flow properties on cleavage toughness Gao et al. [22] developed Weibull stress based procedure enabling correction for constraint loss. Based on detailed three dimensional finite element analyses they derived non-dimensional function that depend on the material flow properties and the Weibull modulus m, but remains identical for all geometrically similar specimens regardless of their absolute size.

Recent studies have proved that the local parameters m and σ_u need to be calibrated against appropriate material data. Although current guidance [26] still allows for the use of such data, it is now recommended, where it is possible, to calibrate local approach models against the pre-cracked data [19].

The Gao et al. method [22] presuppose that, in order to determine calibrated values m and σ_u, it is necessary to calibrate Beremin model against fracture toughness data from specimens with at least two different crack tip constraint conditions. Two data sets data are thus required. One set having the low constraint configuration, the other having high constraint configuration (preferably close to SSY condition). Analyses of both configurations using 3D finite element calculation according to Eq. (5) provide variations of σ_W vs. J, family of curves for various values of shape parameter m.

$$\sigma_w = \left\{ \frac{1}{V_0} \int_\Omega \sigma_1{}^m d\Omega \right\}^{1/m} \tag{5}$$

In Eq. (5) V_0 is characteristic volume (the constant used = $(100 \ \mu m)^3$), m is shape parameter (Weibull modulus), σ_1 is maximum principal stress, and Ω represents plastic zone ahead the crack-tip. Example of the results of such calculations can be seen in *Figure 9 and 10*.

For SSY condition the boundary layer method was used. This method simplifies the generation of numerical solution for stationary and growing cracks with varying level of constraint. The plane-strain FE model has been used for an infinite domain. Mode I loading of far field permit very similar analysis.

For the P-CVN specimen the finite element model included small notch above sharp fatigue crack. This specimen has a square cross-section with B = W = 10 mm and a span of 40 mm. A conventional mesh having 35 focused rings of elements surrounding the crack front is used with small-blunted crack tip (radius of 1μm). The mesh had 15 thickness layers and quarter-symmetric 3D model for P-CVN contains 17 000 nodes.

Using results of FEM the diagram relating the Weibull stress against loading parameter has been plotted as shown in *Figure 9*.

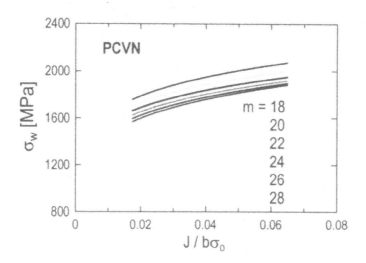

Figure 9. Diagram of Weibull stress determined for P-CVN specimen geometry

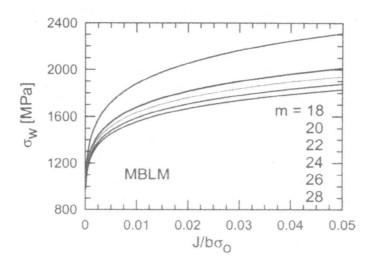

Figure 10. Diagram of Weibull stress determined by
modified boundary layer method (MBLM).

Than by combining both diagrams (based on the same value of Weibull stress σ_w) and relating the loading parameter for the followed geometry to reference stress field

208

the scaling diagram can be obtained. Thus for the P-CVN specimens the transformation diagram was determined and is presented in *Figure 11*.

The only problem that needs to be solved is selection of the value of the scatter parameter. It has been found that direct transmission from the other specimen geometry, in particular from notched tensile specimen etc [26] doesn't supply good solution.

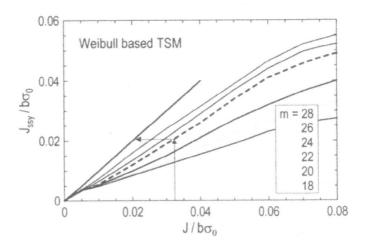

Figure 11. Toughness scaling diagram based on Weibull stress with varying Weibull modulus

For calibration of m parameter a procedure suggested by Gao et al. [22] has been applied for the steel and test specimens investigated here.

From discussion presented in [3], macroscopic values of cleavage fracture toughness measured using high-constraint specimens follow a Weibull distribution with the failure probability given by

$$P_f(J) = 1 - \exp\left[-(\frac{J}{\beta})^2 \right].$$ (6)

where β defines the toughness value at a 63.2 percent failure probability. Relating the Eq. (6) to basic equation for failure probability determination the following equation can be obtained

$$(\frac{J}{\beta})^2 = (\frac{\sigma_W}{\sigma_u})^m$$ (7)

For SSY conditions the value of $\beta_{SSY} = 0.064$ MPam the calibrated value of scatter parameter was possible to read as equal to 24.1. The procedure is obvious from *Figure 12*. Fore more details to this procedure applied in our investigation see the work [28].

The curve corresponding to m = 24 was applied for the constraint loss corrections (dashed line).

The results of data correction is shown in *Figure 13*. There are shown the raw data obtained with P-CVN specimens (filled rhombi) as well as the data from the standard specimens corrected for statistical size effects only (filled circles). The P-CVN data corrected for constraint loss are shown by empty triangles. The rank probability diagram shown in figure proves very good correlation of size corrected data from 1T SE(B) specimens and those corrected for constraint loss from P-CVN specimens.

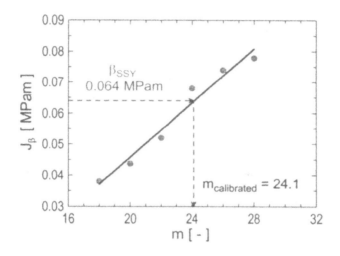

Figure 12. Calibration of Weibull stress parameters using fracture
toughness data based on Gao approach

The second method that can be used for calibration of local parameters [29] accepted the need to reference the fracture toughness data relating to different constraint condition. This method also permits the determination of temperature dependence in σ_u parameter, to enable the rapid upswing in cleavage toughness in the upper transition regime. A random sampling approach is used to derive sets of data specific to different temperature and/or constraint condition against which the local parameters m and σ_u are calibrated using a maximum likelihood method.

It has been found [29] that both mentioned methods provide similar values of m and σ_u for the same data set.

To model damage (microcrack) statistics and pronounced effects on scatter of measured Jc values in the transition region, a local approach arising from Weibull stress appears to be very promising. This local parameter reflects the different rates at which the crack tip stresses increase with applied load (J value) as well as the changes in process zone dimensions caused by loss of constrain in different specimen – crack configurations. For the right use it is essential however the parameters characterising the micromechanics model used to be really representative to actual fracture process. For

210

these purposes the calibration procedure applied seems to be satisfactory from the material properties point of view.

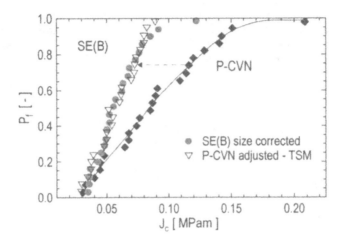

Figure 13. Rank probability diagram for 1T size corrected and P-CVN J_c values adjusted for constraint loss by Weibull strss based TSM

4. Concluding Remarks

The fracture resistance of cast ferritic steel intended for production of container casks for spent nuclear fuel has been analysed combining several approaches. The capability of the pre-cracked Charpy type specimens tested statically for prediction of fracture toughness characteristics of full scale (1T) specimens has been tested.

The toughness scaling model based on equivalency of critically stressed volume (introduced and calculated by Dodds, Anderson and Nevalainen) has been applied for correction of fracture toughness data in constraint dependent regime in transition region of steel investigated. The corrected data have correlated well with scatter band of valid data.

The same approach has been applied for constraint loss corrections of fracture toughness values from pre-cracked Charpy type specimens. Applying the adjustment for statistical size effects the predicted values correlated well with actually measured fracture toughness values for standard SE(B) specimens. For prediction of 90 % probability scatter band the master curve methodology has been successfully applied.

The effect of specimen thickness in terms of out-of-plane constraint loss has been followed. It has been found that for the steel investigated even the standard 1T SE(B) specimens obeys the out-of-plane constraint loss and for exact evaluation the correction has to be taken into account.

The toughness scaling diagram arising from Weibull stress and modulus determination was calculated and tested supplying good correlation of predicted data from pre-cracked Charpy specimens with real fracture behaviour.

According to the results obtained the procedures involving constraint adjustment seem to be very promising and reproducible for transfer of crack resistance characteristics from small specimen to full scale specimen/components.

Acknowledgements

The research was financially supported by grant No. A2041003 of the Grant Agency of the Czech Republic and project No. 972655 within NATO Science for Peace program.

References

1. O'Dowd, N.P. and Shih, C.F. (1991) Family of crack tip fields characterised by a triaxiality parameter, Part I and II, *Journal of the Mechanics and Physics of Solids*, Vol. 39, No. 8, pp. 989-1015 and Vol. 40, pp. 939-963.
2. Betegon, C. and Hancock, J.W. (1991) Two parameter Characterisation of elastic plastic crack tip field, *Journal of Applied Mechanics*, Vol. 58, pp. 104-113.
3. Pineau, A, (1992) Global and local approaches to fracture transferability of laboratory test results to component In *Topics in Fracture and Fatigue*, A.S. Argon, Ed., Springer Verlag, pp. 197/234
4. Anderson, T.L, Dodds, R.H., (1991) Specimen size requirements for fracture toughness testing in the transition region, *Journal of Testing and Eval.*, Vol. 19, pp.123-134.
5. Dodds, R.H., Anderson, T.H, Kirk, M.T. (1991) A framework to correlate a/W ration effects on elastic-plastic fracture toughness (Jc), *International J. of Fracture*, 48, pp. 1-22.
6. Anderson T.L., R.H. Dodds Jr. (1993) Simple constraint corrections for subsized fracture toughness specimens, ASTM STP 1204, *Small Specimen Test Techniques Applied to Nuclear Reactor Vessel Thermal Annealing and Plant Life Extension*, pp. 93-105,
7. Parks, D.M. (1992) Advances in characterisation of elastic-plastic crack tip fields, *Topics in Fracture and Fatigue*, A.S. Argon, ed., Springer Verlag, pp. 59-98.
8. Chlup, Z., Dlouhý, I. (2002) Micromechanical aspects of constraint effect at brittle fracture initiation, *The Transferability of Fracture Mechanical Characteristics*, paper in this volume.
9. Chao, Y. J. and Sutton, M.A. (1994) On the fracture of solids characterised by one or two parameters, *Journal of Mechanics and Physics of Solids*, Vol. 42, pp. 629-647.
10. Sorem, W.A., Dodds, R.H.Jr, Rolfe, S.R. (1991) Effects of crack depth on elastic plastic fracture toughness, *International Journal of Fracture*, 47, pp. 105-126.
11. Dodds, R.H.Jr, Ruggieri, C., Koppenhoefer, K. (1997), 3D constraint effects on models for transferability of cleavage fracture toughness, *Fatigue and Fracture Mechanics: 28th volume*, ASTM STP 1321, J. H. Underwood and B.D. MacDonald, M.R. Mitchell, Eds., pp. 179-197.
12. Dlouhý I., Kozák V., Válka L., Holzmann M. (1996) The susceptibility of local parameters on steel microstructure evaluated using Charpy type specimen. Proc. of Conf- EUROMECH - MECAMAT '96, Fontainebleau.
13. Anderson, T.L., Vanaparthy, N.M.R. and Dodds, R.H. Jr. (1993) Predictions of specimen size dependence on fracture toughness for cleavage and ductile tearing, *Constraint Effects in Fracture, ASTM STP 1171*, Hackett, Schwalbe and Dodds eds. pp. 473-491.
14. Nevalainen, M., Dodds, R.H. (1995) Numerical investigation of 3D constraint effects on brittle fracture in SE(B) and C(T) specimens, *International Journal of Fracture*, 74, pp.131-161.
15. Beremin, F. M., (1983) A local criterion for cleavage fracture of a nuclear pressure vessel steel, *Metal. Trans. A*, Vol. 14A, pp. 2277-2287.

16. Mudry F. (1987) A local approach to cleavage fracture, *Nuclear Engineering and design*, Vol. 105, pp. 65-76.
17. Minami,F. Brückner Foit, A., Munz, D., Trolldenier, B. (1992) Estimation procedure for the Weibull parameters used in the local approach, *International Journal of Fracture*, Vol. 54, pp. 197-210.
18. Ruggieri, C., Dodds, R.H., (1996) A transferability model for brittle fracture including constraint and ductile tearing effects: A probabilistic approach, *International Journal of Fracture*, Vol. 79, pp. 309-340.
19. Ruggieri, C., Dodds, R.H., Wallin, K., (1998) Constraint effects on reference temperature T0, *Engineering Fracture Mechanics*, Vol. 60, pp. 14-36.
20. Bakker, A. and Koers, R.W.J. (1991) Prediction of cleavage fracture events in the brittle ductile transition region of a ferritic steel. *Defect Assessment in Components. Fundamentals and Applications*, ESIS/EG9 Blauel and Schwalbe Eds., Mechanical Engineering Publication, London, pp. 613-632.
21. Koppenhoefer, K.C., Dodds, R.H. (1997) Constraint effects on fracture toughness of impact loaded, pre-cracked Charpy specimens, *Nuclear Engineering and Design*, Vol. 162, pp. 145-158.
22. Gao, X., Ruggieri, C., Dodds, R.H. (1998) Calibration of Weibull stress parameters using fracture toughness data. *International Journal of Fracture*, 92, pp. 175-200.
23. Gao X., Dodds, R.H. Jr. (2001) An engineering approach to asses constraint effects on cleavage fracture toughness, *Engineering Fracture Mechanics*, 68, pp. 263-283.
24. Dlouhý, I., Holzmann, M., Chlup, Z. (2002) Fracture resistance of cast ferritic C-Mn steel for container of spent nuclear fuel, *Transferability of fracture mechanical characteristics*, Kluwer, contribution in this volume.
25. Kozák V., Janík A. (2002) The use of the local approach for the brittle fracture prediction, *The Transferability of Fracture Mechanical Characteristics*, paper in this volume.
26. ESIS P6/98 (1998) Procedure to measure and Calculate Local Approach Criteria Using Notched Tensile specimens, *ESIS document*
27. Standard Test Method For the Determination of Reference Temperature T_0 for Ferritic Steels in the Transition range, ASTM, E1921-97.
28. Kozák V., Holzmann M., Dlouhý I (2001) The transferability of brittle fracture toughness characteristics, *Structural Mechanics in Reactor Technology; SMIRT 16*, Washington DC, Proc. on CD ROM.
29. Sherry A.H., Lidbury, D.P.G., Bass, B.. Williams, P.T. (2001) Development in local approach methodology with application to analysis / re-analysis of the NESC-1 PTS benchmark experiments, *Int. J. of Pressure Vessel and Piping*, 78, pp. 237-249

RELATION OF FRACTURE ENERGY OF SUB-SIZED CHARPY SPECIMENS TO STANDARD CHARPY ENERGY AND FRACTURE TOUGHNESS

H. J. SCHINDLER AND P. BERTSCHINGER
Mat-Tec SA, Unterer Graben 27, CH-8401 Winterthur, Switzerland
schindler@mat-tec.ch

Abstract: The total fracture energy of edge-cracked beams under bending load is strongly dependent on specimen size, so the Charpy energy can only be measured on standard specimens. By means of a simplistic mechanical model a mathematical relation between the total fracture energy of an edge-cracked beam under bending and the fracture toughness is derived, from which a mathematical relation between the fracture energy and specimen size is obtained. It can be used to scale-up the fracture energy of sub-sized tests, and then use the evaluation procedure for standard Charpy specimens mentioned above. Unlike the commonly used empirical correlation formulas, the presented scaling law is applicable to any elastic-plastic material. It holds for the upper-shelf regime, and as a lower bound also in the brittle-to-ductile transition regime. The results are compared with experimental data obtained from different specimen sizes.

Keywords: Charpy-V-notch test, sub-size specimens, fracture toughness, fracture energy, scaling laws, J-Integral, J-R-curve, impact testing, J-resistance curve, analytical relation

1. Introduction

Because of its simplicity the standard Charpy test is still a very popular toughness test, although the measured quantity, the fracture energy KV, is known to be strongly dependent on specimen geometry, size, notch sharpness and loading rate. Therefore, the fracture energy of the standard specimen has neither a direct relation to the fracture energy of a structural part nor to fracture toughness in terms of K_{Ic} or J_{Ic}. The physical processes involved are rather different: In KV, there are contributions from processes like notch root blunting, ductile crack initiation, and, mainly, tearing crack growth, shear-lip formation and plastic deformation of the ligament, whereas J_{Ic} or K_{Ic} represent critical parameters of the corresponding local crack tip load and characterize just the resistance against crack extension. For these reasons KV rather serves as an "indicating" than a "quantifying" toughness parameter. Nevertheless, fracture toughness is often estimated from KV. A number of corresponding empirical correlations are offered in the literature [1-4]. However, for the reasons mentioned above, these correlations are

213

I. Dlouhý (ed.), Transferability of Fracture Mechanical Characteristics, 213–224.
© 2002 *Kluwer Academic Publishers. Printed in the Netherlands.*

considered to be rather unreliable and valid only with restriction to certain classes of materials.

Nevertheless, by means of some simplistic mechanical models, Schindler [5, 6, 7] derived a mathematical relation between the J-R-curve and the fracture energy in bending. Application of this general relation to the special case of the standard Charpy test lead to an analytical relation between KV and J_{Ic}, giving some theoretical justification to the empirical correlation formulas. Furthermore, the analytical relation shows additional influencing material parameters like flow stress and hardening exponent. It has been successfully applied to steels of different strength, aluminium and – as shown in the present paper – to several types of bronze.

In practical material testing there sometimes are situations where sub-sized specimens have to be used instead of standard ones. A typical miniature specimen is the KLST-specimen according to the German standard DIN 50 115, which has the size 4x3x22 mm, or the half-Charpy according to ESIS. In order to interpret and classify the obtained fracture energy, there often is a need to compare them with standard data. However, the relation of the fracture energy of sub-sized specimens to the one of standard specimens is complex, including a significant, material-dependent temperature shift due to the reduced constraints, and a strongly non-linear size-dependence of the upper shelf energy. There are several empirical correlation formulas to relate the upper shelf fracture energy of sub-sized specimens to the one of standard Charpy energy [8, 9, 10]. The relation appears to be not only size- but also material-dependent.

In the present paper an analytical relation between the upper-shelf fracture energy of different specimen sizes is presented which can serve as a scaling law of Charpy-type fracture energy. It is experimentally confirmed by comparison of up-scaled fracture energy with directly measured ones for different materials.

2. Estimation of J-R Curve From Bend Test

As shown by the first author in [11, 12, 13] the J-R-curve can be estimated from the continuous force-displacement-diagram of a single, uninterrupted, static or dynamic bending test (*Figure 1 and 2*) by

$$J(\Delta a) = C \cdot \Delta a^p \qquad \text{for } \Delta a < (W-a_0)/10 \qquad (1)$$

where

$$C = \left(\frac{2}{p}\right)^p \cdot \frac{\eta(a_0)}{B\ (W-a_0)^{1+p}} \cdot W_t^{\ p} \cdot W_{mp}^{\ 1-p} \qquad (2)$$

$$p = \left(1 + \frac{W_{mp}}{2W_t}\right)^{-1} \qquad (3)$$

W_{mp} and W_t are the dissipated energy at maximum force and the total fracture energy, respectively, that can be obtained from the load-displacement diagram (*Figure 2*). The factor η is the well known η-parameter for the edge-cracked 3-point bending specimen, which is according to [8]:

$$\eta = 13.81 \cdot \frac{a}{W} - 25.12 \cdot \left(\frac{a}{W}\right)^2 \qquad \text{for } 0<a/W<0.275 \qquad (4a)$$

$$\eta = 1.859 + 0.03/(1-a/W) \qquad \text{for } a>0.275W \qquad (4b)$$

The crack extension at maximum force F_m was obtained to be

$$\Delta a_m = \frac{W_{mp} \cdot p \cdot b_0}{2W_t} \qquad (5)$$

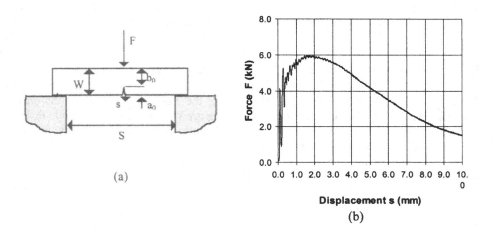

(a)

(b)

Figure 1. Mechanical system (a) and the corresponding force-displacement diagram of a bending test with an edge-cracked specimen in the upper-shelf range (b).

216

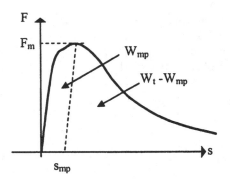

Figure 2. Quasistatic Force-deflection diagram of an instrumented Charpy test (schematic)

According to eqs. (1-3) the J-R curve is determined by only two experimental parameters, W_{mp} and W_t, which can be easily determined from the force-displacement diagram even if it is disturbed by of by dynamic oscillations, as long as the behaviour is essentially quasistatic. Therefore this evaluation procedure is well suited to be applied to testing at increased loading rates like typically Charpy-type testing.

3. Estimation of J-R Curve from a Non-Instrumented Test

In the case of a non-instrumented impact test like the classical Charpy test, the only available experimental value is the total fracture energy W_t. As shown in [6] and [7], by using the additional mathematical condition of maximum force, another equation for the crack extension at maximum force is obtained:

$$\Delta a_m = \frac{A_g \cdot p \cdot b_0}{2} \qquad (6)$$

where A_g denotes the standard uniform fracture strain to be measured on a uniaxial tensile test as the plastic strain at maximum force. By comparison of (6) with (5), W_{mp} can be eliminated from (2) and (3), resulting in the following expressions

$$C = \left(\frac{2}{p}\right)^p \cdot \frac{\eta(a_0)}{B \cdot (W - a_0)^{1+p}} \cdot W_t \cdot A_g^{1-p} \qquad (7a)$$

$$p = \left(1 + \frac{A_g}{2}\right)^{-1} \qquad (7b)$$

Eq. (1), (7a) and (7b) enable the J-R curve to be determined just from the total fracture energy and the uniform fracture strain. By inserting the corresponding parameters of the standard Charpy tests, i.e. $W_t = KV$, $B = W = 10$ mm, $a_0 = 2$ mm and, according to eq. 4, $\eta = 1.76$, the J-R-curve is obtained from a single Charpy test in the upper shelf or upper transition range. The effect of the finite notch root radius is discussed in [6], and the corresponding correction is used in section 5 later on.

4. Scaling Law for Fracture Energy

Although the $J(\Delta a)$-curve determined by eqs. (1), (7a) and (7b) is just an extrapolation of the ductile tearing phase towards the blunting regime and not necessarily equivalent to the near-initiation J-R-curve of the material (see discussion in next section), it is expected to be size-independent in the range of J-controlled crack-tip-loading, i.e. $\Delta a < W/10$. Hence, from two specimens with different sets of geometrical parameters (W, B, a_0 for one specimen, W', B', a_0' for the other), the same value of the factor C as given in eq. (7a) should result. From this condition, one obtains from (7a) the following relation:

$$\frac{\eta(a_0 / W)}{B \cdot (W - a_0)^{1+p}} \cdot W_t = \frac{\eta(a_0'/W')}{B' \cdot (W'-a_0')^{1+p}} \cdot W_t' \qquad (8)$$

From the fracture energy W_t measured on a specimen with the geometrical parameters W, B, and a_0, the fracture energy W_t' of a specimen of a different size and shape (W', B', a_0') can be calculated as

$$W_t' = \frac{\eta(a_0 / W) \cdot B' \cdot (W'-a_0')^{1+p}}{\eta(a_0'/W') \cdot B \cdot (W - a_0)^{1+p}} \cdot W \qquad (9)$$

The parameter p is given in (7b). With some adequate simplification the above given scaling law applied to standard Charpy specimen results in

$$KV' = \frac{5.68}{B} \cdot \eta(a_0 / W) \cdot \left[\frac{8}{(W - a_0)}\right]^{2-A_g/2} \cdot W_t \qquad (10)$$

In this equation, KV' denotes the upper shelf or upper transition Charpy energy estimated from a Charpy-type test using a sub-sized (or over-sized, respectively) specimen with the dimensions B, W and a_0, which have to be inserted in millimetres. Some experimental validation of (10) is given in section 6.

In a similar way, using the corresponding relation given in section 2, scaling laws for further parameters of instrumented bending tests were obtained [13].

5. Determination of Fracture Toughness

KV' values determined from sub-sized specimens by means of (10) can serve to estimate fracture toughness in similar ways as from Charpy fracture energy KV. For this purpose, we suggest to use the semi-analytical relation between KV and J derived in [6, 7], rather than the common empirical correlation formulas. In the following this estimation procedure is roughly outlined.

In principle, the J-R curve estimated by (1), (7a) and (7b) can be used to determine approximate fracture toughness properties like the J_{Ic} or $J_{0.2BI}$ from the J-R-curve, as defined in the corresponding standards like ASTM E 1820 or ISO/DIS 12135. However, one should keep in mind that the J(Δa)-curve according to the above derivation follows from an analytical treatment of the ductile tearing phase of crack extension, i.e. in the range $\Delta a > \Delta a_m$. The analysis is based on a mechanical model in which ductile tearing is assumed to be governed by a constant crack-tip opening displacement (see [6] for derivation). There from the J(Δa)-curve in the range $\Delta a < \Delta a_m$ is obtained by extrapolation. For this reason it does not reflect the J(Δa)- behaviour in presence of the initially higher crack-tip constraints, which depend on the thickness and the initial crack length. For this reason it is expected that the real J-R-curve measured directly exhibits some deviation from the calculated one.

To account for the effect of the initial crack-length a_0, a constraint factor is introduced in [7] such that

$$J_R(\Delta a) = \frac{C}{\kappa\left(\dfrac{a_0}{W}\right)} \Delta a^p \qquad (11)$$

According to [7] a conservative empirical estimation of the constraint correction factor is

$$\kappa(a_0/W) = 1 + 9 \cdot (0.5 - \frac{a_0}{W})^2 \qquad \text{for } a_0 < 0.5W \qquad (12a)$$

$$\kappa(a_0/W) = 1 \qquad \text{for } a_0 > 0.5W \qquad (12b)$$

Comparison with low-blow-data [10, 11, 14] revealed that the shape of the J-R-curve in the high-constraint initiation phase is improved by modifying the exponent p to

$$p = \frac{3}{4} \cdot \left(1 + \frac{W_{mp}}{W_t}\right)^{-1} \qquad (13a)$$

for instrumented tests, or to

$$p = \frac{3}{4} \cdot \left(1 + A_g\right)^{-1} \tag{13b}$$

for non-instrumented tests, respectively.

When dealing with methods to estimate fracture toughness, it is advisable to use conservative definitions of the key parameters. For this reason, $J_{0.2t}$ as defined in *Figure 3* is suggested to be used instead of the standard $J_{0.2Bl}$ according to ISO 12135 (or J_Q according to ASTM E 1820, respectively). The former has a clearer physical meaning than $J_{0.2/Bl}$ or J_Q and can be represented in closed form in terms of C, as given below in (16). The blunting line, which is required to determine $J_{0.2t}$ as well as $J_{0.2/Bl}$ or J_Q, is defined according to the draft international standard ISO/DIS 12135 by

$$J = s_1 \cdot \Delta a \tag{14}$$

where

$$s_1 = 3.75 \cdot R_m \tag{15a}$$

with R_m denoting the ultimate tensile strength. Alternatively, the less conservative s_1 according to ASTM E1820 can be used, i.e.

$$s_1 = 2.0 \cdot \sigma_f \tag{15b}$$

where $\sigma_f = (R_p + R_m)/2$ represents the flow stress. For impact tests we suggest to use rather (15a) than (15b), since the former tends to account for the increased flow stress due to high strain rates[1]. In mathematical terms, $J_{0.2t}$ as defined in *Figure 3* reads

$$J_{0.2t} = \frac{C}{\kappa(a_0/W)} \cdot \left[\left(\frac{C}{s_1}\right)^{\frac{1}{1-p}} + 0.2mm\right]^p + \frac{K_I^2(F_m)}{E}(1 - v^2) \tag{16}$$

The second term in (16) represents the elastic component of $J_{0.2t}$, which is, for the sake of simplicity, assumed to correspond to the elastic component of J at maximum load. In the case of classical (non-instrumented) Charpy tests, the latter is not known, so it is proposed in [6, 7] to estimate it from the plastic collapse condition. Based on ASTM E1820 this results in

$$F_m \cong F_0 \cong \frac{4 \cdot \sigma_f \cdot B \cdot (W - a_0)^2}{3S} \tag{17}$$

---

[1] In the case of impact tests, the strength values like R_m or σ_f should be the ones at the corresponding strain rate. Nevertheless, since the derived formulas are just approximations, we suggest to use the quasistatic values.

220

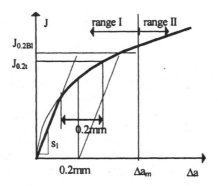

Figure 3. J-R curve and definition of near-initiation values $J_{0.2Bl}$ and $J_{0.2t}$

Using the well-known SIF for a deeply cracked beam under bending [15], the SIF for an edge crack $a_0 > W/3$ under 3-point bending due to the force given in (17) is approximated by

$$K_I(F_m) \cong \frac{4 \cdot \sigma_f \cdot \sqrt{W - a_0}}{3} \qquad (18)$$

From J according to (16) a representative (conservative) fracture toughness value K_{JWt} can be derived by the well-known relation

$$K_{JWt} = (J_{0.2t} \cdot E / (1 - v^2))^{1/2} \qquad (19)$$

If the fracture energy W_t corresponds to KV determined by a standard Charpy test, the estimated fracture toughness K_{JWt} is suggested to be denoted as K_{JKV}. Introducing the values of the parameters corresponding to the standard Charpy specimen, the following equation to determine fracture toughness K_{JKV} from KV or KV' is obtained:

$$K_{JKV} = \sqrt{\frac{C \cdot E}{(1 - v^2)} \cdot \left[\left(\frac{C}{3.75 \cdot R_m} \right)^{1/1-p} + 0.2mm \right]^p + 7.15 \cdot \sigma_f^2} \qquad (20)$$

where

$$C = \frac{0.0250}{8^p} \cdot (KV - \frac{7.97 \cdot \sigma_f \cdot Z}{1 - Z}) \cdot A_g^{1-p} \qquad (21)$$

$$p = \frac{3}{4} \cdot \left(1 + A_g\right)^{-1} \qquad (22)$$

The second term in the bracket of (29) accounts for the effect of the finite notch root radius of the Charpy specimen, as shown in [6]. Therein, Z denotes the reduction in area of a uniaxial tensile test. Like A_g, Z has to be inserted as a non-dimensional number (not as a percentage), and KV in N·mm.

6. Experimental Validation

6.1. SCALING OF FRACTURE ENERGY

To verify the formulas derived above the fracture energy measured on sub-sized specimens is scaled-up to standard sized specimens by means of (9) and compared with directly measured standard Charpy fracture energy for materials of different toughness. From the latter, fracture toughness is estimated by means of (19) and compared with directly measured fracture toughness. The materials and specimen sizes used for these comparisons are shown in TABLE I and II. The two considered types of bronze had the same chemical decomposition, but differing mechanical properties due to different heat treatment. The measured fracture energy W_t (mean values from 3 – 5 specimen each) are given in TABLE III.

TABLE I. Geometry of the used specimens (see *Figure 1*).

	B [mm]	W [mm]	b_0 [mm]	S [mm]	H [-]
Standard Charpy	10	10	8	40	1.76
KLST	3	4	3	22	1.88
Half-size Charpy	5	5	4	22	1.76

TABLE IV shows the W_t values given in TABLE I scaled up to the size of standard Charpy specimens obtained by using formula (10). Compared with the directly measured Charpy fracture energy given in TABLE III, the deviations given in percentages in TABLE IV are obtained. Regarding the facts that these formulas are purely theoretically derived, without any adjustable factor, and that the (natural) scatter of the Charpy energy is usually as much as up to ±5%, the agreement between measured and scaled fracture energy is very satisfying.

TABLE II. Material properties of the tested materials

	R_m [N/mm^2]	R_p [N/mm^2]	A_g [-]	Z [-]	E [N/mm^2]
Steel A533 B	640	470	0.11	0.55	210000
Bronze GZ-CuSn12Ni (Type 1)	299	178	0.11	0.12	120000
Bronze GZ-CuSn12Ni (Type 2)	477	208	0.5	0.43	130000

TABLE III. Experimental W_t-values measured on standard and sub-sized specimens

	Standard Charpy	KLST	Half-size
Steel A355	215 J	8.42 J	29.4 J
Bronze Type 1	5.9 J	0.264 J	-

TABLE IV. Fracture energy KV' estimated from W_t of sub-sized specimens by eq. (10)

	KLST		Half-size	
	W_t'	Deviation	W_t'	Deviation
Steel	203 J	-3.62%	227 J	+5.5%
Bronze Type 1	6.35 J	+7.6%	-	-

6. 2. ESTIMATION OF FRACTURE TOUGHNESS

Applied to steel, the reliability of eqs. (20) – (22) to estimate fracture toughness was shown elsewhere [6, 7]. Being derived semi-analytically, the applicability of (20) – (22) is not restricted to steel, but in principle open to any elastic-plastic material. In the following we show some comparison of measured and predicted fracture toughness for two heat-treatments of bronze GZ-CuSn12Ni, denoted by type 1 (coarse grained) and type 2 (fine grained), where the former is similar to the type considered in the previous section (TABLE II). In TABLE V the fracture toughness measured according to ASTM E1820 is compared with the one estimated from the Charpy fracture energy KV by means of (20) – (22). The agreement is surprisingly good, indicating that the relation between KV and fracture toughness, which forms the basis of the scaling law derived above, is indeed applicable to any type of elastic-plastic materials.

TABLE V. Comparison of fracture toughness predicted by means of eq. (20) – (22)
with standard values K_{Jc} for the considered two types of bronze.

Material	KV [J]	K_{Jc} Measured [N/mm$^{1.5}$]	K_{JCV} Predicted [N/mm$^{1.5}$]
Bronze Serie 292	6.1	1084	1031
Bronze Serie 200	32	3842	3671

7. Discussion and Conclusions

By the presented formula it is possible to estimate the J-R-curve or fracture toughness from the fracture energy of an edge-cracked or sharply notched beam of any size under static or impact bending. A typical application is the estimation of fracture toughness from standard or sub-sized Charpy specimens. However, to estimate fracture toughness from a sub-sized Charpy-type test, it is advisable to proceed in two steps: First, the measured fracture energy should be transferred to the standard size. For this purpose, a scaling law for the total fracture energy has been derived analytically based on simple mechanical models. Second, the obtained Charpy energy shall be used to estimate the fracture toughness. This can be done using the semi-analytical relation derived in [6, 7]. Estimation of $J_{0.2t}$ directly from the sub-sized test is not recommended, because the empirical modification of p as given by (13a) or (13b) corresponds to the standard specimen. It is expected to be somewhat different for other specimen sizes.

The underlying mechanical model is based on the assumption that the involved fracture processes are predominately ductile tearing. Thus, the J-R-curve represented by (1) or (7a) as well as the scaling law (10) are restricted to the upper shelf regime from a theoretical point of view. Nevertheless, they ca be formally applied in the ductile-to-brittle transition (DBT) range and on the lower shelf as well. Although lacking a physical basis the near initiation toughness defined in (16) or (20) represent mathematically lower-bound values in these toughness ranges, as shown in [6], so conservative approximations of fracture toughness are expected to result. However, one has to be aware, that in general there is a significant shift of DBT-temperature due to specimen size, which has to be accounted for. Some empirical data on this subject are given in [8, 9]. Furthermore, the J-R-curve and fracture toughness is in general rate dependent, so the loading rate in terms of dJ/dt or dK_I/dt should be also provided in the test report.

Of course, analytical relations are approximations in any case, because they are based on simplifying analytical models. Nevertheless, in combination with experimental data, they can be used to establish semi-empirical relations that are much more general and reliable than purely empirical relations.

Nomenclature

a	Crack length
A	Non-dimensional constant
a_0	Initial crack length or notch depth, respectively
A_g	Uniform fracture strain (strain at maximum force of a uniaxial tensile test)
B	Specimen thickness
b_0	Initial ligament width, $W-a_0$
C	Constant
Δa	Stable tearing crack extension
Δa_m	Crack extension at maximum load
F_m	Maximum load
J	J integral
$J_{0.2t}$	Near-initiation J as defined in *Figure 3*
J_{Ic}	Critical J integral
K_{JKV}	Critical SIF calculated from $J_{0.2t}$ obtained from the KV-correlation
KV	Charpy fracture energy
KV'	Charpy fracture energy obtained by scaling-up a sub-sized test
p	Exponent
R_m	Ultimate tensile strength
R_p	Yield stress (stress at 0.2% plastic strain)
σ_f	Flow stress, $\sigma_f = (R_p+R_m)/2$

224

σ_{fd}	Flow stress σ_f at increased strain rate (dynamic flow stress)
SIF	Stress intensity factor
W	Specimen width
W_m	Energy consumed by the test specimen up to maximum load
W_{mp}	Plastic (nonrecoverable) part of W_m
W_t	Total fracture energy
E	Young's modulus
Z	Standard reduction of area of a uniaxial tensile test (nondimensional)
ν	Poisson's ratio
( )'	Quantity corresponding to another specimen size

References

1. Server, W.L., (1979), Static and Dynamic fibrous initiation toughness results for nine pressure vessel materials, Elastic-Plastic Fracture, ASTM STP 668, J.D. Landes, et al., Eds., American Society for Testing and Materials, West Conshohocken, PA, 493-514.
2. Norris, D.M., Reaugh, J.E. and Server, W.L., (1981), A fracture toughness correlation based on Charpy initiation energy, Fracture Mechanics: 13th Conference, ASTM STP 743, R. Roberts, Ed., American Society for Testing and Materials, West Conshohocken, PA, 207-217.
3. Barsom, J.M. and Rolfe, S.T., (1987), Fracture and fatigue Control in Structures – Application of Fracture Mechanics, 2nd ed., Prentice-Hall. Englewood Cliffs, NJ.
4. Sailors, R.H. and Corten, H.T., (1972), "Relationship Between Material Fracture Toughness Using Fracture Mechanics and Transition Temperature Tests, Proc. of the 1971 National Symposium on Fracture, ASTM STP 514, ASTM, West Conshohocken, PA, 164-191.
5. Schindler, H.J., (1998), The correlation between Charpy fracture energy and fracture toughness from a theoretical point of view, Proc. 12th European Conference on Fracture, Sheffield, 841-847.
6. Schindler, H.J., (1999), "Relation Between Fracture Toughness and Charpy Fracture Energy - An Analytical Approach", Pendulum Impact Testing: A Century of Progress, ASTM STP 1380, T. Siewert and M. P. Manahan, Sr., Eds., ASTM, West Conshohocken, PA, 337-353.
7. Schindler, H.J., (2001), Abschätzung von Bruchzähigkeitskennwerten aus der Bruch- oder Kerbschlagarbeit, Materialwissenschaft und Werkstofftechnik, Vol. 32, No. 6.
8. Lucon, E., et al., (1999), Characterizing Material Properties by the Use of Full-Size and Sub-Size Charpy Tests, Pendulum Impact Testing: A Century of Progress, ASTM STP 1380, T. Siewert and M. P. Manahan, Sr., Eds., ASTM, West Conshohocken, PA, 146 – 163.
9. Lucon, E., (1999), European Activity on Instrumented Impact Testing of Sub-Size Charpy V-Notch Specimens, Pendulum Impact Testing: A Century of Progress, ASTM STP 1380, T. Siewert and M. P. Manahan, Sr., Eds., American Society for Testing and Materials, West Conshohocken, PA, 242-252.
10. Corowin, W.R., and Houghland, A.M., (1986), Effect of specimen size and material condition on the Charpy impact properties of 9Cr-1Mo-V-Nb steel, The Use of Small-Scale Specimens for Testing Irradiated Material, ASTM STP 888, Philadelphia, 325-338.
11. Schindler, H.J., (1996), Estimation of the dynamic J-R curve from a single impact bending test, Proc. 11th European Conf. on Fracture, Poitiers, EMAS, London, 2007-2012.
12. Schindler, H.J., et al, (2000), Ageing an Irradiation Surveillance by Means of Impact Testing of Pre-Cracked Charpy Specimens, Proc. Int. Symposium on Materials Ageing and Life Management, Ed. B. Raj, et al., Kalpakkam, India, 837-846.
13. Schindler, H.J, Veidt, (1998), M., Fracture Toughness Evaluation from Instrumented Sub-size Charpy-Type Tests, in: Small Specimen Test Techniques, American Society for Testing and Materials STP 1329, W.R. Corowin, et al., Eds., 48 – 62.
14. Böhme, W. and Schindler, H.J., (1999), Application of Single Specimen Methods on Instrumented Charpy Tests: Results of DVM Round Robin Exercises, Pendulum Impact Testing: A Century of Progress, ASTM STP 1380, T. Siewert and M.P. Manahan, Eds., American Society for Testing and Materials, West Conshocken.
15. Tada, H., Paris, P.C. and Irwin, G.R., (1973), The Stress Analysis of Cracks Handbook, Del Research

MASTER CURVE METHODOLOGY AND DATA TRANSFER FROM SMALL ON STANDARD SPECIMENS

M. HOLZMANN, L. JURÁŠEK, I. DLOUHÝ
Brittle Fracture Group
Institute of Physics of Materials ASCR, Brno, Czech Republic

Abstract: In the present paper MC methodology has been used for evaluating of the fracture toughness transition behaviour of the C-Mn cast steel intended for fabrication of the large container for spent nuclear fuel. The PCVN specimens and the standard 1T SENB specimens have been used to establish the fracture toughness values in the ductile–to–brittle transition regime. Based on the experimental results obtained the following problems have been addressed: suggestion and verification of transferability approach for transferring the fracture toughness data from the small PCVN specimens to the 1T standard ones; the determination of the reference temperature T_0 based on the transformed PCVN fracture toughness data; the construction of MC and fracture probability tolerance bounds utilising the average reference temperature T_0 from the transformed PCVN data; the comparison of the toughness K_{Jc} values measured with the standard 1T SENB specimens with the toughness values predicted using small PCVN specimens.

Keywords: fracture toughness, master curve, specimen size effect, data transferability, constraint effect, reference temperature, ductile-brittle transition

1. Introduction

The methodology of master curve [1] is currently widely used for the evaluation of fracture toughness for ferritic steels in the ductile to brittle transition regime. The main material's characteristics in this concept are the reference temperature T_0, the median fracture toughness and associated failure probability bounds. The verification of this concept has been performed for pressure vessel steels, cast steel for radwaste container and weldments [2-9]. For determining the reference temperature T_0, which localizes the master curve (MC) on the temperature axis, large 1T specimens are required. But there are structures (components) under operation for which transition behaviour of the fracture toughness is of great interest (reactor pressure vessels, turbine rotors etc.) and the application of MC concept would be very useful here. However, for these components only small specimens (Charpy V-notch) may be used for the assessment of the degradation (embrittlement).

One of objective in developing mentioned standard has been that it should if possible allow to use of precracked Charpy-size (PCVN) specimen to measure T_0 and associated fracture performance. However, the supporting technical basis document for E1921 [10] presents no experimental evidence, which demonstrates the accuracy of T_0 estimates from precracked Charpy specimens.

225

I. Dlouhý (ed.), Transferability of Fracture Mechanical Characteristics, 225–242.
© 2002 *Kluwer Academic Publishers. Printed in the Netherlands.*

Hence the effort is now concentrated on application of precracked small specimen for these purposes [11]. Some works, mainly of Wallin [7, 12] have shown that the small PCVN specimens can be used in determining the reference temperature T_0 and thereby making possible to apply MC concept for the integrity assessment procedure of these components.

However, at the same time, it has been proved that the fulfilling of the provisions in the standard E1921 for evaluating T_0 using Charpy specimens is very often difficult to satisfy, as will be shown later. Some authors [13, 15] have demonstrated that reference temperature T_0 determined, using PCVN specimens and single-temperature method, is biased and provides non-conservative estimation. These findings have very relevant consequences with respect of utilizing the PCVN specimens for determining MC for degraded conditions. This problem has been discussed in Session 3 "Master curve testing" at symposium held under auspices IAEA [14].

At the present paper MC concept has been used for an assessment of the fracture toughness transition behaviour of C-Mn cast steel intended for fabrication of large container for spent nuclear fuel. The precracked Charpy-size specimens and 1T SENB specimens have been used to establish fracture toughness values in the ductile-to-brittle transition regime.

Based on the experimental results obtained the following tasks have been addressed:
- suggestion and verification of transferability approach for transferring small fracture toughness data to the standard 1T ones. Local approach [16-18] and the constraint adjustment had to be incorporated into the transferability procedure because most of the fracture toughness values from PCVN specimens have not met the validity requirements of the Standard E1921
- the determination of the reference temperature T_0 based on the transformed Charpy data to the 1T standard ones. Various methods (single- and multi-temperature) have been used for determining T_0s. Verification of the established T_0 values by comparing them with reference temperature T_0 which was evaluated using large set of 1T SENB specimens and procedure of the standard E1921
- the construction of MC using the average reference temperature T_0 from PCVN specimens. Verification of MC concept using all the transformed fracture toughness data from PCVN specimens and valid fracture toughness values for 1T standard specimens
- the comparison of toughness values K_{Jc} measured with 1T standard specimens with toughness values predicted using the small PCVN specimens.

2. The Determination of T_0 Reference Temperature.

2.1. SINGLE-TEMPERATURE METHOD (EVALUATION OF T_0 ACCORDING TO ASTM STANDARD E1921)

The critical values of J_I - integral J_c are measured for a set of 1T thick samples at the selected temperature in the ductile-to-brittle transition regime. The J_c values are then converted to their K_{Jc} equivalents. The Weibull scale parameter, K_0, is then calculated from the relationship

$$K_0 = \left[\frac{\sum\limits_{i=1}^{N} (K_{Jc(i)} - K_{min})^4}{(r - 0.3068)} \right]^{1/4} + K_{min} \tag{1}$$

Here r is the number of uncensored results, N, the total number of specimens tested, and $K_{min} = 20$ MPa.m$^{1/2}$. K_0 is then converted to the median toughness of the sample distribution, $K_{Jc(med)}$

$$K_{Jc(med)1T} = (K_0 - K_{min})[\ln(2)]^{1/4} + K_{min} \tag{2}$$

If specimens of various thickness have been used for fracture toughness measurement statistical size correction is then applied to convert the valid K_{Jc} data to 1T equivalence. Data equivalent to that for a 1T specimens, B_{1T} ($B_{1T} = 25$ mm) can be calculated by the following equation:

$$K_{Jc(1T)} = 20 + [K_{Jc(x)} - 20] \left(\frac{B_x}{B_{1T}} \right)^{1/4} \tag{3}$$

These data may be included to dataset of 1T specimens used for the calculation of K_0.

This procedure is based on the assumption that the cleavage initiation toughness distribution follows a three-parameter Weibull distribution with a slope of 4 and threshold stress intensity K_{min} of 20 MPa.m$^{1/2}$. If it is assumed that the fracture toughness versus temperature dependence is universal for a steel with ferritic matrix, having yield stress ranging from 275 MPa to 825 MPa, a value of T_0, the temperature at which $K_{Jc(med)1T} = 100$ MPa.m$^{1/2}$ can be determined. This "master curve" shape is defined within E1921 by equation

$$K_{Jc(med)1T} = 30 + 70 \exp[0,019(T - T_0)] \tag{4}$$

from which T_0 can be expressed as

$$T_0 = T - \frac{1}{0,019} \ln \left[\frac{K_{Jc(med)1T} - 30}{70} \right] \tag{5}$$

where T is the test temperature of data set.

The size of deformation criterion specified within E1921 defines a limiting toughness value $K_{Jc(limit)}$ which prescribes the maximum, allowable measured toughness. This limit is set as

$$K_{Jc(\text{limit})} = (Eb_0\sigma_{ys}/30)^{1/2} \tag{6}$$

where σ_{ys} is the 0,2% offset yield stress of steel at the test temperature and b_0 is the remaining ligament length of the specimen with a crack. Toughness values measured below this value should be independent of in-plane specimen dimensions. Test values which exceed $K_{Jc(limit)}$ are not discarded, but "censored" as described within E1921. Two statistical properties should be considered in determining the scatter for master curve method, i. e. tolerance interval and confidence band on the estimate of T_0. Hour and Yoon [3], Lucon et.al. [13]. For single temperature method the standard deviation for T_0 is given by formula in E1921 standard

$$\Delta T_0 = \frac{\beta}{\sqrt{N}} Z \tag{7}$$

where β factor is given in this standard and N is number of valid data for establishing the value of T_0. Z is a standard two-tail normal deviate, which should be taken from statistical handbook tabulations.

Lucon et.al. [13] calculated the coefficient β for the reference temperature using the Monte Carlo method in the hypothesis of Gaussian distribution. The resulting equation for β is given by

$$\beta = 13,9 + 1178,5 \left(K_{Jc(med)} - 20\right)^{-1,54} \tag{8}$$

2 .2. MULTI-TEMPERATURE EVALUATION OF T_0

If cleavage fracture toughness data are measured for different specimen sizes and test temperatures, a multi-temperature approach can be used to determine the reference temperature T_0

Two methods may be used:
a) Wallin`s multi-temperature maximum likelihood approach [12]

$$\sum_{i=1}^{n} \frac{\delta_i.\exp\{c.[T_i - T_0]\}}{a - K_{\min} + b.\exp\{c.[T_i - T_0]\}} - \sum_{i=1}^{n} \frac{\left(K_{IC_i} - K_{\min}\right)^4.\exp\{c.[T_i - T_0]\}}{\left(a - K_{\min} + b.\exp\{c.[T_i - T_0]\}\right)^5} = 0 \tag{9}$$

This method requires an iterative solution for T_0. In this equation $\delta_i = 0$ (test terminated without cleavage), or 1 (test terminates with cleavage), $a = 31$ MPa.m$^{1/2}$, $b = 77$ MPa.m$^{1/2}$ and $c = 0,019°C$. Here a, b, and c are the master curve versus temperature coefficients for K_0, the mean toughness value of the distribution. K_{Jc} values substituted into Eq. (9) are first converted to their 1T equivalent values using Eq. (3).

b) Chaouadi approach [19]

This method is based on analysing the reference temperatures as derived from each sample. From equation (4) T_0 is determined as

$$T_0^{(i)} = T - \frac{\ln\frac{\left(K_{Jc}^{(i)} - 30\right)}{70}}{0,019} \tag{10}$$

The reference temperature is then derived from data set tested at various temperatures by considering the mean or median value

$$T_0 = mean \ or \ median \ of \left(\sum_i T_0^{(i)}\right) \tag{11}$$

The multi-temperature method is utilized herein to compare the temperatures T_0 determined with T_0 evaluated by single temperature technique. In paper Lucon et al. [13] the following expression is given for deriving an engineering estimate of the equivalent 1T median toughness for a multi-T data set:

$$K_{Jc(med)1T}^{eq} = 30 + 70\left[0,019\left(\frac{\sum_i^N}{N} - T_{0(MT)}\right)\right] \tag{12}$$

where $T_{0(MT)}$ is the reference temperature evaluated by multi-temperature method. The values of $K_{Jc(med)}^{eq}$ can be used to determine the value β. Substituting the β into Eq. (7) the standard derivation of multi-T_0 reference temperature can be calculated and using Eq. (7) the confidence band of $T_{0(MT)}$ may be estimated.

2.3. CONFIDENCE INTERVALS OF T_0 COMPARED TO MONTE CARLO SIMULATIONS OF T_0 EVALUATION

Figure 1 shows the influence of the number N of valid toughness data, measured at the temperature $T-T_0 = -50°C$ and $T-T_0 = 50°C$, on the upper limit $+\Delta T_0$ of the 95% confidence interval for T_0 estimate.

Various curves are presented including ASTM E1921 procedure and Lucon et al. [13] computation of β, eq. (6), compared to Monte Carlo simulations. The Monte Carlo curves have been computed during our investigation using three parameter Weibull distribution with $m = 4$ and 10^5 iterations for individual data set. As seen, the ASTM E1921 procedure yields the most conservative values of $+\Delta T_0$.

In *Figure 2* it is shown the minimum number N of specimens of 1T thickness needed to ensure T_0 estimate in the interval $T_0 \pm 10°C$ with confidence 90% or 95%. Again, the curves in *Figure 2* have been computed using the Monte Carlo simulation. The computation has been performed for both single (STM) and multi-temperature method

(MTM) for T_0 evaluation. For multi-temperature method simulation (eq. 9) the range ±50°C around the temperature T has been chosen but the results are independent on the measured temperature range. As seen the minimum number N is a function of testing temperature. The result are shown for valid test temperature range $T\text{-}T_0 = -50°C$ to $T\text{-}T_0 = 50°C$ (ASTM E1921) and are valid when all measured data meet validity limit, eq. 6

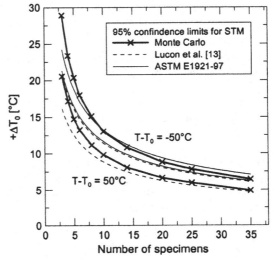

Figure 1. Influence of the number of specimens on the upper limit $+\Delta T_0$ of the 95% confidence interval

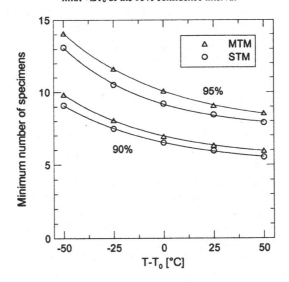

Figure 2. Minimum number of specimens needed to ensure T_0 estimate in the interval $T_0 \pm 10°C$ with confidence 90% or 95%

3. Material, Experimental Technique

The low carbon-manganese cast steel has been used for the investigation. The chemistries of cast steel (weight %) is in TABLE I.

The thick plate of the dimensions 700 mm × 750 mm × 270 mm has been cast (Melt I). The bars of the dimensions 55 mm × 90 mm × 250 mm were cut from this plate in the transverse direction. To guarantee the microstructure of these bars to be the same as that of an inner part of the cast body the computer simulation of heat treatment of these bars was performed. The bars were supplied by Škoda Company for the manufacture of the specimens.

TABLE I. Carbon-manganese cast steel (weight %)

C	Mn	Si	Cr	Ni	Cu	Mo	S	P	Al
0,09	1.18	0.37	0.12	0.29	0.29	0.03	0.025	0.01	0.028

The following specimens have been prepared for mechanical testing:
– cylindrical specimens, diameter 6 mm, for static rate tensile tests over the temperature range −196°C to 20°C for measuring of the temperature dependence of the yield stress and tensile strength
– single-edge notched bend (SENB) specimens of the thickness 1T (25 mm), width 50 mm and the length 220 mm. The fatigue precracking was done in accordance with ASTM standard E399-90. The specimens were loaded by static three point bending using bend test fixture described in mentioned standard (the span support $S = 4W = 200$ mm). The tests were conducted over wide temperature range in cooling chamber the specimens being cooled by liquid nitrogen vapour. The accuracy of pre-set temperature was ± 1,5°C. The cross head velocity was 2 mm/min (displacement control). Force versus force-point displacement was recorded and used for determining J-integral at the onset of the cleavage fracture. The calculation of critical values of J-integral was performed as per standard E1921. These critical J_I - values were converted to their elastic-plastic equivalent stress intensity factor K_J depending on the failure mode.
– precracked Charpy-size specimens (a/W ratio ≈0,5) loaded by static three point bending in special bending fixture in cooling chamber over temperature range from - 160°C to −60°C. This testing set-up was placed into tensile testing machine. The cross head velocity was 2 mm/min. The force-force point displacement traces were used for calculation of critical values of J_I - integral, J_c, at the onset of fracture event. The J_c values were again converted to their elastic-plastic equivalent K_{Jc}-values.

4. Results and Discussion

4.1. TENSILE TESTS

The basic mechanical characteristics at room temperature: the yield stress $R_e = 270$ MPa; the tensile strength $Rm = 435$ MPa; elongation $A5 = 32,5$ %; reduction in area $Z = 74,7$ %

232

The temperature dependence of the upper and lower yield stress and tensile strength is illustrated in *Figure 3*. As seen the cast steel investigated exhibits relatively low values of the upper and lower yield stress and with decreasing temperature the increase of these characteristics is very slow (e.g. at 100°C R_{eL} is only 380 MPa). This resulted in the necessity to perform static loading tests of the small PCVN specimens at very low temperatures to evaluate the reference temperature, T_0, as per guidelines of standard E1921.

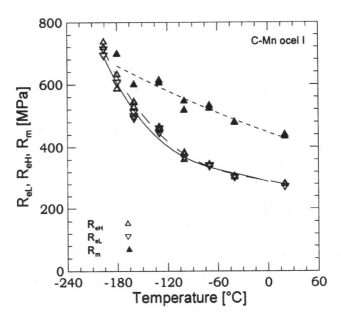

Figure 3. Upper R_{eH} and lower R_{eL} yield stress and tensile strength respectively as a function of temperature

4.2. TRANSITION BEHAVIOUR OF THE FRACTURE TOUGHNESS – 1T SENB SPECIMENS

The measured values of the fracture toughness as a function of temperature are plotted in *Figure 4*.

The following fracture toughness values have been measured:

K_{Jc} : fracture toughness values meeting the validity condition, Eq. (6), which is represented in figure by line marked $K_{Jc(limit)}$.

K_{QJc} : fracture toughness values for tests terminated by cleavage event, but the high deformation level resulted in the loss of constraint at the onset of unstable fracture.

K_{Ju} : fracture toughness values for unstable cleavage fracture occurring after some ductile crack extension.

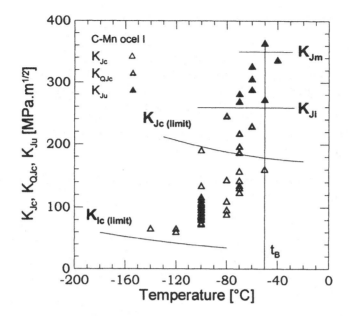

Figure 4. Fracture toughness temperature
diagram - standard 1T SENB specimens

The $K_{Ic(limit)}$ line marks the maximum values of valid plane-strain fracture toughness K_{Ic}. The K_{Ji} line represents the estimated fracture toughness value for an initiation of ductile fracture, since the results below this line are fracture toughness initiated by the cleavage without prior ductile tearing and those ones above this line already exhibit some ductile crack extension prior unstable failure.

Finally, K_{Jm}-line is K_J value for maximum force in the force versus force-displacement record.

The temperature T_B is the brittleness transition temperature since at and below this temperature the cleavage initiated fractures without prior ductile tearing may already be observed. As may be seen in temperature region −70°C to −50°C the overlapping of transition range with K_{Ju}-values and K_{Jc}-ones occurs.

For correct determining of the reference temperature T_0 single temperature method as per the standard E1921 has been employed. At the temperature −100°C data set of twenty three specimens has been tested. All the measured fracture toughness values, as seen, fall below the validity limit.

By utilizing the Eqs. (1), (2) and (5) the reference temperature T_0 has been calculated

$$T_0 = -92°C$$

Taking from the standard E1921 the parameter β for $K_{Jc(med)} = 90$ MPa.m$^{1/2}$, $\beta = 18°C$ and substituting this value in the Eq. (7) the standard deviation on T_0 ($\sigma(T_0)$)

can be calculated. Choosing the 85% confidence on T_0 estimate one gets the confidence limit

$$\Delta T_0 = \pm 5°C$$

Note, that Lucon et al. [13] when comparing reference temperatures T_0 evaluated using data set of specimens with various dimensions have chosen for the confidence band two standard deviation $\pm 2\sigma$ (T_0), i.e. 95% confidence.

4.3. TRANSITION FRACTURE BEHAVIOUR OF PRECRACKED CHARPY SIZE SPECIMENS

The measured toughness values in temperature range from $-160°C$ to $-65°C$ are plotted as a function of temperature in *Figure 5*.

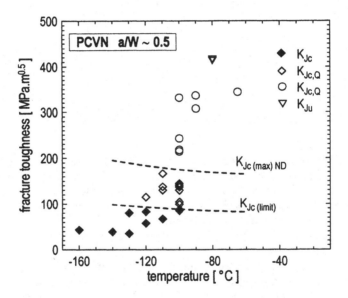

Figure 5. Fracture toughness temperature diagram - PCVN specimens

The following fracture toughness values are plotted:

K_{Jc} : valid values of K_{Jc} lying under the validity line (Eq. (6)) denoted $K_{Jc(limit)}$.

$K_{Jc,Q}$: constraint dependent fracture toughness values, that are divided into two groups, one of which lying above $K_{Jc(max)}$ N-D line (details see later). Despite the high values of $K_{Jc,Q}$, tests always terminated by cleavage event, demonstrating high resistance of cast steel examined against ductile fracture initiation.

K_{Ju} : fracture toughness for unstable cleavage failure occurring after some ductile crack growth.

As seen from figure the procedure of determining T_0 as per the standard E1921 cannot be effectively used for PCVN specimens due to relatively low yield stress of cast steel examined even at low temperatures. Only five valid K_{Jc}-values larger than 50 MPa.m$^{1/2}$ namely at different temperatures could be measured.

Therefore, Dodds–Anderson model [16, 17] is referred here that enables the measured unvalid, due to constraint loss, toughness values under large scale yielding deformation state (LSY) $J_{c,Q}$ (stress intensity equivalent $K_{Jc,Q}$) to transfer to the material toughness J_0 (stress intensity equivalent K_{J0}) under 2D-SSY deformation state (the toughness scaling diagrams). Model is based on the equivalency of the critically stressed volume of the material ahead of the crack tip between these two deformation states. Using 2D model (plane–strain) for both SSY and LSY deformation states, the equivalence is based on the equivalency of an critically stressed areas enclosed within the principal stress counter σ_1/σ_0 between these two deformation states. The computing study under 2D-LSY conditions of various fracture specimens revealed an independence of the $J_{c,Q}/J_0$ ratio on the selected values of σ_1/σ_0 (σ_0 is the yield stress).

However, 3D-FEM analysis of various fracture mechanics specimens performed by Nevalainen and Dodd (N-D) [18] have proved that the ratio $J_{c,Q}/J_0$ is the in-plane constraint component which is strong dependent on principal stress counter σ_1/σ_0.

The toughness scaling diagram used herein is therefore that one published by Nevalainen and Dodds [18] for specimen geometry having $W = B$; $a/W = 0,5$, loaded by three-point bending; material flow properties: $E/\sigma_0 = 500$; $n = 10$ (*Figure 10a* of N-D work). The critical stress counter of stressed area in midplane with normalized value of $\sigma_1/\sigma_0 = 3,4$ has been chosen for in-plane constraint adjustment.

The reference temperature T_0 determination has been performed in two phases. At the initial stage of investigation only small number of PCVN specimens were available for measurement of fracture toughness versus temperature diagram (*Figure 5*) [20].

The reference temperature T_0 has been determined as follows.

4.3.1. *Single Temperature Method (STM)*:
- testing temperature chosen: -100°C
- six fracture toughness values were measured, from which only two values met the validity condition, Eq. (6). The four values have been in-plane constraint adjusted using D-A model with toughness scaling diagram of N-D
- all the fracture toughness values have been now statistical size converted to 1T equivalence
- the reference temperature T_0 has been then calculated using Eqs. (1), (2) and (5):

$T_0 = -89°C$

The confidence interval on T_0 estimate for $N = 6$ and 85% confidence (Eqs. (7)):

$\Delta T_0 = \pm 10°C$

4.3.2. *Multi-temperature Method (MTM)*:
The set of 14 fracture toughness values in temperature region $\langle$-130; -100$\rangle$°C ,*Figure 5*, [20] have been used for determining T_0. Only eight fracture toughness values met the validity condition, Eq. (6). Six values have been in-plane constraint adjusted as mentioned above. Then, all values have been statistical size converted to 1T equivalence.

The evaluation of reference temperature T_0:
- maximum likelihood method, Eq. (9):

$$T_{0(MT)} = -91°C$$

- Chaouadi approach, Eqs. (10) and (11):

$$T_{0(MT)} = -82°C .$$

After calculating $K^{eq}_{Jc(med)1T}$ Eq. (12), the corresponding β parameter, Eq. (7), may be taken from the standard E1921 and for 85% confidence one gets the confidence band $\Delta T_0 = \pm 7°C$.

During further phase of the investigation large number of specimens have been tested at $-100°C$ (TABLE II) [21]. But only for the fracture toughness values falling under the line denoted $K_{Jc(max)ND}$ an in-plane constraint adjustment procedure can be utilized. The line $K_{Jc(max)ND}$ demonstrates the maximum values of toughness to which N-D toughness scaling diagram was computed.

TABLE II.   Measured $K_{Jc,Q}$ toughness values and constraint adjusted and size corrected $K_{Jo(1T)}$ values

$K_{Jc.Q}$	84.5	86.4	94.9	100.2	100.3	103.9	120.5	123.5
$K_{Jo(1T)}$	67.2	72.8	73.2	76.4	76.5	77.9	86.2	87.8
$K_{Jc.Q}$	123.5	129.2	131.2	138.4	139.8	141.8	142.9	156.7
$K_{Jo(1T)}$	88.9	91	92.8	94.2	94.8	95.1	96.2	100.3
$K_{Jc.Q}$	160.3	161.7	164.2	170.1	177.9	178.9	184.4	
$K_{Jo(1T)}$	101	102	103.3	104.4	106.3	107.5	109.5	

The set of twenty three fracture toughness values (TABLE II) was accounted for to determine T_0. From this set only two values fulfilled the validity condition, Eq. (6), the other have been an in-plane constraint adjusted, as described above. Then, all fracture toughness values were statistical size converted to 1T equivalent. Single temperature method has been utilized to determine T_0. The resulting value for T_0 is

$$T_0 = -90°C$$

and for 85% confidence band one gets

$$\Delta T_0 = \pm 5°C$$

The reference temperatures T_0 evaluated by means of various methods and with specimens of different geometry and size are plotted, with confidence bands, in *Figure 6* as a function of the remaining ligament length.

As can be seen there is relatively good agreement among reference temperatures T_0 determined from precracked Charpy-size specimens and reference temperature T_0 from standard 1T SENB specimens having been determined using rigorously the guidelines of the standard E1921. This T_0 may be regarded as a basic material characteristic of cast steel examined. The reference temperatures T_0 from PCVN specimens fall either into the confidence band of T_0 of the standard 1T SENB specimen or lie on the conservative side of this band.

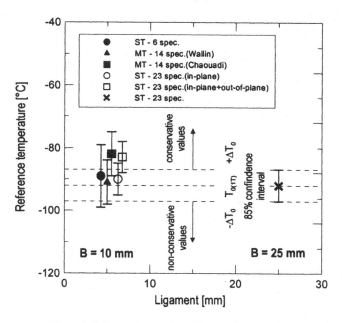

Figure 6. Reference temperatures T_0 versus ligament size

4.4. THE MASTER CURVE CONCEPT BASED ON PRECRACKED CHARPY SIZE SPECIMEN DATA.

Master curve (MC) and associated fracture probability bounds for $P_f = 5\%$ a 95% have been constructed utilizing the average reference temperature T_0 evaluated by using PCVN specimens. The average median fracture toughness curve and the fracture probability bounds are drawn in *Figure 7* using the equations

$$K_{Jc(med)1T} = 30 + 70\exp[0,019(T + 85°C)] \tag{14}$$

$$K_{Jc(0,05)} = 25,4 + 37,8\exp[0,019(T + 85°C)] \tag{15}$$

238

$$K_{Jc(0,95)} = 34,6 + 102,2 \exp[0,019(T + 85°C)] \qquad (16)$$

In *Figure 7* the margin-adjusted 5% fracture probability bound is drawn as well using average $\Delta T_0 = +7°C$ (85% confidence) from PCVN data. This margin-adjusted 5% tolerance bound is revised to:

$$K_{Jc(0,05;0,85)} = 25,4 + 37,8 \exp[0,019(T + 78°C)] \qquad (17)$$

Now, all measured fracture toughness data for standard 1T SENB specimens and those obtained from PCVN specimens constraint adjusted and size corrected are plotted in the diagram. Practically all data fall in the scatter band limited by 5% and 95% tolerance bounds. These results validate the use of PCVN data, constraint adjusted and size corrected, for determining T_0 and the E1921 master curve concept for fracture toughness behaviour of C-Mn cast steel in the lower-end of the ductile-to-brittle transition regime.

These conclusions are in coincidence with the results of more recently published paper of Joyce and Tregoning [22].

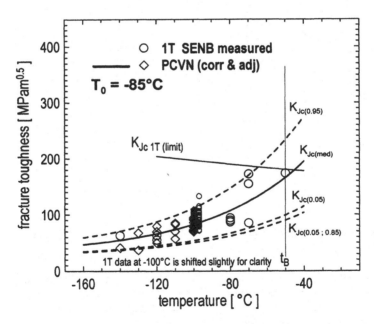

Figure 7. The E1921 Master curve based on the reference temperature T_0
from PCVN constraint adjusted and size corrected data
and 1T SENB fracture toughness data

4.5. THE COMPARISON OF FRACTURE TOUGHNESS K_{Jc} VALUES FOR STANDARD 1T-SENB SPECIMENS WITH PREDICTED VALUES USING PRECRACKED CHARPY DATA

The fracture toughness K_{Jc} values of 1T-SENB specimens converted from the measured J_c-integral at the onset of cleavage fracture at the temperature $-100°C$ are given in TABLE. III. In this table there are further presented following data:

- constraint dependent fracture toughness $K_{Jc,Q}$ values converted from $J_{c,Q}$-integral measured at the onset of unstable cleavage fracture of PCVN specimens tested at $-100°C$ (taken from TABLE II).
- in-plane constraint adjusted toughness values K_{Jo}, converted from material toughness values J_o under 2D-SSY deformation state determined from measured $J_{c,Q}$ values using toughness scaling diagram of N-D for critical stress counter $\sigma_l/\sigma_0 = 3,4$.
- fracture toughness $K_{Jo(1T)}$ values, which are the K_{Jo} values statistical size corrected to the thickness 1T using eq. 6 (taken from TABLE II).
- in-plane and out-of-plane constraint adjusted $K_{\bar{J}o}$ toughness values using N-D model for determination of the out-of-plane constraint components.
- $K_{\bar{J}o(1T)}$ - fracture toughness $K_{\bar{J}o}$ statistical size converted to the thickness 1T. $K_{\bar{J}o(1T)}$ represents inherent material fracture toughness for 3D-idealized SSY deformation state (Joyce and Tregoning [22]).

Based on data in TABLE III the following conclusion may be drawn:

- the differences between data in second and third column represent the magnitude of the in-plane constraint component.
- the differences between data in the third column and the fifth column represent the out-of-plane constraint component at LSY deformation of PCVN specimens.
- the data in the first and the fourth column, respectively, enable the comparison of the measured fracture toughness values for the standard 1T-SENB specimens and the predicted ones from PCVN data. As may be seen the coincidence is relatively close. This fact demonstrates that constraint loss at LSY deformation of PCVN specimens is mainly due to in-plane constraint component.
- the comparison of data in the first and the sixth column, respectively, reveals that with 1T-SENB specimens some out-of-plane constraint has occured mainly for higher values of K_{Jc}.

Finally, the reference temperature T_0 is evaluated for 3D-idealized SSY deformation state

$$T_0 = -83°C$$

and confidence band for 85% confidence $\Delta T_0 = \pm 5°C$. This value of T_0 is also plotted in *Figure 6*. As seen it is about 9°C higher than T_0 for 1T specimens.

TABLE III. Fracture toughness data of 1T-SENB specimens and PCVN specimens

K_{Jc}	$K_{Jc.Q}$	K_{Jo}	$K_{Jo(1T)}$	$K_{\bar{Jo}}$	$K_{\bar{Jo}(1T)}$
71.8	84.5	79.4	67.2	73.3	62.4
72.6	86.4	86.4	72.8	74.8	63.6
72.8	94.9	86.9	73.2	79.9	67.6
72.8	100.2	90.9	76.4	83.4	70.4
78.9	100.3	91	76.5	83.5	70.5
81.2	103.9	92.8	77.9	84.9	71.6
83.6	120.5	103.2	86.2	94.1	78.9
84.8	123.5	105.3	87.8	95.8	80.3
85.6	123.5	106.6	88.9	97.8	81.9
86.9	129.2	109.3	91	99.2	83
86.9	131.2	111.5	92.8	101.2	84.6
89.7	138.4	113.3	94.2	102.7	85.8
91.1	139.8	114.1	94.8	103.4	86.3
94.8	141.8	114.5	95.1	103.2	86.1
95.4	142.9	115.8	96.2	105.7	88.2
98.7	156.7	121	100.3	109.2	90.9
99.8	160.3	121.9	101	109.9	91.5
103.1	161.7	123.1	102	110.9	92.3
104.8	164.2	124.7	103.3	112.3	93.4
106.4	170.1	126.1	104.4	113.4	94.3
111.3	177.9	128.5	106.3	115.6	96
114.9	178.9	130	107.5	116.8	97
132.7	184.4	132.5	109.5	119.1	98.8

5. Conclusions

- Transferability procedure of fracture toughness values from PCVN specimens to the standard 1T ones and the prediction of master curve and associated failure probability bounds for cast C-Mn steel have been analysed
- Various methods for determining the E1921 T_0 reference temperature have been tested
- The master curve and associated failure probability bounds based on the reference temperature from PCVN specimens were validated using the fracture toughness data of standard 1T specimens and constraint adjusted and size corrected data of PCVN specimens

- The comparison of fracture toughness K_{Jc} values determined from standard 1T-SENB specimens with predicted ones being established from constraint adjusted and size corrected PCVN data, has demonstrated very good coincidence of both toughness values. Therefore, the small PCVN specimens may be used to determine fracture toughness for SSY deformation state if constraint loss and statistical size effect of PCVN specimens are properly evaluated.

Acknowledgements

The research and work on contribution has been financially supported by grant No. 101/00/0170 of the Grant Agency of the Czech Republic and project Nr. 972655 within NATO Science for Peace program.

References

1. Standard ASTM E1921 – 97 Standard Test Method for Determination of Reference Temperature, T_0, for Ferritic Steels in the Transition Range.
2. Yoon, K. K.: Alternative Method of RT_{NDT} Determination for Some Reactor Vessel Weld Metals Validated by Fracture Toughness Data, Journ. Pressure Vessel Technology, 1995; 117; pp. 378-382.
3. Hour, K. Y. and Yoon, K. K.: "Fracture Toughness Test on Precracked Charpy Specimens in the Transition Range for Linde 80 Weld Metals", Small Specimen Test Techniques, ASTM STP 1329, pp. 173–195, W. R. Corwin, S. T. Rosinski and E. van Walle Eds., Amer. Soc. for Testing and Materials, West Conshocken, P. A.
4. Holzmann, M., Dlouhý, I. and Brumovský, M.: Measurement of Fracture Toughness Transition Behaviour of Cr-Ni-Mo-V Pressure Vessel Steel using Precracked Charpy specimens, J. Pressure Vessel and Piping 1999, 76, pp. 591-598.
5. Aurich, D., Jaenicke, B. and Veith, H.: Safety and Reliability of Plant Technology, 22 MPa Seminar, 1996, paper 21.
6. Mc Cabe, D. E., Sokolov, M. A. and Nanstad R.K., In Proc. of Structural Mechanics in Reactor Technology, 14. Intern. Conf., Vol. 4, division G, Lyon, France 1997, pp. 349-356.
7. Wallin, K.: Small Specimen Fracture Toughness Characterization – State of the Art and Beyond, in Advances in Fracture Research, ICF 9, Sydney, April 1997, B. L. Karihaloo, Y-W. Mai and M. I. Ritchie Eds., PERGAMON Amsterdam, 1997, pp. 2333-2344.
8. Link, R. E. and Joyce, J. A. "Experimental Investigation of Fracture Toughness Scaling Models "Constraint Effect in Fracture, ASTM STP 1244, 1995, pp. 286-315, M. Kirk and Ad Bakker Eds., ASTM, West Conshohocken, P. A.
9. Dlouhý, I., Chlup, Z. and Holzmann, M.: "Master curve evaluation at static and dynamic loading (for cast steel and pressure vessel steels), IAEA SPECIALIST' MEETING, Master Curve Testing and Application, Prague, September 2001, Paper 16.
10. NUREG/GR-5540, US Nuclear Regulatory Commision, Washington, DC, November 1998, cited in [22]
11. Mayfield, M. E., Vassilaros, M. G., Hacket, E. M., Mitchel, M. A., Wichman, K. R., Strosmider, J. R. and Shao, L. G., Aplication of revised fracture toughness curves in pressure vessel integrity analysis, pp. 13–20
12. Wallin, K. "Validity of small Specimen Fracture Toughness Estimates Neglecting Constraint Correction, Ibid as [8], pp.519-537.
13. Lucon, E., Scibetta, M. and Eric van Walle "Assessment of the Master Curve Approach on Three Rector Pressure Vessel Steels (JRQ, JSPS, 22 NiMoCr37), Ibid as [9], paper No.7.
14. IAEA SPECIALIST' MEETING, Master Curve Testing and Application, Prague, 2001.
15. Chomic, E., Liendo, M. and Iorio, A. F. "Master curve testing on 22 NiMoCr 37 material", Ibid as [14], paper No.9.
16. Anderson, T. L. and Dodds, R. H. "Specimen Size Requirements for Fracture Toughness Testing in the Transition Region "Journal of Testing and Evaluation, JTEVA. ASTM 1991, 19, pp. 123-134.

17. Anderson, T. L., Vanaparthy, N. M. R. and Dodds, J. R. "Predictions of specimen size dependence on fracture toughness for cleavage and ductile tearing "Constraint Effect in Fracture, ASTM STP 1171, pp. 473-491, E. M. Hackett, K.-H. Schwalbe and R. H. Dodds Jr. Eds., ASTM Philadelphia, PA 19103.
18. Nevalainen, M. and Dodds, R. H. "Numerical Investigation of 3D Constraint Effects on Brittle Fracture in SE(B) and C(T) Specimens, Inter. J. of Fracture, 1995, 74, pp. 131-161.
19. Chaouadi, R. "Fracture toughness measurements in the transition regime using small size samples" Ibid as [3], pp. 214-237.
20. Holzmann, M., Kozák, V. And Dlouhý, I. „Master curve methodology and fracture behaviour of the cast steel" ECF 12 – Fracture from defects, Sheffield, 1998, Vol. II, M.W. Brown, E.R. de los Rios and K.J. Miller Eds., pp. 823-828, EMAS PUBLISHING
21. Dlouhý, I., Holzmann, M., Kozák, V., Possible use of small specimens in fracture behaviour evaluation of CMn cast steel, Welding 2000, 49, pp. 247-251
22. Joyce, J. A. and Tregoning, R. L. Development of the reference temperature from precracked Charpy specimens, Eng. Fract. Mechanics 2001, 68, pp. 861-894.

MASTER CURVE VALIDITY FOR DYNAMIC FRACTURE TOUGHNESS CHARACTERISTICS

I. DLOUHÝ [1]), G. B. LENKEY [2]), M. HOLZMANN [1])

[1])*Institute of Physics of Materials, Academy of Sciences, Žižkova 22, 61662 Brno, Czech Republic*

[2])*Bay Zoltán Foundation for Applied Research, Institute for Logistic and Production Systems, H-3519 Miskolctapolca, Hungary*

Abstract: Selected aspects of the master curve methodology have been investigated experimentally based on dynamic fracture toughness data. A manganese cast steel has been utilised for experiments the specimens being cut form surface location and midthickness of the thick walled plate. Drop weight tower, standard pre-cracked bend specimens and two procedures – quasistatic and dynamic key curve method were applied for dynamic fracture toughness determination. Pre-cracked Charpy type specimens have been also followed to investigate the statistical size effects. The master curve concept has been found to be fully valid for dynamic fracture data and capable for prediction of standard specimens behaviour from the small specimens.

Keywords: dynamic fracture toughness, cast ferritic steel, master curve, pre-cracked Charpy specimens, drop weight test

1. Introduction

The master curve (MC) approach has been developed to quantify fracture toughness of low-alloyed ferritic steels in the brittle to ductile transition region [1-8]. This method has been based on a more than 15-year research of Wallin's group [1-3 etc.]; its methodology is recently covered by ASTM standard E1921-97 [4] for data obtained at quasistatic loading conditions. The MC approach was applied to numerous fracture toughness data sets of ferritic steels, in nuclear power industry in particular, supplying very good results and proving capability of this concept for fracture toughness quantification and prediction in lower shelf and transition region [4-8].

Joyce, Viehrig [5,7] applied this method to dynamic fracture toughness data and showed the applicability of this approach in this field. The MC appears to be an applicable for the description of brittle (cleavage) fracture initiation at higher loading rates [5]. According to the main results the reference temperature obtained from dynamic fracture toughness data exhibits a systematic shift when comparing the T_0 for quasistatic and dynamic test data. Wallin analysed the dynamic fracture toughness data using the master curve approach and calculated the T_0 shifts from static to dynamic data

I. Dlouhý (ed.), Transferability of Fracture Mechanical Characteristics, 243–254.

in terms of loading rates [3]. According to results obtained by Viehrig et al. with nuclear pressure vessel steel A 533 B Cl 1, the K_{Jd} values determined under dynamic loading do not follow the course of the MC designed for quasistatic conditions and arguments have been supplied against a simple, unmodified transfer of quasistatic to dynamic condition. The results stimulated to carry out further investigation with a broader spectrum of materials.

For determining the reference temperature, T_0 that is taken as a basic material characteristic localising the MC on the temperature axis the standard large (1T) specimens are usually required. But there are structures under operation for which the transition behaviour of fracture toughness is of great interest (reactor pressure vessels, rotors etc.) and application of MC concept would be very useful. For these components only small specimens (e.g. pre-cracked Charpy type specimens, P-CVN) can be used for assessment of degradation, however. It has been shown [2,5] that the P-CVN specimens can be used in determining the reference temperature T_0 and thereby making possible to apply the MC concept for the integrity assessment procedure of these components. Specimen size and loading rate effects on brittle fracture of ferritic steels tested in the ductile to brittle transition region remain key topic for the application of pre-cracked Charpy specimens.

As proved by Dodds et al. and others [10-12] the P-CVN specimens lose constraint well before the onset of unstable fracture. The test results obtained provide nonconservative and invalid estimate of conditions required for the onset of unstable fracture in full-scale component. It has been found however, that if the P-CVN specimens were loaded in impact the additional constraint would be present at the crack tip allowing to predict correctly the unstable fracture initiation from the P-CVN data [5]. This allows the factor in the size-deformation limit to be reduced from the range of 50 to 100 to between 25 and 30 and the P-CVN results to be qualified for prediction of unstable fracture initiation in 1T specimens or full-scale components [13] without any adjustment for crack tip constraint loss.

The aim of this work has been experimentally to investigate the validity of the MC concept for dynamic fracture toughness data obtained for cast ferritic steel in ductile to brittle transition regime with standard specimens. Additionally, the prediction of the fracture toughness scatter of large (1T) specimens from small pre-cracked Charpy ones tested statically and/or dynamically has been also performed. Note that the effect of loading rate on reference temperature has been followed systematically in other contribution in this volume [14].

2. Mechanical Testing and Calculations

The material used in this study was the C-Mn cast ferritic steel supplied by ŠKODA Nuclear Machinery. This material was available as a 270 mm thick plate (Melt II according to nomenclature introduced in paper [15]). The verification of master curve methodology for dynamic fracture toughness data has been carried for specimens cut from two locations: from the surface parts of the plate (E) and from the midthickness (C).

Three point bend specimens of dimensions 23x23x110 mm^3 were used for dynamic fracture toughness determination. These dimensions have been adjusted to maximum possibilities limited by geometrical dimensions of fixtures on drop weight tower [16]. The specimens were tested at drop weight velocities of 1.99 ms^{-1} to 3.00 ms^{-1} in the temperature range from -50 to -10 °C. The dynamic fracture toughness values, K_{Jd}, have been determined by two procedures: (i) the first one was based on standard approach as used for quasistatic conditions of testing (labelled as QST). The only difference was the more careful reading of unstable (fracture) load and necessity to apply smoothed traces before the determination of plastic work to fracture. (ii) The other method was based on dynamic key curve approach, a method that is predetermined for high strain rate loading (labelled as DKC). Detailed analysis of the test methodology has been presented in work of Lenkey [17] in this volume. The results of both assessment procedures have been compared and applied for master curve evaluation.

Pre-cracked Charpy type specimens (P-CVN) have been tested dynamically by low blow method on instrumented impact tester (loading rate of about 1 ms^{-1}, according to estimations similar strain rate at the crack tip as in case of large specimens was supposed). The same specimen geometry has been tested at quasistatic loading conditions (1 $mm.min^{-1}$) with this material [15].

For master curve analyses the standardised approach was applied according to standard ASTM E 1921 –97 [4]. The results presented here represent data evaluated with multitemperature method and procedure including the maximum likelihood method. For more details see [6].

From supporting procedures applied it is necessary to mention the FEM calculations and fractographic analysis. The standard MARC [21] and ABAQUS 5.7 FEM codes were used to model elastic-plastic behaviour (almost in 3D) for the test specimen geometries investigated.

3. Results Description and Discussion

3.1. COMPARISON OF DKC AND QST METHOD (BASED ON THE MASTER CURVE METHODOLOGY)

As generally known for dynamic tests the load - deflection traces are associated with oscillations that complicate the assessment procedures of fracture toughness determination. For purposes of this investigation two methods have been followed and applied.

In case the fracture load can be read exactly from load deflection trace the quasistatic approach can be still applied. Standard 1T SENB specimens and the P-CVN ones have been assessed by this method.

For evaluation tests at higher loading rates the dynamic key-curve method [18] may be applied supposing the onset of unstable fracture is determined by some supporting method. Magnetic emission method was used for this purpose in our investigation. This approach was applied for specimens tested with drop weight tower [16]. The methodological aspects of the methods and full analysis of results obtained have been described in other contribution in this volume [17].

All the data obtained by both methods applied (quasistatic – QST and dynamic curve method – DKC) are summarised in *Figures 1 and 2*. Both data sets are plotted in

246

these figures commonly with the results of reference temperature determination (see also TABLE I) and master and bounding curves. In both cases the 90 % probability scatter bands correspond well to the given data sets (QST or DKC correspondingly). For

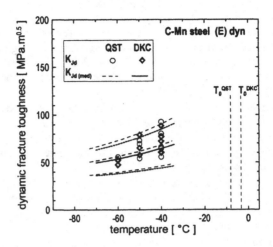

Figure 1. Comparison of K_{Jd} and corresponding master and bounding curves in lower part of transition for location E (K_{Jd} values being determined by two methods)

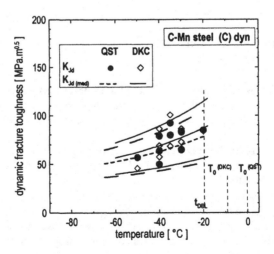

Figure 2. Comparison of K_{Jd} and corresponding master and bounding curves in lower part of transition for location C (K_{Jd} values being determined by two methods)

central part of the plate (location C) remarkable difference between both methods of K_{Jd} determination has been found. For surface location (E) more consistent results have been obtained. Although differences have been found between separate specimens assessed either by QST or by DKC method, the resulting properties of data set are practically the same, i.e. the master and bounding curves are not affected by procedure of K_{Jd} determination. This can be seen in the very near values of reference temperature (TABLE I) and in the same localisation of master curve on temperature axis.

When arising from the primary results gained the following knowledge may be obtained:

Independently of difference between both methods applied at dynamic fracture toughness determination (QST and DKC) the median curve and 90 % probability scatter band correspond well to the data used for the reference temperature calculation. From this point of view the MC concept is suitable methodology as for description of data in transition region so for possible prediction of transition behaviour based on specimen of non-standard geometry.

At dynamic loading the onset of unstable fracture is usually associated with adiabatic heating localised into plastic zone at crack tip. As the result of this specimen behaviour a sharp, discontinuous, increase of fracture toughness values is usually observed instead of continuous curve. This is the typical material behaviour in transition region that specimen fails either completely in brittle manner or completely by ductile mechanism. Generation of data for purposes of master curve evaluation or similar purposes is thus complicated due to this material behaviour.

The highest temperatures that supplied valid experimental fracture toughness values and data for master curve determination are well below the reference temperature obtained. The difference is more than 20 °C for location C and about 40 °C for location E. Although good correlation has been found as for mean values so for scatter characteristics, the highest temperature displaying the unstable fracture initiation limits also the validity of master curve.

TABLE I: Reference temperatures obtained for cast ferritic steel, for both locations C and E, standard SENB and pre-cracked Charpy specimens, respectively and for two loading rates

Specimen location	1T SENB statically	P-CVN statically	SENB dynamically QST	SENB dynamically DKC	P-CVN low blow QST
C	-126	-127	0	- 9	9
E	-144	-144	- 8	- 3	- 5

3.2. THE EFFECT OF SPECIMENS LOCATION

In addition to the first two figures the direct comparison of the effect of specimens location is available from the *Figure 3*. Here not only the transition region but also the upper shelf data are informatively shown (for the location E) to characterise fully the material fracture behaviour in both brittle and ductile regime. Because the quasistatic approach (QST) to dynamic fracture toughness determination has been found to give

248

more consistent data the data obtained by this methodology have been applied only for purposes of the comparisons in the *Figure 3*.

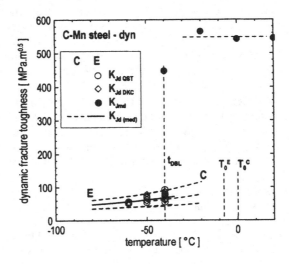

Figure 3. Fracture toughness K$_{Jd}$ values, master and bounding curves in lower part of transition region for location E and comparison to location C

For the midthickness location C of the thick walled plate the master curve (by dashed line) with 90 % probability scatter band is introduced as the representative characteristics of all experimental values in *Figure 3*. For surface location E also all valid experimental values are introduced as well as the master curve including the 90 % probability scatter band (full line and dashed lines being overlapped by scatter band of data from location C).

In addition to this direct comparison of the master curves locations on temperature axis the reference temperatures are introduce in the right side of the figure, showing relatively small difference in fracture behaviour at dynamic loading between surface and midthickness of thick walled plate.

The only difference between both specimen locations can be seen in the location of the same curve on the temperature axis, i.e. that data from both specimen locations are lying in the same scatter band.

3.3. EFFECT OF LOADING RATE ON REFERENCE TEMPERATURE T$_0$

Supposing the master curve has the same shape independently of loading rate, all the data plotted against temperature difference T-T$_0$ (where T represent the test temperature) has to be located in the same probability scatter band. For the steel investigated, for static fracture toughness values (data introduced in paper [15] in this volume) and the dynamic fracture toughness values obtained in this investigation by using standard SENB specimens and P-CVN ones this has been proved as it is shown in

Figure 4. It follows from the figure that for the steel followed and range of loading rates applied the transition curves and corresponding scatter bands have, more or less, the same shape when plotted against the value of T-T$_0$.

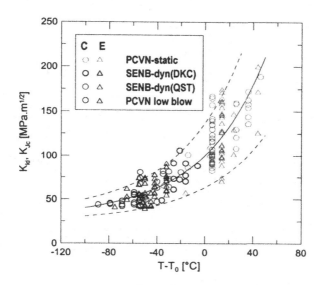

Figure 4. Fracture toughness data K$_{Jd}$, K$_{Jc}$ obtained at different loading rates (at 1 mm.min^{-1} and 1 m.s^{-1}) plotted against T-T$_0$

From the material point of view one of the main findings is that steel is strongly susceptible to loading rate. Remarkable shift of reference temperature (and/or other transition temperatures in corresponding way) is observed being on level of more than 120 °C for location C and 130 °C for location E when comparing these temperatures with values obtained at static loading (dK/dt about 10^5 MPam$^{0.5}$s^{-1} against 1 MPam$^{0.5}$s^{-1}) as it is evident from the TABLE I.

This shift can be also observed from the *Figure 5* where all data obtained at static and dynamic conditions of loading are introduced (for Melt II of C-Mn steel investigated here and for Cr-Mo steel [19]).

The character of this dependence (reference temperature vs. loading rate) has been studied. Firstly more detailed study has been carried out for Cr-Mo pressure vessel steel [19] tested statically and dynamically at 5 different loading rates. Experimental evidence on the linear dependence of reference temperature on loading rate dK/dt (or shift of this temperature when comparing it with the static loading) in co-ordinates with abscise in logarithmic scale has been proved [14]. It has been also shown that the higher slope of this dependence is more or less material characteristics and most probably may correspond to higher susceptibility of the cast ferritic steel to loading rate.

It is necessary to underline that the reference temperature determination is connected with low accuracy; this is caused by comparably higher value above the highest test temperature displaying the unstable fracture. The shift of reference temperature caused

250

by different loading rates is evident from the right horizontal axis. Ones having this shift available it can be applied for prediction of static behaviour based on data from dynamic test and vice versa as it will be shown later.

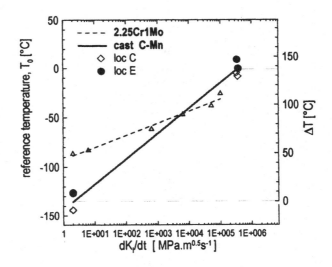

Figure 5. The effect of loading rate on reference temperature T_0 and increase of this temperature ΔT comparing to static loading.

3.4. FRACTURE TOUGHNESS PREDICTIONS BY USING P-CVN SPECIMENS

Pre-cracked Charpy type specimen (P-CVN) is the most suitable geometry for the assessment of radiation and elevated temperature ageing of container cask and RPV steels, as well as for analysis of strain rate effect. As mentioned, for P-CVN specimens the similar data sets of fracture toughness values as for standard SENB specimens have been generated (the raw data are presented in other contribution in this volume [15]).

The data from P-CVN specimens tested dynamically that met the size deformation validity limit

$$K_{Jd(limit)} = [(Eb\,R_{e\,d})/50]^{1/2} \qquad (1)$$

have been corrected for statistical size effects according to [14] by using the equation

$$K_{Jc(1T)} = 20 + (K_{Jc(10)} - 20)(\frac{B_{10}}{B_{1T}})^{1/4} \qquad (2)$$

and the reference temperature and master curves have been obtained. In equation mentioned the values B_{10} and B_{1T} represent the specimen thickness of P-CVN and standard SENB specimen respectively. The value of b represents the ligament length. R_{ed} is dynamic yield stress of the material corresponding to loading rate applied on P-CVN specimens. The reference temperatures obtained by multitemperature method are obvious from TABLE I.

For the data from dynamically loaded P-CVN specimens and corrected additionally for statistical size effect (for the cast ferritic steel investigated, surface location E) the predicted master curve and 90 % probability scatter band corresponding to standard specimens thickness is shown in *Figure 6*. This characteristic can be compared to standard SENB specimen data determined by the same methodology, i.e. by quasistatic determination procedure of K_{Jd}. The master curve and 90 % probability scatter band obtained is lying slightly in conservative side of experimental fracture toughness data; nevertheless very good correlation to this data can be observed. For location C (midthickness of the experimental plate) of the same steel very similar results have been obtained [6].

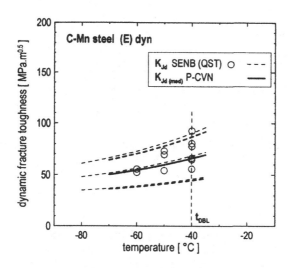

Figure 6. Prediction of 1T SENB specimen behaviour using the P-CVN specimen tested dynamically, cast ferritic steel - location E

The above-mentioned approach is applicable in case enough number of valid K_{Jd} values is available, i.e. the fracture toughness values are meeting the validity condition. The data from P-CVN specimen not meeting the $K_{Jd(limit)}$ may be adjusted for constraint as it has been shown in [20] applying the toughness-scaling model [12]. Thus by combining adjustment according to the toughness scaling model [12] and size correction the master curve methodology is possible to apply for prediction of 1T specimen behaviour using data obtained from the pre-cracked Charpy type specimens.

Having the valid data from P-CVN specimens tested by low blow method a prediction for standard 1T SENB specimens behaviour at static loading could be carried out. The reference temperature $T_{0(dyn)}$ from P-CVN data can be determined according to standard procedure as shown above. Than the shift of reference temperature ΔT can be read from *Figure 5* and reference temperature for static fracture behaviour $T_{0(st)}$ can be obtained subtracting from $T_{0(dyn)}$ the shift ΔT. Such procedure was applied for the data from Melt I applying the shift measured for Melt II, see *Figure 7*. After application of the reference temperature shift (by ΔT_0) the transition behaviour for static fracture toughness could be predicted as it was discussed also in work [6].

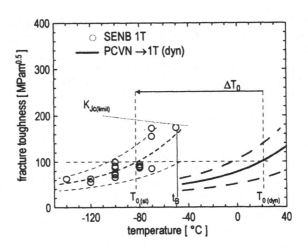

Figure 7. Prediction of 1T SENB specimen behaviour using the P-CVN specimen tested dynamically, cast ferritic steel (Melt I)

4. Conclusions

The fracture resistance of cast ferritic steel applying the master curve methodology was followed. The special attention has been paid to the application of this methodology for data obtained at dynamic conditions of loading. Additionally, the prediction of the fracture toughness temperature dependence and scatter for large (1T) specimens from pre-cracked Charpy ones tested statically and/or dynamically has been also performed.

Two procedures of dynamic fracture toughness determination have tested. For surface location the same results (fracture toughness temperature diagram) has been obtained as for quasistatic so for dynamic key curve method.

The effect of loading rate on reference temperature has been followed based on data obtained for cast ferritic steel and CrMo pressure vessel steel.

It has been found that dependence of reference temperature on loading rate (ln dK/dt) has more or less linear character and can be used for prediction of the effect of loading rate on shift of transition region.

The master curve concept has been shown to be valid in the lower part of transition range for C-Mn cast steel for both specimen locations investigated, i.e. for the central part and near surface layer of thick walled plate.

Reference temperature T_0 determined using size corrected P-CVN data obtained from static or dynamic testing (applying the shift of reference temperature) was only slightly different from T_0 evaluated using 1T SENB specimens.

Although extensive engineering applications of master curve approach can be found the physical basis to this methodology is still absent. This could be one of the aims for the future investigation in this field. Except for these fundamental investigations of fundamental nature the effect of loading rate on reference temperature has to be studied in order to predict this dependence based commonly available mechanical characteristics.

Acknowledgements

The financial supports of Grant Agency of the Academy of Sciences of the Czech Republic under the project Nr. S2041001, the grant OTKA T-030057 of the Hungarian Scientific Research Fund and No. 972655 within NATO Science for Peace Program to this research are gratefully acknowledged.

References

1. Wallin K., (1993), Macroscopic Nature of Brittle Fracture, J. de Physique, Colloq 7, Suppl J. de Physique II, Vol. 3, pp 575-583.
2. Wallin, K. (1995) Validity of Small Specimen Fracture Toughness Estimates Neglecting Constraint Correction, *Constraint Effect in Fracture, ASTM STP 1244*, Eds. M . Kirk, Ad Bakker, pp. 519-537.
3. Wallin K., Planman, T (2001), Effect of strain rate on the fracture toughness of ferritic steels, IAEA Specialist Meeting on Master Curve Testing and Results Application, Prague.
4. ASTM, E1921-97 (1997), *Standard Test Method For the Determination of Reference Temperature T_0 for Ferritic Steels in the Transition Range.*
5. Joyce J. A., (1998) On the Utilization of High Rate Charpy Test Results and the Master Curve to Obtain Accurate Lower Bound Toughness Predictions, *Small specimens test techniques, ASTM STP 1329*, W.R. Corwin, S.T: Rosinski, and E. Van Walle eds., pp. 3-14.
6. Dlouhý I., Lenkey, G.B., Holzmann M. (2001) Master curve evaluation at static and dynamic conditions of loading for cast ferritic steel, *Conf. Structural Mechanics in Reactor Technology, SMiRT 16*, Washington, proc. on CD ROM.
7. Viehrig H.W., Boehmert, J., Dzugan, J. (2002) Use of instrumented Charpy impact tests for master curve determination, IAEA Specialist Meeting on Master Curve Testing and Results Application, Prague
8. Yoon K.K., Van Der Sluys W.A., Hour K. (2000) Effect of Loading Rate on Fracture Toughness of Pressure Vessel Steels", *J. Pressure Vessel Technology*, pp. 125-129.
9. Joyce, J. A., Tregoning R.L. (2001) Development of T0 reference temperature from pre-cracked Charpy specimens, Engineering Fracture Mechanics, 68, pp. 861-894.
10. Dodds, R.H., Anderson, T.H, Kirk, M.T. (1991) A Framework to Correlate a/W Ratio Effects on Elastic-Plastic Fracture Toughness, *International Journal of Fracture,* 48, pp 1-22.
11. Anderson, T.L, Dodds, R.H. (1991) Specimen Size Requirements for Fracture Toughness Testing in the Transition Region, *Journal of Testing and Eval.*, JTEVA, ASTM, 19, pp.123-134.
12. Nevalainen, M., Dodds, R.H. (1995) Numerical Investigation of 3D Constraint Effects on Brittle Fracture in SE(B) and C(T) Specimens", *Int. Journal of Fracture*, pp.131-161.
13. Aurich D., Jaenicke B., Veith H., (1996), Statistical Base of Evaluation of Toughness Properties of Components in ..., *MPA seminar Safety and Reliability of Plant Technology*, p. 21.

254

14. Kohout, J., Jurášek, L. Holzmann, M. (2002) Evaluation of loading rate effects on transition behaviour by applying the master curve methodology, Transferability of fracture mechanical characteristics, Kluwer, contribution in this volume.
15. Dlouhý, I., Holzmann, M., Chlup Z. (2002) Fracture resistance of cast ferritic C-Mn steel for container of spent nuclear fuel, *Transferability of fracture mechanical characteristics*, Kluwer, contribution in this volume.
16. Lenkey G. B: Instrumented Impact Experiments on Pre-cracked Bend Specimens, Bay Zoltan Institute, University of Miskolc, NATO SfP project, Internal rpt., 2001.
17. Lenkey G. B. (2002) Dynamic fracture toughness determination of large SENB specimens, *Transferability of fracture mechanical characteristics*, Kluwer contribution in this volume.
18. Böhme, S.W. (1990) Dynamic Key Curves for Brittle Fracture Impact Tests and Establishment of a Transition Time *Fracture Mechanics: Twenty First Symposium, ASTM STP 1074*, J. P. Gudas., J.A. Joyse, & E.M. Hacket, eds., ASTM Philadelphia, pp. 144-156.
19. Holzmann, M., Dlouhy, I. (2002) To the Effect of Loading Rate on Reference Temperature and Master Curve, manuscript of paper for Int. Journal of P. Vessel and Piping.
20. Dlouhý I., Chlup Z. (2000) Micromechanical Aspects of Constraint Effect in Steel for Containers of Spent Nuclear Fuel, *13th European Conference on Fracture, Fracture Mechanics: Applications and Challenges*, San Sebastian, Spain (CD ROM).
21. Lenkey G. B., Balogh, Z.S. Hegman, N. (2002) Finite element Modelling of Charpy impact testing, *Transferability of fracture mechanical characteristics*, Kluwer, contribution in this volume.

EVALUATION OF STRAIN RATE EFFECTS ON TRANSITION BEHAVIOUR APPLYING THE MASTER CURVE METHODOLOGY

J. KOHOUT [a]), V. JURÁŠEK [b]), M. HOLZMANN [b]), I. DLOUHÝ [b])

[a]) *Military Academy in Brno, Military Technology Faculty, Department of Physics, Kounicova 65, 612 00 Brno, Czech Republic*
[b])*Institute of Physics of Materials, Academy of Sciences, Žižkova 22, 616 62 Brno, Czech Republic*

Abstract: The master curve methodology has been used for an evaluation of strain rate effects on transition behaviour of cast ferritic CrMo steel. The physical aspects of strain rate effect on reference temperature has been analysed as a base for the prediction of this dependence. Statistical aspects of the strain rate effects on the reference temperature and the shift of master curve on temperature axis has been discussed showing capability of the method for the prediction of strain rate susceptibility of steel fracture behaviour.

Keywords: ferritic steel, master curve, reference temperature, strain rate effect, cast steel, rotor steel, statistical aspects

1. Introduction

Master curve (MC) concept is currently widely used for the evaluation of temperature dependence of fracture toughness in lower shelf and transition region. The fracture toughness values applied for the master curve prediction have to be obtained at small scale yielding conditions acting at a crack tip at the moment of unstable fracture initiation. The MC concept includes the weakest link theory describing distribution of fracture toughness values at a given temperature [1,2], methodology for the accounting of the statistical size effects [3] and, in addition, different crack length effect on fracture toughness level [4,5]. Based on the data obtained for low alloyed and low carbon steel with predominantly ferritic microstructure and with yield stress ranging from 275 to 825 MPa, the MC has been verified and the independence of the MC shape itself on various alloying, heat treatment, and radiation embrittlement has been shown. All these factors can only shift the MC in the direction of temperature axis.

Kirk et al. [6] have shown that exponential shape of master curve is affected only by the obstacles producing stress field of short range. Comparing the experimental values of larger data sets they have shown, that the shape of master curve is the same not only for the above-mentioned group of steel but also for all steels with the bcc lattice, i.e. with dominant ferritic (or martensitic) microstructure. All other parameters such as chemical composition, heat treatment and irradiation cause only certain shift of master

255

I. Dlouhý (ed.), Transferability of Fracture Mechanical Characteristics, 255–270.
© 2002 *Kluwer Academic Publishers. Printed in the Netherlands.*

curve along temperature axis. The localisation of the master curve on temperature axis is expressed by reference temperature T_0 which is the temperature corresponding to fracture toughness value (50 %) of 100 MPa m$^{1/2}$.

Independently of certain limitations (e.g. for intergranular fracture), the master curve has been successfully used in many applications such as the assessment of the metallurgical effects, the effects of technological and operational degradation, radiation embrittlement etc. [7-10]. Usually the evaluation has been based on data of quasi-static fracture toughness. For many critical applications the dynamic fracture toughness is necessary to be evaluated including the prediction of transition behaviour. Only a few authors focused their effort on this topic [e.g. 11-15]. This was partly due to a lack of satisfactory number of dynamic fracture toughness data and partly due to unknown behaviour of master curve under the loading rate effect. Summarizing experimental observations [15], the conclusion has been done that dependence of reference temperature on loading rate in terms of $\dot{K}$ has more or less linear character if logarithmic scale of $\dot{K}$ is considered. Similar observations confirmed this finding [14].

A large number of data has been generated at the Institute of Physics of Materials for CrMo reactor pressure vessel steel and for cast ferritic steel regarding the fracture toughness at various loading rates. These data sets enabled verification of master curve concept use for different loading rate. Also, different ways how to predict the loading rate effect on reference temperature and thus on master curve position on temperature axis could be followed on the basis of mentioned data sets.

The aim of this contribution has been to analyse the data obtained by experimental observation and, being based on this analysis, to develop physical background for the quantification of loading rate effects on the reference temperature.

2. Theoretical Background

As the main mechanism preceding to fracture and deciding about the resistance of material against failure, the dislocation movement over various microstructural obstacles, i.e. plastic deformation, is considered. In order to move over the obstacles, minimum activation energy is necessary to reach for the dislocation. The frequency of successful jumps over the microstructural obstacles is usually described by relation that follows from the basic Arrhenius equation

$$\nu_1 = \nu_0 A \exp\left(-\frac{\Delta G}{kT}\right) \qquad (1)$$

where ν_0 is vibration frequency for dislocation, ΔG is free enthalpy, k is the Boltzmann constant, T is temperature in K, and A is constant. Comparing this equation with the Orowan equation quantifying the relation between deformation and dislocation velocities

$$\frac{\mathrm{d}a}{\mathrm{d}t} = \rho b \bar{v} \qquad (2)$$

where a is sheare strain, ρ is dislocation density, b is the Burgers vector, $\bar{v}$ is mean dislocation velocity, then the basic relation for strain rate determined by temperature activated dislocation velocity can be obtained

$$\dot{\varepsilon} = \dot{\varepsilon}_0 \exp\left(-\frac{\Delta G}{kT}\right)$$ (3)

The obstacles that have to be overcome can be divided into two groups:
(i) Obstacles forming the stress field of short range; these obstacles may be overcome by means of temperature activation. Lattice itself supposes such case of obstacles.
(ii) Obstacles producing the stress field of long-range order. The force acting on a dislocation is changed with location of dislocation. Typical examples are precipitates, grain boundaries, other dislocations etc. Any change of temperature has no substantial effect on the dislocation movement over these obstacles.
Eq. (3) can be also written in forms

$$\dot{\varepsilon}_0 = \dot{\varepsilon} \exp\frac{\Delta G}{kT} \equiv Z \quad \text{and} \quad \frac{\Delta G}{k} = T \ln\left(\frac{\dot{\varepsilon}_0}{\dot{\varepsilon}}\right) \equiv \Theta$$ (4)

where Z is temperature-compensated strain rate, usually called the Zener-Hollomon parameter, see e.g. Roberts [16], and Θ is rate-compensated temperature, also sometimes called the Zener-Hollomon parameter by certain authors, see e.g. Wallin and Planman [11]. The importance of these parameters consists in the possibility to replace the couple of quantities $\dot{\varepsilon}, T$ by single quantity Z (or single quantity Θ).

The value of parameter Z (or parameter Θ) is deciding for overcoming the obstacles by the stress fields of short range. The overcoming of other obstacles has to be covered by increasing applied stress. The Zener-Hollomon parameter may be used for description of temperature and strain rate effects on yield stress. It can be used also for description of these effects on fracture toughness (transition behaviour). In the frame of experimental scatter the master curve shape itself is not affected by the strain rate effects. From the application point of view is thus possible to describe the strain rate effect by temperature shift of the master curve.

Generally, fracture toughness is a function of both temperature and strain rate as external factors which can be replaced by single Zener-Hollomon parameter. Fracture toughness is then a function of this parameter (in this case parameter Θ is used)

$$K = K(T, \dot{\varepsilon}) = K\left(T \ln\frac{\dot{\varepsilon}_0}{\dot{\varepsilon}}\right).$$ (5)

The size of the temperature shift can be expressed from the difference of reference temperatures at the stipulated value of fracture toughness. According to master curve approach the value 100 MPa m$^{1/2}$ is used as the stipulated fracture toughness. Generally, for various values of deformation rate $\dot{\varepsilon}_1$ and $\dot{\varepsilon}_2$ different reference temperatures T_1 and T_2 can be obtained and they satisfy following equation

$$K\left(T_1 \ln\frac{\dot{\varepsilon}_0}{\dot{\varepsilon}_1}\right) = K\left(T_2 \ln\frac{\dot{\varepsilon}_0}{\dot{\varepsilon}_2}\right) = 100 \text{ MPa m}^{1/2}.$$ (6)

From the equality of arguments the relation for the computation of the reference temperature T_2 can be derived supposing that the reference temperature T_1 is known

$$T_2 = T_1 \frac{\ln(\dot{\varepsilon}_0/\dot{\varepsilon}_1)}{\ln(\dot{\varepsilon}_0/\dot{\varepsilon}_2)}. \tag{7}$$

To express the deformation rate $\dot{\varepsilon}$ ahead the crack tip is rather complicated. Much more easily the deformation rate (loading rate) can be described by the temporal variation of the stress intensity factor $\dot{K}$. Since the relation between the deformation rate and time variation of stress intensity factor is operative

$$\dot{K} = \frac{\mathrm{d}K}{\mathrm{d}\varepsilon}\dot{\varepsilon} \tag{8}$$

this proportionality can be utilized and one can write

$$K\left(T_1 \ln\frac{\dot{K}_0}{\dot{K}_1}\right) = K\left(T_2 \ln\frac{\dot{K}_0}{\dot{K}_2}\right) = 100 \text{ MPa m}^{1/2}. \tag{9}$$

This equation must be valid not only for two couples $T_1, \dot{K}_1$ and $T_2, \dot{K}_2$ but for arbitrary couple $T, \dot{K}$

$$T \ln\frac{\dot{K}_0}{\dot{K}} = C \tag{10}$$

where C means a constant value independent of quantities $T, \dot{K}$. For the unitary quasi-static loading rate $\dot{K}_I = 1 \text{ MPa m}^{1/2} \text{ s}^{-1}$, corresponding to reference temperature T_s, one can receive

$$T_{01} = \frac{C}{\ln \dot{K}_0}. \tag{11}$$

Now the dependence of the reference temperature on the loading rate can be expressed as

$$T_0 = T_0(\dot{K}_I) = \frac{C}{\ln \dot{K}_0 - \ln \dot{K}_I} = \frac{T_{01} \ln \dot{K}_0}{\ln \dot{K}_0 - \ln \dot{K}_I} \tag{12}$$

and also in following form

$$\frac{1}{T_0} = \frac{\ln \dot{K}_0 - \ln \dot{K}_I}{C} = \frac{1}{T_{01}} - \frac{\ln \dot{K}_I}{C} = \frac{1}{T_{01}} - \frac{\ln \dot{K}_I}{T_{01} \ln \dot{K}_0}. \tag{13}$$

Using reciprocal scale for reference temperature T_s and logarithmic scale for loading rate $\dot{K}_I$, Eq. (13) can be represented by straight line with the slope $-1/C$, where $C = T_{01} \ln \dot{K}_0$, and the value of $1/T_s$ for $\dot{K}_I = 1 \text{ MPa m}^{1/2} \text{ s}^{-1}$. Note that the same equation as Eq. (12) was derived by Wallin and Planman [11] where designation $\Gamma = \ln \dot{K}_0$ was used.

For the case of $\left| \ln \dot{K}_I \right| \ll \ln \dot{K}_0$, which is very often valid, an approximate simplification of Eq. (12) can be done

$$T_0 = \frac{C}{\ln \dot{K}_0} \left(1 - \frac{\ln \dot{K}_I}{\ln \dot{K}_0} \right)^{-1} \approx \frac{C}{\ln \dot{K}_0} \left(1 + \frac{\ln \dot{K}_I}{\ln \dot{K}_0} \right) = \frac{T_{01}}{\ln \dot{K}_0} \ln \dot{K}_I + T_{01} \qquad (14)$$

whereby the linear dependence between reference temperature and logarithm of time variation of stress intensity factor is obtained. The slope of this straight line is $T_{01}/\ln \dot{K}_0$ and the value for $\dot{K}_I = 1\,\mathrm{MPa\ m^{1/2}\ s^{-1}}$ is T_{01}. The linear dependence between reference temperature and logarithm of loading rate was used also by Yoon et al. [15] but they used it as a phenomenological relation without any theoretical grounds or derivation.

Eqs. (12) or (14) describing reference temperature dependence on time variation of stress intensity factor contain two basic parameters T_{01} and $\dot{K}_0$ (or $\Gamma = \ln \dot{K}_0$ in Wallin's approach). The former determines the position of the master curve (reference temperature for $\dot{K}_I = 1\,\mathrm{MPa\ m^{1/2}\ s^{-1}}$), the latter determines the shift of MC with changing loading rate. Unfortunately, physical meaning of $\dot{K}_0$ is quite unclear. Using parameter Z instead of parameter Θ, see Eq. (4), Eqs. (13) and (14) can be rewritten in forms

$$\frac{1}{T_0} = \frac{1}{T_{01}} - \frac{k}{\Delta G} \ln \dot{K}_I \qquad (13')$$

and

$$T_0 \approx \frac{k T_{01}^2}{\Delta G} \ln \dot{K}_I + T_{01}. \qquad (14')$$

Now the shift of transition curve due to various loading rate is determined by activation enthalpy ΔG that is very well defined physical quantity and it can be measured by several mechanical tests (e.g. tensile test or creep test with step change in temperature, stress relaxation, see Lukáč [17]). It means that having the transition curve for defined value of loading rate, its shift for various loading rates can be predicted. This is very valuable result of theoretical considerations; its verification is under way at present.

3. Experimental Part

For the investigation of the loading rate effect on the transition behaviour of fracture toughness the steel 2.25Cr-1Mo [18], used for pressure vessels in petrochemical industry, was selected. A wrought plate with dimensions $4200 \times 1800 \times 30\ \mathrm{mm}^3$ was in normalised and tempered state. The basic mechanical characteristics of the steel at room temperature were as follows: $R_e = 308\ \mathrm{MPa}$, $R_m = 495\ \mathrm{MPa}$, $KV = 192\ \mathrm{J}$.

Two types of test specimens for fracture toughness determination were manufactured from the above mentioned plate:

(1) Specimens for static three-point bending test with dimensions of $B = 25$ mm, $W = 50$ mm and $L = 240$ mm (support span $S = 4\,W = 200$ mm), crack length to

specimens width ratio $a/W \approx 0.5$. They were L-T oriented, i.e. their longitudinal axis was identical with rolling direction (according to ČSN EN ISO 12 737 [19], ASTM E 1820-99a [20]). The specimens were tested in temperature range from -180 ° to 20 °C. During their loading the dependence force vs. loading point displacement was measured by two symmetrically located inductance transducers. Loading rate expressed by stress intensity factor rate $\dot{K}_I$ determined from linear-elastic part of the measured dependence was $2 \text{ MPa m}^{1/2} \text{ s}^{-1}$. The fracture mechanics characteristics K_{IC}, K_{JC} or K_C were determined in agreement with above-mentioned standards.

(2) Specimens for three point bending test having dimensions of $B = W = 15$ mm and $L = 75$ mm in length (support span $S = 4\ W = 60$ mm), crack length $a/W \approx 0.5$ were used for the fracture toughness measurements at various loading rate. They were L-T oriented. For loading of this specimen set the servo-hydraulic test machine ZWICK Rel was used whose piston speed could be controlled in the range from 0 up to 6 m s^{-1}. For three-point bend test a special jig was produced, which was subjected to tensile force as a whole; the deflection of tested specimen was carried out by the piston shift. The tests were carried out in temperature range from -140 to 20 °C. The piston speeds used at tests are given in TABLE I. together with corresponding temporal change of stress intensity factor $\dot{K}_I$, calculated for linear-elastic part of the dependence force vs. displacement (piston path). The quasi-static approach was used for fracture toughness determination.

TABLE I. Loading rates (piston speeds) and corresponding values of $\dot{K}_I$
used in fracture toughness measurements

specimen set	1	2	3	4	5
piston speed [mm s^{-1}]	$5 \cdot 10^{-2}$	$5 \cdot 10^{-1}$	5	$5 \cdot 10$	$5 \cdot 10^{2}$
$\dot{K}_I$ [MPa m$^{1/2}$ s^{-1}]	6.2	64	640	$6.4 \cdot 10^{3}$	$1.0 \cdot 10^{5}$

4. Experimental Results and Discussion

Based on temperature dependences of fracture toughness values two procedures of reference temperature and master curve determination were applied both based on multitemperature approach:
(i) The iteration method based mainly on the relation derived by Wallin [25] exploiting maximum likelihood method (MLM); and
(ii) Using the regression (least square method - LSM) of experimental results in transition region which could be fitted using general equation

$$K_{JC} = K_{min} + A \exp(BT). \tag{15}$$

The methodologies used have been similar to those applied in work [25].

Both mentioned procedures were directly used only when the experimental data K_{JC} were measured on test specimens with thickness 1 T (i.e. 25 mm).

If the measurements were made on specimens with thickness B_x different from thickness 1 T, it was necessary to convert fracture toughness values to 1T equivalent thickness values (correction for statistical size effects). Standard equation was used for these purposes

$$K_{JC(1T)} = 20 + \left[K_{JC(x)} - 20 \right] \left(\frac{B_x}{B_{1T}} \right)^{1/4}. \tag{16}$$

The equation was applied for conversion of fracture toughness values K_{JC} greater than 40 up to 50 MPa m$^{1/2}$ (the weakest link theory is not valid for smaller ones).

Estimated values of T_0 for particular loading rates are shown in TABLE II. Following results could be drawn from the results obtained:

- The reference temperature shifts to higher temperature with increasing strain rate. For loading rate $\dot{K}_I = 10^5$ MPa m$^{1/2}$ s^{-1} temperature shift ΔT_0 is approximately 60 °C according to both methods of T_0 evaluation.

- The difference among T_0 values estimated according to method (i) and (ii) is not large, maximally 9 °C. The T_0 values determined using maximum likelihood method are always higher than those from regression (except one value) and, therefore, the maximum likelihood method seems to be more conservative. The systematic difference between the results of both methods is fully natural because the maximum likelihood method usually leads to biased estimates.

TABLE II. The reference temperatures T_0 determined using maximum likelihood method (i), and using least square method (ii), respectively

$\dot{K}_I$ [MPa m$^{1/2}$ s^{-1}]	2	62	64	640	$6.4 \cdot 10^3$	$1.0 \cdot 10^5$
T_0 [°C] (i)	−87	−83	−62	−47	−38	−26
T_0 [°C] (ii)	−91	−74	−67	−51	−44	−35

Known value of reference temperature T_0 allows the master curve on temperature axis to localise for given loading rate $\dot{K}_I$. At quasi-static loading conditions (i.e. $\dot{K}_I = 2$ MPa m$^{1/2}$ s^{-1}) a value of reference temperature $T_0 = -87$°C was established.

As the master curve keeps the shape independent on loading rate, all fracture toughness data measured at different loading rates $\dot{K}_I$ will align into one scatter band in K_{JC} vs. $(T - T_0)$ coordinates. The scatter band is specified by tolerance limits defined by equations [3]:

$$K_{JC(0.05)} = 25.4 + 37.8 \exp[0.019 (T - T_0)] \tag{17}$$

$$K_{JC(0.95)} = 34.6 + 102.2 \exp[0.019 (T - T_0)] \tag{18}$$

Temperature dependence of fracture toughness is shown in *Figure 1* for the highest loading rate used in tests. Experimental data are presented here together with the master curves obtained from reference temperatures determined according to both above-mentioned procedures.

The significant knowledge follows from the figure:

- Very good agreement of the experimental data with master curve shape including the tolerance bands that can be taken as a good evidence of validity of master curve concept at dynamic loading conditions.
- Despite differences in reference temperature T_0 caused by different methods of its determination good agreement between master curve and experimental data has been observed for both methods followed.
- Simultaneously, the conservative nature of maximum likelihood method is confirmed.

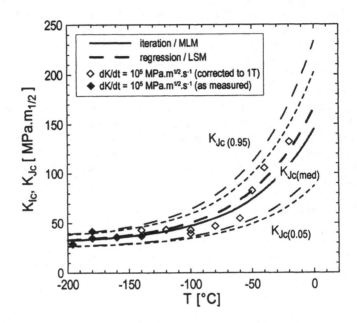

Figure 1. Fracture toughness temperature diagram for the highest loading rate used

In *Figure 2* the fracture toughness diagram versus temperature difference $T–T_0$ is shown. It confirms above-mentioned assumption that the master curve shape itself is not dependent on loading rate. Similar results were presented in paper of Yoon et al. [15].

It is worthy of notice that valid fracture toughness values less than 50 MPa m$^{1/2}$ are localized also in 90 % confidence interval though for these fracture toughness values the master curve concept is not valid. Experimental points in *Figure 2* are distinguished: solid symbols represent original values of fracture toughness measured on specimens with thickness $B = 25$ mm and also $B = 15$ mm without conversion according to Eq. (16) while empty symbols represent values converted using mentioned equation.

In *Figure 3* the dependence between reference temperature shift ΔT_0 and logarithm of loading rate $\dot{K}_I$ is shown as it was published by Yoon [15]. As it can be seen there is a number of experimental points for SA 515 steel investigated by Joyce [13]. There are included also ΔT_0 values determined from T_0 values from TABLE II. The temperature shift is given by selection of loading rate $\dot{K}_I$ at quasi-static tests. Other experimental points in *Figure 3* represent similar dependences for weldments. All these experiments show that the dependence of reference temperature T_0 or temperature shift ΔT_0 on $\ln \dot{K}_I$ can be approximated by the straight line in semi-logarithmic coordinates. This phenomenological approach of Yoon et al. has been supported by our theoretical considerations resulting in Eqs. (14) and (14').

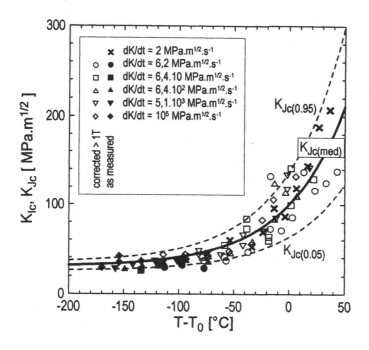

Figure 2. Fracture toughness at different loading rates as a function of temperature difference $T\text{-}T_0$

The importance of the diagram in *Figure 3* can be seen in possible transferability of the data received at the static loading by means of the small size specimens, most frequently pre-cracked Charpy specimen (P-CVN, usually not meeting the EPLM conditions for standard specimens), or real components and details [24]. Further possibilities of the applications can be expected in technical praxis where:

- The selection of steel type is done in the stadium of design proposal in order to ensure the integrity of the structure for given operation conditions even in the case when an undetected technological imperfection (flaws) is present in the structure (see European standards for steel structures [22,23], instruction AASHTO in USA for the bridge projects [23]).

- A defect appears or is discovered after certain operating time (fatigue crack, corrosion crack etc.) and checking calculation of component integrity or of its residual life is necessary to do.

264

To solve such problems the dependence $K_{JC(Bx)} = f(T, \dot{K}_I)$ must be known. To determine it the master curve approach can be used. If reference temperature T_0, determined at static loading (for its determination only very small set of test specimens is sufficient now) is known, temperature shift ΔT_0 for given service conditions (e.g. for the most adverse value of $\dot{K}_I$) can be taken from ΔT_0 diagram. Knowing T_0 and ΔT_0 values for given steel, the master curve position can be determined and the tolerance bands of fracture toughness for required probability of failure with respect to service temperature can be found out. Choice of failure probability level depends on the importance of design node or component in which the defect can be expected or was detected.

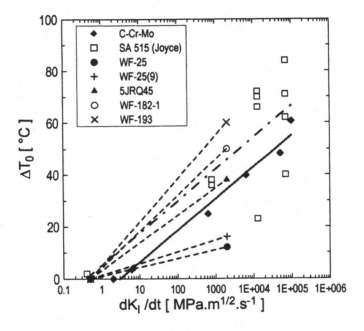

Figure 3. Dependence of temperature difference T-T_0 as on loading rate represented in terms of $\dot{K}_I$

5. The Accuracy of the Reference Temperature T_0 Determination at Different Loading Rates

The prediction of the fracture toughness temperature dependences for different loading rates is also influenced by the accuracy of the reference temperature T_0 estimation procedure. The fracture toughness values exhibit large inherent scatter in the transition region and this fact has significant influence on the accuracy of the reference temperature T_0 determination and, hence, the effect on the temperature shift ΔT_0. Two basic procedures may be usually used for the evaluation of reference temperature T_0: single temperature method (STM) as proposed in standard ASTM E 1921-97 [3], or some other method utilizing for the determination of T_0 fracture toughness data measured at various temperatures - multitemperature method (MTM). Therefore, the influence of the number of tested specimens on the confidence interval of T_0 values has to be followed (as shown also by Wallin [25], for both single temperature and the multi-temperature methods).

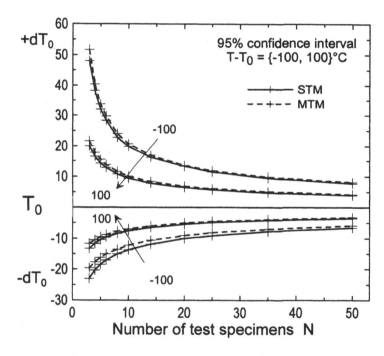

Figure 4. Confidence intervals for reference temperature T_0 for temperature difference T-T$_0$ lying in range of –100 to 100°C

Figure 4 shows the 95% confidence interval dT_0 as a function of specimen number used for evaluating of T_0. The Monte Carlo method (10^5 replicates) and three parameters Weibull distribution with $b = 4$ have been employed for evaluating of dT_0. In the case of MTM method the measurement of T in the relationship $T–T_0$ is represented by the middle of temperature region in which the fracture toughness values have been measured. These computations were done for the case of uniform distribution of measured data in temperature interval. However, it has been found that the width of this temperature region has not affected the resulting confidence interval.

As it may be seen, the accuracy of T_0 determination increases with the number of specimens used for the fracture toughness determination but a noticeable increase of the accuracy cannot be expected when test temperatures lie deep below temperature T_0. Similar situation can be expected for higher loading rates when step (discontinuous

266

[14]) increase of fracture toughness values occurs from lower to upper shelf level. Then only the values from lower shelf region can be used and further course of the master curve has to be extrapolated. To obtain comparable accuracy in this case, the number of fracture toughness data is necessary to increase. However, for very high loading rates also this approach need not be successful.

Confidence intervals of dT_0 evaluated by both mentioned methods are practically identical. Only in the case of MTM method a small shift (bias) to higher temperatures may be seen in *Figure 5* where the dependence of median (average) on the number of specimens is drawn for both methods and for various $(T–T_0)$ values. Although the shift is quite small it confirms the conservative nature of MTM.

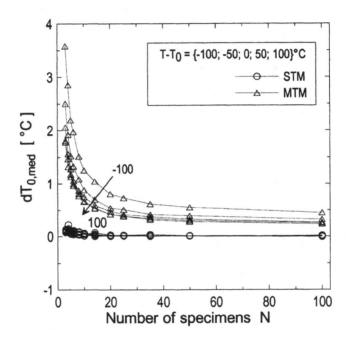

Figure 5. The effect of specimen number on median of reference temperature T_0 for both the STM and MTM method

Based on these computations the 95% confidence limits (bars) of T_0 are plotted in *Figure 6* showing the temperature T_0 as a function of $\ln(\dot{K}_I)$. Straight line dependence is theoretically supported by Eqs. (14) and (14') and regression calculations allow to determine the values of regression parameters T_s and ΔG (or $\Gamma = \ln \dot{K}_0$) together with their standard deviations: $T_s = (-90\pm3)$ °C $= (183\pm3)$ K, $\Delta G = (0.49\pm0.05)$ eV $= (47\pm5)$ kJ/mol, $\Gamma = 31\pm3$. Parameter $\dot{K}_0$ itself is not suitable to be calculated because its standard deviation is higher than its value.

The confidence limits are comparable because approximately the same number of specimens was used for determining T_0 and the master curve had not be extrapolated even for the highest loading rates. These conditions have not been apparently met in measurements of T_0 for steel SA 515 performed by Joyce as a large scatter of dT_0 values can be observed already at $\dot{K}_I = 10^4$ MPa m$^{1/2}$ s^{-1}.

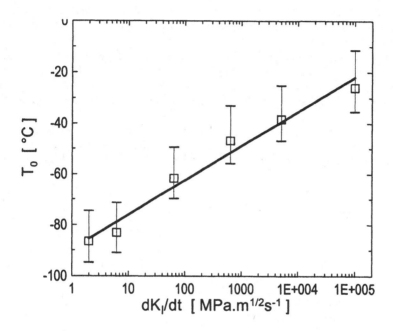

Figure 6. Reference temperatures confidence intervals for different loading rates

In principle, only two T_0 values determined at two different loading rates are sufficient to establish experimentally the dependence of reference temperature T_0 on the loading rate $\dot{K}_I$. However, it is useful to achieve the comparable accuracy of T_0 determination in both cases, which may need larger number of specimens in the case of the higher loading rate.

For determining sufficient number of specimens the plot in *Figure 7* may be utilised providing the minimum number of specimens needed for resulting reference temperature T_0 values that fall with 90% or 95% confidence into the temperature region $T_0 \pm 10$ °C.

6. Conclusions

The effect of loading rate on the change of reference temperature has been followed taking into account physical basis and statistical aspects of the experimentally determined dependence. Main knowledge of the investigation presented can be summarised are as follows:

It has been experimentally proved that the master curve shape versus temperature does not depend within scatter on the loading rate.

The relationship between reference temperature T_0 (or the shift of T_0 with respect to quasi-static loading rate) and loading rate may be based on theory of thermally activated processes and dislocation mechanics. It can be described with linear dependence if logarithmic scale of loading rate is considered. These theoretical findings have been experimentally demonstrated.

For experimental establishing of the dependence of reference temperature on loading rates only two T_0 values as determined at different loading rates are sufficient. However, it is useful to achieve the comparable accuracy of both T_0 determinations which may demand larger number of specimens for the measurement of T_0 at the higher loading rate.

From the engineering point of view, for evaluation of the influence of loading rates on the temperature behaviour of fracture toughness the dependence between reference temperature and loading rate is fully sufficient which can be determined by the values of two parameters.

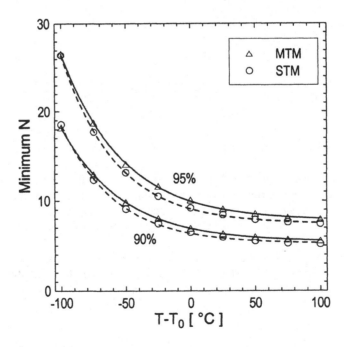

Figure 7. Minimum number of test specimens needed for reference temperature determination in interval $T_0 \pm 10$ °C with 90% and/or 95% reliability

Acknowledgements

The research was financially supported by grant No. A2041003 of the Grant Agency of the Czech Republic and project No. 972655 within NATO Science for Peace program.

References

1. Wallin K., (1993), Macroscopic Nature of Brittle Fracture, J. de Physique, Colloq 7, Suppl J. de Physique II, Vol. 3, pp 575-583.
2. Wallin, K. (1995) Validity of Small Specimen Fracture Toughness Estimates Neglecting Constraint Correction, *Constraint Effect in Fracture, ASTM STP 1244*, Eds. M . Kirk, Ad Bakker, pp. 519-537.
3. ASTM E 1921-97, Standard Test Method for Determination of Reference Temperature, T_0, for Ferritic Steels in the Transition Range.
4. Joyce, J.A, Tregoning, R.L. (2001) Development of the T0 reference temperature from precracked Charpy specimens, Engineering Fracture Mechanics 68, pp. 861-894.
5. Wallin K., Planman, T., Valo, M., Rintamaa R. (2001) Applicability of miniature size bend specimens to determine the master curve reference temperature T0, Engineering Fracture Mechanics 68, 1265-1296.
6. Kirk M. T., Natishan M. E., Wagenhofer M. (2001) Microstructural Limits of Applicability of the Master Curve, ASTM STP 1406, pp. 1-16.
7. Natishan, M., Rosinski S., Wagenhofer, M. (2001) Implementation of a Physics-based, predictive model for fracture toughness transition behaviour, *IAEA Specialists Meeting on Master Curve Testing and Results Application*, Prague, Paper No.4..
7. Holzmann, M., Dlouhý, I., Brumovský, M. (1999) Measurement of fracture toughness transition behaviour Cr-Ni-Mo-V pressure vessel steel using pre-cracked Charpy specimens, Inter, J. Pressure Vessel and Piping, 1999, 76, pp. 591-598.
8. Dlouhý, I., Lenkey, G., Holzmann, M.: Master Curve Evaluation at Static and Dynamic Conditions of Loading for Casts Ferritic Steel, SMIRT 16 – 16th International Conference on Structural Mechanics in Reactor Technology, Washington DC, August 12–17, 2001, Paper #G10/3.
9. Wallin, K. (1993) Irradiation damage effects on the fracture toughness transition curve shape for reactor pressure vessel steels, International Journal of Pressure Vessel and Piping, **55**, pp. 61-79.
10. Kirk, (2001) Shift in toughness transition temperature due to irradiation: T0 vs. T41J, a comparison and rationalisation of differences, *IAEA Specialists Meeting on Master Curve Testing and Results Application*, Prague, Paper No.23.
11. Wallin, K., Planman, T. (2001) Effect of strain rate on the fracture toughness of ferritic steels, *IAEA Specialists meeting on Master Curve Testing and Results Application*, Prague, pp.
12. Viehrig, H.W., Boehmert, J., Dzugan, J. (2001) Use of instrumented Charpy impact tests for master curve determination, *IAEA Specialists Meeting on Master Curve Testing and Results Application*, Prague, pp.
13. Joyce J. A., (1998) On the Utilization of High Rate Charpy Test Results and the Master Curve to Obtain Accurate Lower Bound Toughness Predictions, *Small specimens test techniques, ASTM STP 1329*, W.R. Corwin, S.T: Rosinski, and E. Van Walle eds., pp. 3-14.
14. I. Dlouhý, G. B. Lenkey, M. Holzmann (2002) Master curve validity for dynamic fracture toughness characteristics, *The Transferability of Fracture Mechanical Characteristics*, Kluwer, paper in this Volume.
15. Yoon K.K., Van Der Sluys W.A., Hour K. (2000) Effect of Loading Rate on Fracture Toughness of Pressure Vessel Steels", *J. Pressure Vessel Technology*, pp. 125-129.
16. Roberts W.: (1984) Dynamic changes that occur during hot working and their significance regarding microstructural development and hot workability. In: Krauss G. (ed.): *Deforming, Processing, and Structure*, Metals Park (Ohio, USA), American Society for Metals.
17. Kratochvíl P., Lukáč P., Sprušil B.: Introduction to Metal Physics I (in Czech), SNTL/Alfa 1984.
18. Holzmann M. and Dlouhý I.(2002) The effect of Loading Rate on Reference Temperature and Master Curve, manuscript of paper under preparation for Int. Journal of Pressure Vessel and Piping.
19. ČSN EN ISO 12737 – Metallic Materials – Determination of Plane strain Fracture Toughness, 2001.

20. ASTM E 1820-99a - Standard Test Method for Measurement of Fracture Toughness, 1999.

21. Holzmann, M. , Jurášek L., Dlouhý, I., 920020 Master Curve Metodology and Data Transfer from Small on Standard Specimen,, this volume.

22. Holzmann, M. (1997) Sborník Česko-Slovenské mezinárodní konf. "Ocelové konstrukce a mosty '97", květen 1997, Eds. J. Melcher, J. Skyva, vyd. CENTA, Ltd, Brno, pp.. 4-21 - 4.30 (in Czech).

23. Eurocode 3 (April 1996) Design of Steel Structures, Part 2, Steel Bridges, ENV 1993 - 2Draft.

24. Barsom, J. M. and Rolfe, T. S. (1999) Fracture and Fracture Control in Structures, ASTM, PA 19428-2959.

25. Wallin, K., Validity of Small Specimen Fracture Toughness Estimate Neglecting Constrain, ASTM STP 1244.

DYNAMIC FRACTURE TOUGHNESS DETERMINATION OF LARGE SE(B) SPECIMENS

G. B. LENKEY

Bay Zoltán Foundation for Applied Research, Institute for Logistics and Production Systems,
Iglói u. 2., H-3519 Miskolctapolca, Hungary

Abstract: In this paper the experimental determination of dynamic fracture toughness of large SE(B) specimens by instrumented drop weight testing will be described. Two evaluation methods have been used and compared: force based quasi-static evaluation and dynamic key curve method. For the latter one the time-to-fracture measurement is necessary, which was done by applying the magnetic emission measurement technique. The magnetic emission probe was successfully installed onto a drop weight tower first time in the world. The two evaluation methods delivered very similar K_{Id} values in the lower shelf region.

Keywords: dynamic fracture toughness, instrumented drop weight testing, cast ferritic steel

1. Introduction

In the engineering practice it is of importance to know the effect of loading rate on the material behaviour. The instrumented impact testing technique is widely used for determining dynamic fracture toughness properties of small Charpy type specimens. But this specimen size usually does not deliver valid fracture toughness data. For larger specimens the instrumented drop weight testing is an applicable experimental technique.

Due to dynamic effects for higher impact velocities several special problems are encountered. Therefore, the application of conventional force based analyses is limited up to a certain loading rate, especially in the lower shelf region when brittle fracture can occur after a few hundred microseconds. Over this range, the dynamic effects very often overshadow the true material response and additional measurement techniques must be applied to determine dynamic fracture toughness. The magnetic emission technique (ME) has potential ability for detecting the crack initiation, therefore the real time-to-fracture can be determined for dynamic analyses [1,2]. One particular objective of the present work was to study the applicability of ME technique for time-to-fracture measurement during instrumented drop weight testing.

The main aim of our work was to determine the dynamic fracture toughness values of large SE(B) specimens of cast ferritic steel, and to compare two evaluation methods: the quasi-static force-based evaluation and the dynamic key curve method.

I. Dlouhý (ed.), Transferability of Fracture Mechanical Characteristics, 271–282.

2. Implementation of the Magnetic Emission Measurement Technique for Drop Weight Testing

The magnetic emission technique has been developed especially for impact testing (by Winkler [1,2]), but can be applied for investigating fast fracture processes of ferromagnetic materials as well, like brittle fracture. The principle of this measurement technique is demonstrated in *Figure 1*. Two physical phenomena contribute to the magnetic emission signal: (a) mechanically induced Barkhausen signals appear when the internal magnetic structure changes during loading, and (b) a propagating crack causes the internal magnetic field to emerge from the solid into the gap between the two crack surfaces, thereby changing the external magnetic field. These field variations can be observed locally by a magnetic transducer which basically consists of a coil. The transducer's output voltage is the *magnetic emission (ME)* signal which is proportional to the derivative of the magnetic field (MF).

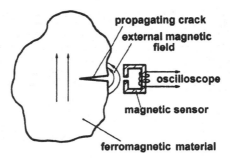

Figure 1. The principle of the magnetic emission(ME) measurement

The magnetic emission probe was successfully installed onto a drop weight tower of 550 J maximum available energy. *Figure 2.* shows the testing machine with the magnetic probe. The position of the probe can be adjusted in order to have it 2-3 mm far from the specimen side, and near to the crack tip. The probe was fixed to the moving weight, so it can follow the movement of the specimen and it is always near to the crack tip until the crack propagation does not start.

The impact velocity can be varied with changing the drop height, and the impact velocity is measured with an optical trigger device fixed onto the frame of the machine. This device makes the data acquisition start. A two pins flag is fixed on the falling weight which goes through the optical device, two pulses are produced and the time interval between these pulses is measured by a clock, and from this time value the impact velocity can be determined. The strain gauges of the tup and the emission probe are connected to a voltage supplies and amplifiers, and their signals are recorded by a TEKTRONIX TDS 420A digital oscilloscope. Then after transferring the data to a PC the evaluation is usually done with spreadsheet procedure.

a)

b)

Figure 2. The instrumented drop weight tower (a) and
the instrumented tup with the magnetic emission probe (b)

3. Experiments and Evaluation Methods

The tested material was cast ferritic steel which can be used for production of spent nuclear fuel containers. Two series of large SE(B) specimens were tested: series E was cut from near surface of a thick cast plate, and series C – from the centre part. The specimen dimensions were 23x23x110 mm which was selected taking into account the width of the instrumented tup. Finally two impact velocities were applied - v_o= 1.94 and 2.94 m/s – in order to eliminate the strong dynamic effects. The test temperature was varied in the range of from -60 to 20 °C.

As the first step, the transition region had to be determined in order to find the lower shelf region for K_{Id} determination.

3.1 UPPER SHELF AND TRANSITION REGION

In the *upper shelf region*, when completely ductile fracture was observed as it is shown in *Figure 3.a* (typically between –40 and 20 °C), the characteristic value of dynamic J-integral can be determined related to the maximum force using eq. 1 [3]:

$$J_{md} = \frac{2 \cdot U_m}{B \cdot (W - a_0)} \tag{1}$$

where U_m was calculated by integrating the force-displacement curve up to the maximum force using eq. 2:

$$U_m = \int_{s=0}^{s_m} F(s)ds \tag{2}$$

The displacement was determined from the measured force-time curve using eq. 3 and 4:

$$s(t) = \int_{\tau=0}^{t} v(\tau)d\tau \tag{3}$$

$$v(t) = v_0 - \frac{1}{m}\int_{\tau=0}^{t} F(\tau)d\tau \tag{4}$$

The mass of the hammer was m = 51.25 kg, and the v_0 impact velocity was measured as it was described earlier.

Then the relevant K_{Jmd} value can be determined for plain strain condition using eq. 5:

$$K_{Jmd} = \sqrt{\frac{E \cdot J_{md}}{(1 - v^2)}} \tag{5}$$

where E = 210000 MPa was taken as the Young's modulus, and ν=0.3 as the Poisson ratio of the investigated steel materials.

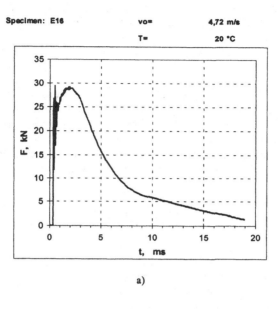

a)

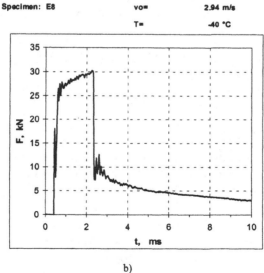

b)

Figure 3. Typical force time curves in the upper shelf (a)
and in the transition (b) region

In the transition region the brittle fracture occurs usually after significant plastic deformation or even after some ductile crack extension (see *Figure 3.b*). When the brittle crack initiation occurred at maximum load, the critical J integral and K values at crack initiation (Jid, Kid) were determined using eq. 1. When significant stable crack extension was observed, the characteristic J integral (Jmd) was determined at maximum load.

3.2 LOWER SHELF REGION

In the lower shelf region brittle crack initiation occurred preceding by no any macroscopic plastic deformation (*Figure 4*). For the higher impact velocity the dynamic effects were more pronounced, and the „3τ" criteria - proposed by Ireland [4] - was usually not fulfilled.

But nevertheless, the quasi-static evaluation method was tried to be applied. For this, the force based analyses was used [5], and the K_{Id} was determined as:

$$K_{Id} = \frac{Y\left(\frac{a_0}{W}\right) \cdot F_{max}}{B\sqrt{W}}, \tag{6}$$

where F_{max} - maximum force, N
       a_0 - initial crack length, mm
       W - specimen width, mm
       B - specimen thickness, mm
       $Y(a_0/W)$ - geometry function for SE(B) specimen [5]:

$$Y\left(\frac{a_0}{W}\right) = 6\sqrt{\frac{a_0}{W}} \frac{\left[1.99 - \frac{a_0}{W}\left(1 - \frac{a_0}{W}\right)\left(2.15 - 3.93\frac{a_0}{W} + 2.7\left(\frac{a_0}{W}\right)^2\right)\right]}{\left(1 + 2\frac{a_0}{W}\right)\left(1 - \frac{a_0}{W}\right)^{3/2}} \tag{7}$$

When the dynamic effects are too strong the force based analyses usually can not be applied. For these cases Böhme developed the dynamic key curve (DKC) method [6]. According to this method the dynamic fracture toughness can be determined on the basis of the measured time-to-fracture value with eq. 8:

$$K_{Id} = \frac{E.Y\left(\frac{a_0}{W}\right)}{\sqrt{W}C_s^*\left(1 + \frac{C_m}{C_s}\right)} v_0 t_F k_{dyn} \left(t = t_F\right) \tag{8}$$

where E - specimen's Young modulus , MPa
      v_0 - impact velocity, m/s
      $k_{dyn}(t_F)$ - dynamic key curve

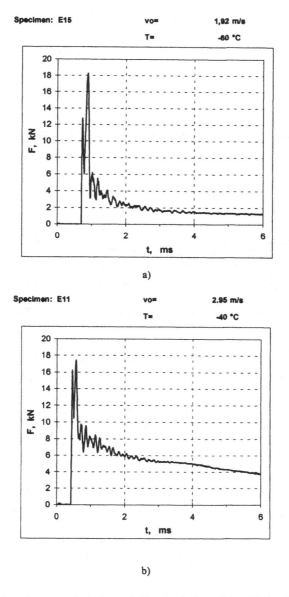

Specimen: E15   vo=   1,92 m/s
              T=   -60 °C

a)

Specimen: E11   vo=   2.95 m/s
              T=   -40 °C

b)

Figure 4. Typical force time curves in the lower shelf region for lower (a) and higher (b) impact velocities

t_F	-	time-to-fracture, s
C_s	-	specimen compliance, m/N
C_m	-	machine compliance, m/N
$C_s^* = EBC_s$	-	dimensionless specimen compliance [6]:

$$C_s^*\left(\frac{a_0}{W}\right) = 20.1 + 135\left(\frac{a_0}{W}\right)^2 \left\{ \begin{array}{l} 1 - 2.11\left(\frac{a_0}{W}\right) + 8.76\left(\frac{a_0}{W}\right)^2 - 19.9\left(\frac{a_0}{W}\right)^3 + 41.4\left(\frac{a_0}{W}\right)^4 \\ -67.7\left(\frac{a_0}{W}\right)^5 + 92.1\left(\frac{a_0}{W}\right)^6 - 76.7\left(\frac{a_0}{W}\right)^7 + 35.6\left(\frac{a_0}{W}\right)^8 \end{array} \right\} \quad (9)$$

$Y(a_0/W)$ - geometry function for SE(B) specimen.

According to the DKC method, the value of k_{dyn} depends on the W/c_1 ratio, where c_1 is the longitudinal wave propagation velocity for plane strain:

$$c_1 = \sqrt{\frac{E}{\rho\left(1 - v^2\right)}} \quad . \quad (10)$$

When $t_f < 39.4 W/c_1$ - for $a_0/W = 0.5$ - (168 µs in our case) we are in the fully dynamic time range. In this time range, the crack tip loading is significantly affected by dynamic effects. In the intermediate time range – if 39.4 $W/c_1 < t_f < 80.9$ W/c_1 (between 168 and 345.2 µs) – the dynamic key curve value is $k_{dyn}=1$. In this range the dynamic effects have decreased significantly, but the externally measured loads are still influenced by them. Above this, the loading rate can be considered as quasi-static. Almost all of the experiments were within the intermediate range as it is shown in *Figure 5*.

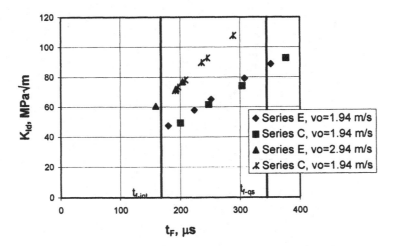

Figure 5. Dynamic fracture toughness values vs. time-to-fracture for two different impact velocities
(int – intermediate time range, qs – quasistatic time range)

The time-to-fracture determination was based on the force and on the magnetic signals. Unstable crack propagation is indicated by a force drop and usually is accompanied by a sharp peak of the magnetic signal according to a rapid crack jump. But due to the strong oscillation of the force signals, it was sometimes difficult to determine the instant of the brittle fracture directly from the force-time curves. *Figure 6* shows one example for the time-to-fracture determination.

The machine compliance was determined with elastic low-blow experiments on un-notched steel specimen as described in [7], and was obtained to $C_m = 1.403e-8$ m/N.

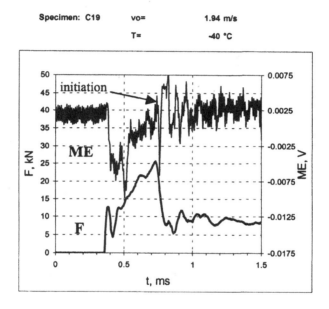

Figure 6. Time-to-fracture determination using magnetic signal

4. Results

The experimental results for the two experimental series are summarised in *Figure7-Figure 10*.

As it can be seen from *Figure 7* and *Figure 8* the series E specimens seem to have sharper transition, while the transition range of the series C specimens seems to be wider. The transition temperature is lower for the series E. (It should be mentioned that the exact determination of the transition temperature was not possible due to the limited number of specimens, and this was not the aim of the experiments.)

Comparing the dynamic fracture toughness values in the lower shelf region (*Figure 9* and *Figure 10*) it can be seen that the two evaluation methods (quasi-static and DKC) delivered very similar results. For series C, the exponential fit gave practically the same curve, and for the series E the DKC method delivered a little bit lower K_{Id} values (with app. 5 MPa√m).

280

Series E

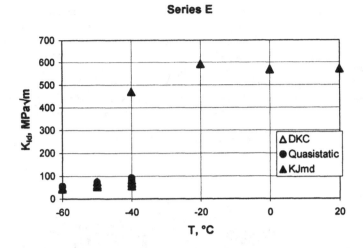

Figure 7. Dynamic fracture toughness vs. temperature for series E

Series C

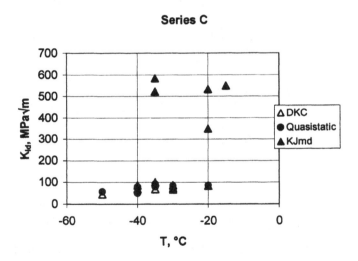

Figure 8. Dynamic fracture toughness vs. temperature for series C

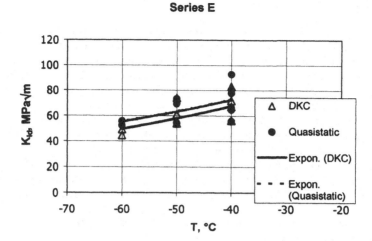

Figure 9. Dynamic fracture toughness vs. temperature for series E in the lower shelf region

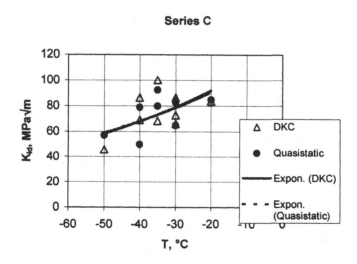

Figure 10. Dynamic fracture toughness vs. temperature for series C in the lower shelf region

5. Summary and Conclusions

Two specimen series of cast ferritic steel were investigated experimentally to determine the temperature dependence of the dynamic fracture toughness. Large SE(B) specimens have been tested with instrumented drop weight testing. Two different evaluation methods were used: the force-based quasi-static evaluation and the dynamic key curve (DKC) method. For the DKC method the time-to-fracture values were determined using the magnetic emission measurement technique.

On the basis of the obtained results the following can be concluded:

1. The magnetic emission probe was installed onto the drop weight tower, and was successfully applied for time-to-fracture determination.
2. The specimens cut from the surface part of a thick cast plate (series E) had a higher transition temperature than of the specimens which were originated from the centre part (series C) of the plate. Series C specimens had higher scatter in the transition range.
3. The two different evaluation methods delivered approximately the same dynamic fracture toughness values as average.

Acknowledgement

The financial support of OTKA T-030057 and NATO SfP 972655 projects is greatly acknowledged.

References

1. Winkler, S. R. (1988) Magnetische Emission, Ein neues Brucherkennungs-verfahren, Fraunhofer-Institut für Werkstoffmechanik Bericht T 3/88, internal report
2. Winkler, S. R. (1990) Magnetic Emission Detection of Crack Initiation, *Fracture Mechanics: Twenty-first Symposium, ASTM STP 1074*, J. P. Gudas, J. A. Joyce and E. M. Hackett, Eds, American Society for Testing and Materials, Philadelphia, 178-192.
3. Blumenauer, H., Pusch, G. (1982) *Technische Bruchmechanik*, Deutscher Verlag für Grundstoffindustrie, Leipzig
4. Ireland, D. R. (1976) Critical Review of Instrumented Impact Testing, *Proceedings of International Conference on Dynamic Fracture Toughness*, London, 47-57.
5. ASTM E-399 (1986) Standard Test Method for Plane-Strain Fracture Toughness of Metallic Materials, American Society for Testing and Materials, Philadelphia
6. Böhme, W. (1990) Dynamic key-curves for brittle fracture impact tests and establishment of a transition time, *Fracture Mechanics: Twenty-First Symposium, ASTM STP 1074*, Gudas, J. P. and Hackett, E. M. (Eds.), American Society for Testing and Materials, Philadelphia, 144-156.
7. Ireland, D. R. (1974) Procedures and Problems Associated with Reliable Control of the Instrumented Impact Test, ASTM STP 563, American Society for Testing and Materials, Philadelphia, 3-29.

FRACTURE TOUGHNESS BEHAVIOUR OF THICK-WALLED NODULAR CAST IRON AT ELEVATED LOADING RATES

K. MÜLLER, P. WOSSIDLO and W. BAER
Department of Materials Engineering, Federal Institute for Materials Research and Testing (BAM),D-12205 Berlin, Germany

Abstract: In consideration of the specific demands required for heavy-section ductile cast iron (DCI) for transport and storage casks the fracture mechanics research programme of BAM is focused on a systematic material characterization under static and dynamic loading conditions with respect to such parameters as microstructure, test temperature, sample size and loading rate. In the present study, ductile cast iron from an original DCI container with a wide variety of microstructure was investigated in order to determine the materials fracture toughness under dynamic loading conditions in the temperature range from -50 °C to +22 °C using three-point bending specimens with thicknesses of 140 mm and 15 mm, respectively. In contrast to static fracture behaviour, the fracture toughness values of thick-walled DCI at higher loading rates show a remarkable reduction with decreasing temperature between +22 °C and -50 °C and a significant shift of the transition range from -40 °C up to +22 °C. On the other hand, the lower bound fracture toughness value used in the design code for transport and storage casks of DCI in Germany was confirmed for dynamic loading conditions by these first investigations using large specimens. Furthermore, at present, BAM works on a research program which comprises systematic investigations of the mechanical and fracture mechanical behaviour of heavy section DCI at elevated loading rates.

Keywords: ductile cast iron, fracture mechanics, specimen size effect, loading rate effect

1. Introduction

For more than twenty years ductile cast iron has been used very successfully for spent fuel casks in Germany. During this period, the mechanical behaviour of hundreds of containers has been investigated. Based on the BAM acceptance criteria for transport and storage containers, material specifications for DCI were established [1-3]. Now, new developments in cask design as well as efforts to extend the application limits require further investigations, especially in the field of fracture mechanical assessment of DCI at elevated loading rates.

The materials toughness of DCI in terms of fracture mechanics characteristics is strongly influenced by microstructural parameters, especially the pearlite content as well as size, morphology and distribution of graphite nodules in the ferritic matrix. Furthermore, it has to be taken into account that in large DCI castings the micro-structure is often not homogeneously distributed over the wall thickness.

The International Atomic Energy Agency (IAEA) standard requires that containers certified for transport have to withstand hypothetical accidents, for example, a 9 meter free drop onto an essentially unyielding target at the lowest service temperature of –

I. Dlouhý (ed.), Transferability of Fracture Mechanical Characteristics, 283–290.
© 2002 *Kluwer Academic Publishers. Printed in the Netherlands.*

40 °C without the loss of radioactive contents [4]. Therefore, under these accident conditions, the influence of strain rate at the crack tip on the material behaviour has to be considered within design and safety analysis of containers and relevant material characteristics have to be provided.

2. Investigated Nodular Cast Iron Material

The nodular cast iron investigated in this study, EN-GJS-400-15 according to the German material standardization, was taken from a container, which had not totally fullfilled the required material specifications. Due to the cooling conditions in this large casting the microstructure was not homogeneously distributed over the wall thickness. The investigated specimens represent a wide variety of microstructure in terms of pearlite content of the matrix as well as size and distribution of the graphite nodules. Therefore, the mechanical properties vary either, as it can be seen from *Figure 1*. The materials strength, given by the 0,2 %-offset yield strength, $R_{p0,2}$, seems to be on a comparable level for pearlite contents not exceeding 20 %. However, an increasing amount of pearlite results in reduced ductility values. As expected, an increase in loading rate leads to higher strength values. Concerning the ductility, the limited available data does not indicate a clear influence of the loading rate.

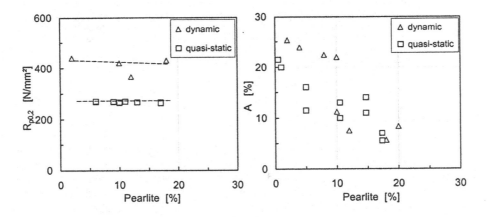

Figure 1. Yield strength, $R_{p0,2}$, and ultimate elongation, A, of DCI as a function of pearlitecontent at T = 22 °C under quasi-static and dynamic ($\dot{\varepsilon} \approx 1 \cdot 10^2 \, s^{-1}$) loadingconditions, respectively

3. Experimental Procedure

The fracture mechanical investigations included the testing of large (thickness 140 mm) and small (thickness 15 mm) single edge bend specimens (SE(B)) at elevated loading rates. A schematic outline and characteristic dimensions of the specimens are given in *Figure 2*. Prior to testing, all specimens were fatigue precracked providing initial a_0/W

ratios of 0,5. The experimental determination of fracture toughness values using SE(B)140 bend specimens was carried out in a test stand for shock loading, *Figure 3*.

Figure 2. Schematic SE(B) specimen and its dimensions

TABLE I. Test specimen dimensions

Specimen dimensions	SE(B)15 specimen	SE(B)140 specimen
Thickness B [mm]	15	140
Width W [mm]	30	280
Length L [mm]	160	1350
Mechanical crack starter notch M [mm]	10	112
Initial crack length a_0 [mm]	15	140

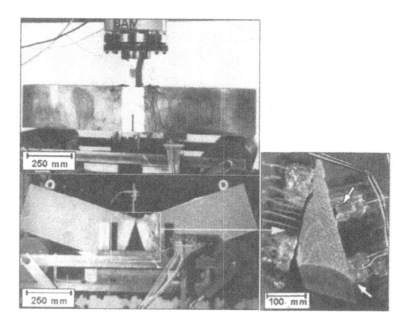

Figure 3. Test arrangement for experimental determination of dynamic fracture toughness values of DCI materials using SE(B)140 specimens (arrows mark strain gages)

The test stand is operated by a servohydraulic test cylinder of 1000 kN maximum force and 4 m/s maximum speed. In the fracture mechanical tests a maximum force of sensors in the ligament area. By means of this instrumentation, the crack initiation could reliably be deduced and dynamic fracture toughness values were determined. Within the analysis of the results the requirements of ASTM E 1820 standard [5] for rapid loading K_{Ic} determination were adopted and met so that valid dynamic fracture toughness values K_{Id} were obtained. The average stress intensity rate, $\dot{K}$, was about $5 \cdot 10^4$ MPa$\sqrt{m}$/s.

The experiments for the determination of dynamic crack resistance curves on SE(B)15 specimens were performed with a 750 J Charpy impact testing machine using an instrumented 150 J hammer and by applying the low-blow multiple specimen technique, *Figure 4*. Within these configuration impact speeds in the range of 1 to 2 m/s were realised. By analysis of the registered force - deflection curves dynamic crack resistance curves were deduced by application of the J-integral concept. Dynamic crack initiation toughness values J_{Ic} were determined from these dynamic J-R curves using the procedure of ASTM E 1820 standard.

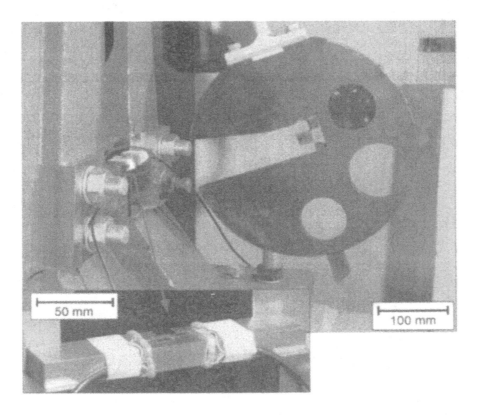

Figure 4. Experimental determination of dynamic crack resistance curves
of DCI materials using SE(B) 15 specimens

4. Results and Discussion

The fracture toughness values of ductile cast iron at elevated loading rates show a remarkable decrease with decreasing temperature between +22 °C and -50 °C in *Figure 5*. This material response describes the transition behaviour of dynamic fracture toughness of DCI in dependence on the test temperature. In the upper transition range of fracture toughness towards ambient temperature elastic-plastic material behaviour gains growing influence. At test temperatures of -40 °C and -50 °C the lower shelf of fracture toughness - characterised by fully linear-elastic material behaviour and brittle fracture - is almost reached. It should be mentioned that in the investigated temperature range all fracture mechanics characteristics of the SE(B)140 specimen met the requirements for valid K_{Ic} or K_{Id} values, respectively.

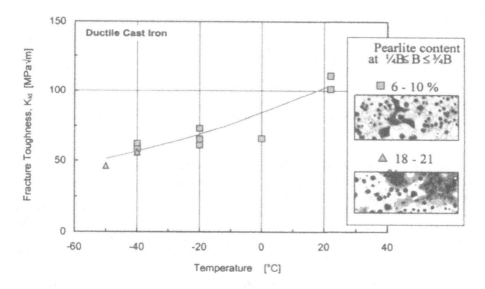

Figure 5. Dynamic fracture toughness of DCI as a function of temperature and pearlite content at test temperatures from –50 °C to +22 °C and loading rate $\dot{K} \approx 5 \cdot 10^4 \, \text{MPa}\sqrt{\text{m}}/\text{s}$, SE(B)140 specimens.

Under static loading conditions, the transition region where ductile fracture changes to brittle fracture is supposed to be between -80 °C and -100 °C for small specimens [6], and for larger thicknesses between -40 °C and -80 °C as shown in *Figure 6*. The results show that in comparison to quasi-static loading conditions the transition range is shifted to higher temperatures between about -40 °C and +22 °C due to the elevated loading rates. The levels of fracture toughness values in the lower and upper shelf are nearly equal to those of static loading conditions. Nevertheless, there should be a certain increase in toughness in the upper shelf region as a result of increased material strength (see *Figure 1*.) in the case of dynamic loading.

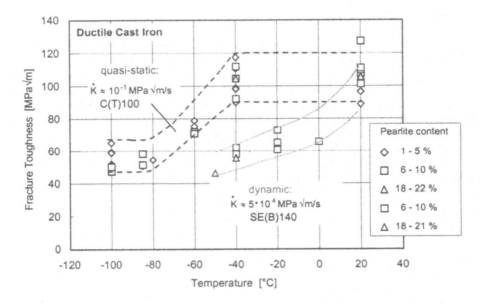

Figure 6. Fracture toughness behaviour of ductile cast iron as a function of test temperature and loading rate (specimen dimensions and loading rate are noted in the figure), [2, 3].

Fracture mechanics characteristics of DCI are required for structural safety analysis within the container design and - with respect to the relevant material specification in case of irregularities of the cast iron quality - for production control and quality assurance programs. In the latter case, the fracture mechanical evaluation procedure is restricted to the results of relatively small specimens which can be machined from samples taken directly from the container without totally destroying the component. Therefore, smaller bend type specimens of SE(B)15 geometry (*Figure 2.*) were investigated in the present study.

Due to the elastic-plastic fracture behaviour of these specimens dynamic crack resistance curves could be determined and crack initiation toughness values J_{Id} could be deduced. If required, these data can be converted to toughness values in terms of fracture toughness, K, according to ASTM E 1820, for instance. *Figure 7.* shows that both the crack initiation values, J_{Id}, as well as the fracture resistance curves, J_d-Δa, are strongly affected by the pearlite content. Increasing pearlite content leads to lower crack resistance. The same way, a decrease of the test temperature results in lower fracture toughness as indicated in *Figure 8.* for ferritic DCI.

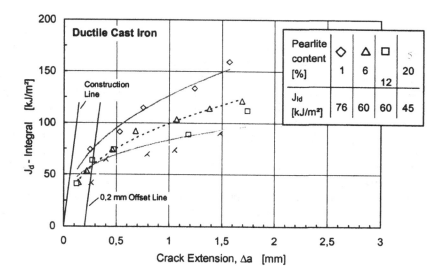

Figure 7. Crack resistance behaviour of DCI under impact loading conditions as a functionof pearlite content at ambient temperature: Low-blow test with $v_0 \approx 1$ m/s,SE(B)15 specimens, regression according to ASTM E 1820 (definition of J_{Id}similar to J_{Ic} at static loading)

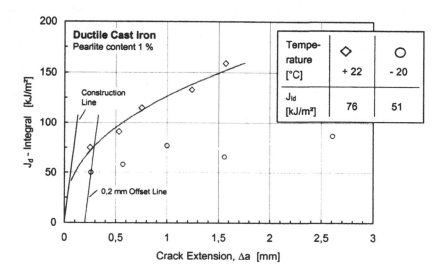

Figure 8. Crack resistance behaviour of ferritic DCI under impact loading conditions as a function of test temperature: Low-blow test with $v_0 \approx 1$ m/s, SE(B)15 specimens, regression according to ASTM E 1820 (definition of J_{Id} similar to J_{Ic}, at static loading)

5. Summary and Outlook

In comparision to static test results an increasing loading rate is predominantly responsible for a higher transition temperature and the change from elastic-plastic to linear-elastic material behaviour of ductile cast iron. The lower bound fracture toughness value of 50 MPa√m used for DCI in the design code for transport and storage casks in Germany was confirmed by the first investigations at elevated loading rates.

However, further investigations are required to determine the dynamic fracture toughness values especially on small-size specimens and a statistical assessment procedure according to the materials behaviour. At present, BAM works on a research programme which comprises systematic investigations of the mechanical and fracture mechanical behaviour of heavy section DCI at elevated loading rates taking into account parameters like variation of microstructure, test temperature, sample and component size as well as loading rate.

References

1. D. Aurich, R. Helms and K. Wieser, (1987), Das sicherheitstechnische Konzept der BAM für Spärogußbehälter (The BAM safety concept for nodular cast iron containers), *Amts- und Mitteilungsblatt der Bundesanstalt für Materialforschung und -prüfung (BAM)* 17, Nr. 4, 657-663.
2. B. Rehmer, H.D. Kühn, S. Weidlich, H. Frenz, (1995), BAM production control programme for containers for transport and storage of nuclear materials, *RAMTRANS* 6, Nos. 2/3, 205-219.
3. K. Müller et al. (2001), Bruchmechanische Untersuchungen von Gusseisen mit Kugelgraphit bei Stoß- und Schlagbiegebeanspruchung. In: *DVM-Bericht 223*, 33. Tagung des DVM-Arbeitskreises Bruchvorgänge, Deutscher Verband für Materialforschung und -prüfung, Berlin, 267-274
4. International Atomic Energy Agency (1996), Regulations for the safe transport of radioactive material, *IAEA-Safety Standard Series No. ST-1: Requirements*.
5. ASTM E 1820-99, Standard Test Method for Measurement of Fracture Toughness, In: *Annual Book of ASTM standards*, Vol. 03.01, American Society for Testing and Materials.
6. J.A. Smith, D. Salzbrenner, K. Sorenson and P. McConnel (1988), *Fracture mechanics based design for radioactive material transport packagings - Historical review*, Sandia Report SAND-98-0764, Sandia National Laboratories, Albuquerque, New Mexico.

FINITE ELEMENT MODELLING OF CHARPY IMPACT TESTING

G. B. LENKEY, ZS. BALOGH, N. HEGMAN
Bay Zoltán Foundation for Applied Research, Institute for Logistics and Production Systems
Iglói u. 2., H-3519 Miskolctapolca, Hungary

Abstract: In this paper static and dynamic loading of Charpy V-notched and pre-cracked specimens was simulated by FEM analysis using elastic-plastic material equation of a cast ferritic steel. The results of 2D plane strain, plane stress and 3D modelling are presented and discussed here. Different impact velocities were applied in the calculations. Contact situation was considered between the hammer and the specimen, as well as between the specimen and the anvils. The calculation results were analysed and compared with experimental load-deflection curves. In the case of dynamic loading not only the calculated load-deflection curve, but also the magnitude of the inertia peak, the load oscillations, as well as the loss of contact effect showed good agreement with the experimental results. The possibility for predicting the dynamic fracture toughness values on the basis of the FEM results is also discussed.

Keywords: finite element modelling, Charpy impact testing, cast ferritic steel

1. Introduction

Despite the fact that the Charpy impact testing was introduced a century ago during the International Conference on Material Testing held in Budapest in 1901 [1], this testing method is widely used even nowadays. With applying the instrumented version of the Charpy test much more information can be obtained about the fracture behaviour of the materials at intermediate loading rate, and dynamic fracture mechanics characteristics can also be determined. With the great development in micro-mechanical description of fracture processes and the local approach of fracture the significance of the Charpy impact testing has increased. Especially for the application of local approach of fracture [2-3] (e.g. the Beremin-model) the finite element modelling (FEM) of the impact test is of great importance. The FEM offers us a more precise investigation of fracture processes. Several approximation methods can be found in the literature for the finite element simulation of the impact tests [4-6].

The aim of our work was to develop 2D and 3D models for FEM simulation of the impact test for different loading conditions, to check the calculated results using experimental data, and to investigate the possibility of the dynamic fracture toughness prediction using the numerical calculation results.

I. Dlouhý (ed.), Transferability of Fracture Mechanical Characteristics, 291–302.
© 2002 *Kluwer Academic Publishers. Printed in the Netherlands.*

2. Geometrical Models and Material Properties

Charpy V-notched specimen was modelled and two loading conditions (quasi-static and dynamic) were considered. The impact test arrangement and the mesh for 2D and 3D modelling are shown in *Figure 1*. Standard Charpy V-notched and pre-cracked specimens were also modelled. For 2D calculations a half model was used, and a quarter model for 3D modelling. The specimen is put on a rigid anvil and is loaded in three points bending statically or dynamically.

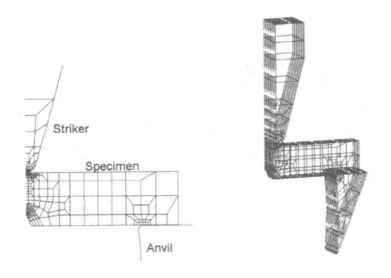

Figure 1. 2D and 3D FEM models of the Charpy impact test arrangement

    Contact situation was considered between the striker and the specimen, as well as between the specimen and the anvils introducing a Coulomb friction with a small friction coefficient of 0.05. A fixed rigid contour line represents the anvil in the 2D model, and an elastically deformed body in the 3D model. An averaged experimental stress-strain curve of a cast ferritic steel determined at −160 °C was used for the specimen (strain rate and temperature dependence were not considered in the calculations). The applied material parameters are shown in *Figure 2* and TABLE I.

    The striker and the anvil were considered as elastic material with the same Young modulus as of the specimen. 2D plane strain and plane stress, as well as 3D modelling of V-notched and pre-cracked Charpy specimens have been performed for quasistatic and dynamic loading.

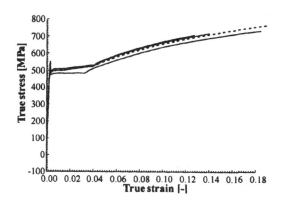

Figure 2. Experimental quasistatic stress-strain curves of cast manganese steel at –160 °C
(solid line – three different specimens' results, dotted line – average curve)

TABLE I. Material parameters for the FEM analysis

Material behaviour	Specimen	Hammer
	Elasto-plastic	Elastic
E (MPa)	205 e3	205 e3
Material equation	$\sigma_y = 498$ if $\varepsilon <= 2.43e\text{-}3$	
	$= 695.6*\varepsilon + 496.3$ if $2.43e\text{-}3 < \varepsilon <= 3.98e\text{-}2$	-----
	$\sigma = 1151* \varepsilon^{0.2436}$ if $3.98e\text{-}2 < \varepsilon < 0.18$	
Poisson ratio, ν	0.3	0.3
Mass density, ρ (kg/m³)		6.9 e6
	7.8 e3	(due to zoom)

TABLE II. Chemical composition and mechanical properties (at room temperature)
of the cast manganese steel

Material	C %	Mn %	Si %	P %	S %	Cr %	Ni %	Cu %	Mo %	Al %
Cast manganese steel	0.09	1.18	0.37	0.010	0.025	0.12	0.29	0.29	0.03	0.028

Yield strength, R_y, MPa	340	Tensile strength, R_m, MPa	480	Elongation, A_5, %	35	Reduction of area, Z, %	70

3. Quasi-Static Modelling

For the FEM modelling 8 node parabolic elements were used in 2D, and 20 node parabolic elements in 3D. As the first step, a V-notched specimen loaded quasi-statically was modelled. This loading condition was defined by *Y(t)=constant* displacement of the hammer. The loading rate in this case was 3 mm/180 s calculated in 120 increments. For computational analysis a two-dimensional plain-strain arrangement and the updated Langrange method of the MARC FEM code was applied. 4-node isoparametric elements were used for discretization with four Gauss points. Due to the large displacements and the non-linear material equation, the Newton-Raphson iteration method was applied using the following equation [7]:

$$K(u)\delta u = F - R(u),$$ (1)

where u is the vector of the nodal displacements, F is the vector of the external nodal loading, R is the vector of the internal nodal loading, K is the tangent stiffness matrix.

Thus R can be expressed as:

$$R = \sum_{element} \int_V \beta^T \sigma \, dv,$$ (2)

where β^T is the transponent of the strain tensor, σ is the stress tensor. In time step $n+1$, on the basis of the displacement calculated in iteration $i-1$, the equation for the iteration i is:

$$K\left(u_{n+1}^{i-1}\right)\delta u^i = F - R\left(u_{n+1}^{i-1}\right)$$ (3)

Solving the equation (3) for δu^i, the following equations can be obtained:

$$\Delta u^i = \Delta u^{i-1} + \delta u^i$$ (4)

and

$$u_{n+1}^i = u_{n+1}^{i+1} + \delta u^i \ .$$ (5)

This iteration is continuously repeated by the program until the given convergence condition is fulfilled, however maximum 20 repetitions were allowed. A convergence condition for the ratio of the infinite norms of the maximal residual load and the maximal reaction load was prescribed:

$$\frac{\left\| F_{max}^{residual} \right\|_\infty}{\left\| F_{max}^{reaction} \right\|_\infty} \leq 0.1.$$ (6)

4. Dynamic Modelling

In the case of dynamic loading the initial velocity of the striker was in the range of $1\div5$ m/s with a reduced mass of 20 kg. This mass was considered in the calculation by concentrating it into a smaller volume according to the geometry by proper modification of the density of the tup. The striker was not controlled during dynamic impact, only its initial velocity was given according to experimental data. In the course of the dynamic modelling of the V-notched test specimen, the contact situation was to analyse in time between the hammer and the specimen, and the diagram of the stress transferred by the hammer to the specimen was to determine. The geometric model of the test arrangement was the same as in the quasi-static analysis, and the same material parameters were used. Time steps should be chosen so that the oscillation wave forming in time step n should not return to its origin in time step $n+1$, i.e. the distance that the wave travels during one time step should be less then twice the thickness of the specimen. The time resolution of the dynamic simulation between the increments was 2 μs. This time scale was suitable to exhibit the inertia peak and further vibrations of the loaded specimen, but it is too large to observe the elementary stress wave propagation between the boundaries of the specimen. Speed of the oscillation wave in the material, which can be characterised by density ρ, and by elastic modulus E can be written as follows:

$$c = \sqrt{\frac{E}{\rho}} \ . \tag{7}$$

On the basis of the equation (7) the speed of the oscillation wave in the test specimen is 5126 m/s.

Thus in the test specimen of 0.01 m in thickness, the wave will return to its origin within 4 x 10^{-6} seconds. So the simulation lasting for 200 μs should be divided at least ≈ 50 time steps. The limited hard disk capacity allowed us to separate 100 steps.

During modelling the damping properties of the hammer and of the specimen were specified by the Rayleigh damping factors. The damping matrix can be written as:

$$D = \sum \alpha_i M_i + \beta_i K_i + \frac{\gamma_i K_i \Delta t}{\pi} \ , \tag{8}$$

where α_i is the mass factor, β_i is the stiffness factor, γ_i is the numerical factor, M_i is the mass matrix, K_i is the stiffness matrix.

In the case of dynamic problem, the following equation can be formulated:

$$M\ddot{u} + D\dot{u} + Ku = R \ , \tag{9}$$

where D is the damping matrix, R is the external load vector, $\ddot{u}$, $\dot{u}$ and u are the acceleration, speed and displacement vectors respectively.

The computer code solves this equation at equilibrium with the Newmark direct integration method. General form of the Newmark operators is the following [7]:

$$u^{t+\Delta t} = u^t + \dot{u}^t \Delta t + [0.5 - \beta] \Delta t^2 \ddot{u}^t + \beta \Delta t^2 \ddot{u}^{t+\Delta t} \tag{10}$$

$$\dot{u}^{t+\Delta t} = \dot{u}^t + [1 - \gamma] \Delta t \ddot{u}^t + \gamma \Delta t \ddot{u}^{t+\Delta t} \ , \tag{11}$$

where t indicates the time step, β, γ are the parameters determining the precision of the integration.

According to Newmark's suggestion [7], γ should be set as 1/2 and β should be set as 1/4, the acceleration values will be approximated within Δt with the arithmetic mean of the acceleration values determined in time step t and $t+\Delta t$. Applying equation (9) to time step $t+\Delta t$, the following equation will be obtained:

$$M^{t+\Delta t}\ddot{u} + D^{t+\Delta t}\dot{u} + K^{t+\Delta t}u = R^{t+\Delta t} \, ,$$

from which the displacement and its derivatives can be calculated.

5. Modelling Results of V-Notched Specimen

Figure 3 shows the static and dynamic stresses in the vicinity of the notch, at the symmetry axis. The static curve time scales were calculated on the basis of the corresponding bending state of the dynamic study. Load-deflection curves together with an experimentally measured one are shown in *Figure 4*. The dynamic curves slightly oscillate around the static results and are shifted with time due to time relaxation of the transient impact event. Although this wavy behaviour on the dynamic stress curves is not so pronounced as it is on the dynamic load-deflection curve. This effect was also experimentally observed that the impact event causes larger force oscillation in the hammer than in the specimen in the vicinity of the notch [8-9].

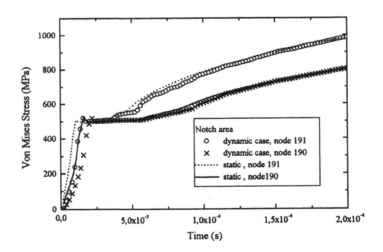

Figure 3. Dynamic and static stress states in the vicinity of the notch. (Nodes 190 and 191 are on the symmetry axis of the specimen within 0.1 mm from the notch root.

The FEM results of dynamic modelling were in good agreement with the experimental observations (*Figure 4* - experiment at T = -80 °C). Static curve gives a rough average of dynamic trends. Good agreement can be seen between the experiment and dynamic simulation comparing the main features: as inertia peak magnitude and location, as well as further vibrations. The magnitude of the inertia peak (approximately 9 kN) and its time period (15 μs) were also realistic. Changing the impact velocity, the inertia peak magnitude varied proportionally with the velocity, as it was expected (see *Figure 5*). It was clearly detected in the simulation that during the increasing part of the inertia peak the nodes in the centre part of the specimen reached the speed of the striker, and the striker speed remains constant due to its relatively large mass. Comparing the dynamic and static loading results, the static curve practically smoothes the dynamic curve paths since the static deflection takes place without sample acceleration and vibration.

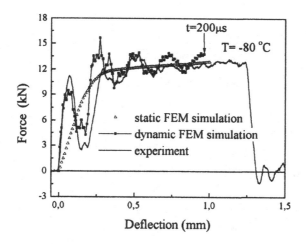

Figure 4. Comparative representation of the experimental results with static and dynamic modelling (2D plane strain model)

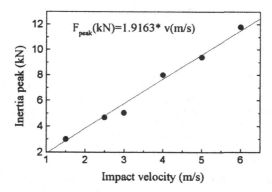

Figure 5. Inertia peak magnitude for V-notched specimen as a function of impact velocity (2D plane strain model)

298

A pronounced loss of contact effect was found between the specimen and the anvil up to 16 μs when forces appeared first time on the anvil and increased afterwards. (Naturally in the case of static loading the force on the anvil appears immediately.) This effect was experimentally proved [10] and has been analysed by several analytical models [5, 11-12]. A change of some parameters during dynamic loading can be seen in *Figure 6a* as a function of time at different locations of the specimen and at different impact velocity. This figure shows the coincidence between the inertia peak and the acceleration of the notched area up to the hammer speed. *Figure 6b* magnifies the specimen loss of contact effect taking place at a node close to the anvil-sample interface in 20 μs window. The positive movement of the chosen node corresponds to the loss of contact, afterwards negative domain stresses are initiated. The first 5μs - until the influence of collision is delayed (showed by arrow) - is the travelling time of stress waves to reach the anvil interface.

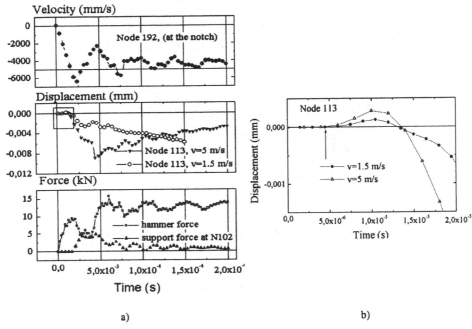

a)   b)

Figure 6. Characteristic features obtained vs. time in the dynamic 2D simulation.
a) Upper diagram: velocity variation of the point just at the intersection of the notch and centre line (node 192). The hammer velocity is –5 m/s. Middle: Displacement of one point in the sample, just above the sample-anvil interface (0.5mm), at 1.5 and 5 m/s impact velocity. Bottom: Hammer force and the force of the point (node 102) at the support-sample interface next to the anvil edge
b) The insert section of the part a) (middle).

After the inertia peak very similar oscillations appeared both in the dynamic FEM results and experimental curve due to the vibration of the specimen and the hammer [5]. In the experimental curve (see *Figure 4*) this load oscillation strongly attenuates before the failure due to damping of the stress waves and crack initiation. In the FEM model numerical damping factor of 0.2 was used. Increasing this factor from zero up to 0.2, the original high frequency noise, likely numerical scattering was smoothed out to approach the real periodic signal. This calculation was stopped at t=200 μs, before the fracture took place in the experiment, since the crack initiation and propagation processes was not included in this state of simulation.

Comparison of the 2D plane strain, plane stress and 3D modelling results (*Figure 7* and *Figure 8*) shows that there is a significant difference between the plane strain and plane stress condition, and the 3D calculation gives more realistic force values in between.

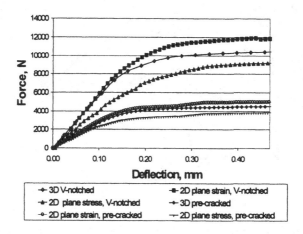

Figure 7. Comparison of the 2D and 3D modelling results for V-notched and pre-cracked specimen

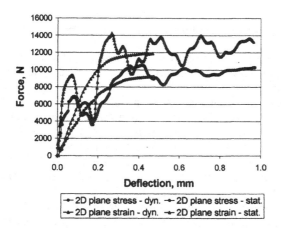

Figure 8. Comparison of the 2D static and dynamic modelling results for V-notched specimen

300

6. Modelling Results of Pre-Cracked Specimen

For the pre-cracked specimen a sharp crack was considered with relative crack depth of a/W=0.5. Both for static and for dynamic loading conditions the maximum load values were much smaller than for the V-notched specimen (*Figure 7* and *Figure 9*). The 3D and 2D plane strain dynamic modelling gave very similar results which were in good agreement with the experiments concerning the magnitude and the oscillation of the load (*Figure 9*).

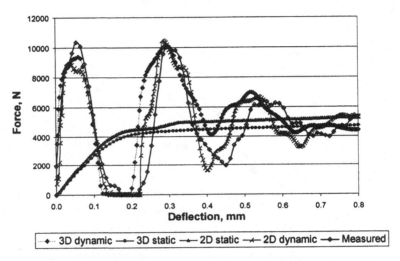

Figure 9. Comparison of the 2D (plane strain) and 3D dynamic and static modelling results for pre-cracked specimen

7. Possibility for Dynamic Fracture Toughness Determination

The principles of both the impact response curve method [9] and the dynamic key-curve method [10] are based on the assumption that in the elastic region the changes of the stress intensity factor depends only on the elastic behaviour of the specimen and of the testing machine, as well as on the specimen geometry. Therefore the dynamic fracture toughness is somehow proportional to the time-to-fracture value. So if an accurate FEM model can describe well the specimen behaviour, the dynamic fracture toughness can be determined on the basis of the calculated K_I values. This is shown in *Figure 10* and *11* for 2D and 3D dynamic modelling. The black triangles represent the experimentally measured K_{Id} values on pre-cracked Charpy specimens (at different temperatures in the range of $-100 \div -40$ °C) applying the impact response curve method. As it can be seen, the 3D model gives a very good agreement with the experimentally determined values. In this way it is possible to determine the K_{Id} values only by measuring the time-to-fracture values (e.g. with magnetic emission measurement or on-specimen strain gauges), without any force measurement (the force measurement is usually not accurate in the case of higher impact velocities and short time-to-fracture values anyway).

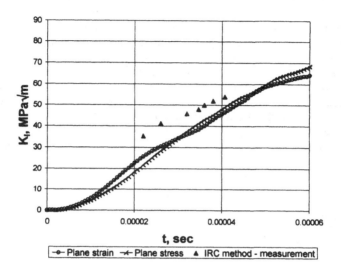

Figure 10. Comparison of the 2D dynamic modelling results for K_I with experimentally measured K_{Id} values

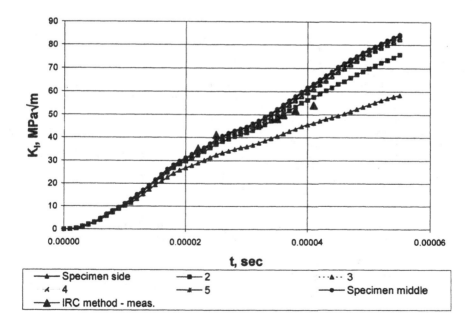

Figure 11. Comparison of the 3D dynamic modelling results in the elastic region for K_I with experimentally measured K_{Id} values

302

8. Summary and Conclusions

In the present work static and dynamic loading of Charpy V-notched and pre-cracked specimens was simulated by FEM analysis using static elastic-plastic material equation of a cast ferritic steel. Applicable and accurate FEM models have been developed for both loading conditions. The developed 2D and 3D models made it possible to describe quite well the real processes, so the effects during the impact can be examined from several points of view. Good agreement was found between the calculations and the experiments. 3D modelling usually delivered more realistic results especially for V-notched specimen. The developed 3D model for pre-cracked specimens gives the possibility to determine the dynamic fracture toughness values on the basis of time-to-fracture measurements.

Further investigations are necessary to take into account the strain rate and temperature dependence of the material behaviour.

Acknowledgement

The financial support of OTKA T-030057 and NATO SfP 972655 projects is greatly acknowledged.

References

1. Charpy, M. G. (1901) Note sur l'essai des métaux á la flexion par choc de barreaux entaillés, *Int. Conf. on Material Testing, Budapest*, 1-30.
2. Beremin, F. M. (1983) A Local Criterion for Cleavage Fracture of a Nuclear Pressure Vessel Steel, *Metallurgical Transactions A*, 14A, 2277-2287.
3. Minami, F., Brückner-Foit, A., Munz, D. and Trolldenier, B. (1992) Estimation procedure for the Weibull parameters used in the local approach, *Int. Journal of Fracture*, 54, 197-210.
4. Miyazaki, N. (1991) Application of line-spring model to dynamic stress intensity factor analysis of pre-cracked bending specimen, *Engineering Fracture Mechanics*, 38, 321-326.
5. Sahraoui, S. and Lataillade, J. L. (1998) Analysis of load oscillations in instrumented impact testing, *Engineering Fracture Mechanics*, 60, 437-446.
6. Marur, P. R., Simha, K.R.Y. and Nair, P.S. (1994) Dynamic analysis of three point bend specimens under impact, *Int. J. of Fracture*, 68, 261-273.
7. MARC Analysis Research Corp., Theory and User information, Vol A, (1997), Par. 11.
8. Ireland, D. R. (1976) Critical review of instrumented impact testing, *Conf. on Dynamic Fracture Toughness, London*, 47-61.
9. Kalthoff, J. F., Winkler, S. and Böhme, W. (1985) A novel procedure for measuring the impact fracture toughness K_{Id} with precracked Charpy specimens, *Journal de Physique*, C5 No. 8, 279-186.
10. Böhme, W. and Kalthoff, J. F. (1982) The Behavior of Notched Bend Specimens in Impact Testing, *Int. J. of Fracture*, 20, R139-R143.
11. Orynak, I. V. and Krasowsky, A. Ja. (1998) The modeling of elastic response of a three-point bend specimen under impact loading, *Engineering Fracture Mechanics*, 60, 563-575.
12. Marur, P. R. (1996) Numerical simulation of anvil interactions in the impact testing of notched bend specimen, Int. J. of Fracture, 81, pp. 27-37.

NOTCH SENSITIVITY ANALYSIS ON FRACTURE TOUGHNESS

G. PLUVINAGE[1] and A. DHIAB[2]

[1]*Laboratoire de Fiabilité MécaniqueUniversité de Metz (France)*

[2]*ENIS Sfax (Tunisie)*

Abstract This paper is devoted to the following problems: definition of notch effects, physical basis of notch effect, notch effects on brittle-ductile transition, notch sensitivity of static fracture toughness of a steel, fracture toughness from Charpy shelf plateau, and notch sensitivity in mixed mode fracture.

Keywords: notch effects, notch fracture mechanics, notch sensitivity, brittle-ductile transition

1. Notch Effects

Fracture load for a notched structure is lowered comparing to a smooth one. Fracture Load reduction occurs for two reasons: ligament size bearing load is reduced and stress concentration at notch tip appears. Reduction of fracture load relative to identical ligament area is considered as "notch effect".

If we plot critical gross stress versus notch radius (*Figure 1*), it can been seen that the evolution can be divided into three stages:

(a) for notch radius less than a critical one, critical gross stress is constant and equal to those obtained from a cracked specimen,
(b) this step is followed by a linear evolution,
(c) at last, critical gross stress reaches asymptotically ultimate strength of material Rm.

Such a kind of evolution has been mentioned by Yokobori et al. [1] for steel. It is shifted towards higher values of notch radius for decreasing temperature.

Notch effect can be defined as the relative difference of ultimate strength and critical gross stress.

303

I. Dlouhý (ed.), Transferability of Fracture Mechanical Characteristics, 303–322.
© 2002 *Kluwer Academic Publishers. Printed in the Netherlands.*

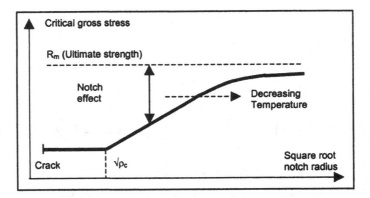

Figure 1. Schematic evolution of critical gross stress versus square root of notch radius.

This simple scheme does not take into account possible modifications of fracture mode due to loss of constraint when notch radius increases.

Fracture emanating from stress concentrators like notches can be classified as belonging to three limit states: plastic collapse, brittle fracture or elastoplastic fracture. Increasing notch radius moves limit state to a more ductile one. However failure by plastic collapse for sharp and long notches is relatively seldom. These limit states are sensitive to geometrical parameters and particularly non dimensional ligament size b* which affect net stress and constraint factor values. Relationships between limit stress σ_L critical gross stress σ_g^c and non dimensional ligament size are different from each limit state and are presented in table1.

TABLE I. Relationships between limit stress σ_L critical gross stress σ_g^c
and non dimensional ligament size

Plastic collapse	brittle fracture	elastoplastic fracture
$\sigma_L = Rmf_1(1-b^*)$	$\sigma_g^c = K_{IC} /\left(\sqrt{\pi(1-b^*)}f_2(1-b^*)\right)$	$\sigma_g^c = K_{\rho,c} /\left[\left(\pi(1-b^*)\right)^\alpha f_3(1-b)*\right]$

f_1, f_2 and f_3 are geometrical function, K_{Ic} fracture toughness measured with cracked specimens, $K_{\rho,c}$ fracture toughness measured with notched specimens and α stress distribution parameter. For a given value of non dimensional ligament size, critical gross stress is maximum for plastic collapse, minimum for brittle fracture and has intermediate value for elastoplastic situation.

Relative notch effect Θ can be defined for a given notch radius as the ratio of the difference between limit stress and critical net stress for elastic or elastoplastic fracture and limit stress.

$$\theta = \frac{\sigma_L - \sigma_N^c}{\sigma_L} \qquad (1)$$

A schematic evolution of Θ is shown in *Figure 2*. Relative notch effect is zero for very short or very long notches which fail by plastic collapse. Its increases when notch radius decreases and is maximum for a crack for which notch radius is considered as equal to zero. Evolution exhibits a maximum which depends of notch radius.

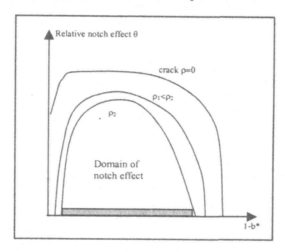

Figure 2. Schematic evolution of notch effects with relative ligament size and notch radius.

2. Physical Basis of Notch Effect

Notch effect is associated with the presence of stress concentrator, which leads to a typical stress distribution. An example of elastoplastic stress distribution is presented in *Figure 3* where the non dimensional stress (i.e. opening stress σ_{yy} divided by the gross stress) is plotted versus non dimensional distance (i.e. distance r divided by notch depth a). This distribution is computed by finite element method for a CT specimen. The dimensions of the specimen are: thickness B = 15 mm ; width W = 80 mm ; height L = 96 mm ; ligament size b = W - a = 36 ; notch angle $\psi = 40°$. The material is steel called E 550 according to French standard. Mechanical properties are as follows: Yield stress (Re = 572 MPa); ultimate strength (Rm = 684 MPa).

This distribution is characterised by maximum stress σ_{max} and high stressed region at notch tip. The maximum stress is not at notch tip but at a little distance behind due to plastic relaxation at the free surface. Notch tip is followed by a high stressed region with limits which are difficult to define.

The fundamental question is to know if fracture process obeys a local or semi local criterion. Formerly, it was assumed that maximum stress governs the fracture process. This assumption suffers the following objections:

306

(a) for a crack in an elastic body, stress is infinite at crack tip due to singularity,

(b) for a notch in an elastic body, maximum stress is higher than ultimate strength corrected by constrain factor,

(c) for a notch in an elastoplastic body, place of maximum stress is different from place to fracture initiation,

(d) for different loading modes with the same maximum stress, fracture doesn't occurs in each case,

(e) for loading mode leading to high stress gradient, a significant scale effect occurs.

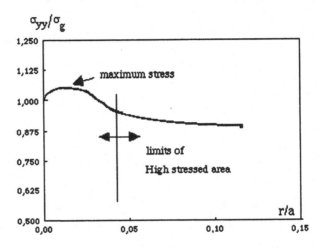

Figure 3. Elastoplastic stress distribution at the notch tip of a CT specimen.

To overcome these critiques, several authors [2] have based a criterion upon the assumption of micro crack coalescence with a fracture nucleation site located at some distance. They located it between notch or crack tip and peak stress position.

Obviously, this distance called the effective distance is depending on notch geometry and loading conditions, but is also depending on material structure. Fracture process occurs in a high-stress volume, which is different in size from the notch plastic zone but similar in shape. Notch plastic zone has a practically circular shape [3]. For this reason, is natural to assume a cylindrical fracture process volume of diameter equal to effective distance and height equal to the thickness of the structure.

For determination of effective distance, a graphical procedure is used. It has been seen that the effective distance is connected to the minimum of relative stress gradient χ defined by:

$$\chi = \frac{1}{\sigma_{yy}} \frac{d\sigma_{yy}}{dr} \tag{2}$$

It can be easily shown that this distance corresponds to the beginning of the pseudo stress singularity of the notch tip stress distribution.

Charpy V notched specimens made in a CrMoV steel (yield stress 771 MPa) were tested under quasi-static bending at one selected temperature in lower shelf region. The tensile stress distribution at the notch has been calculated using FEM. A 2D model under plane strain conditions was used for the elastic-plastic analysis.

The effective distance X_{ef} has been determined using normal stress distributions below notch root plotted in bilogarithmic scale. Relative stress gradient (equation 2) has been plotted on the same graph in order to obtain the exact effective distance. For a fracture load equal to 13,1 kN, the effective distance has been found as $X_{ef} = 0.380$ mm.

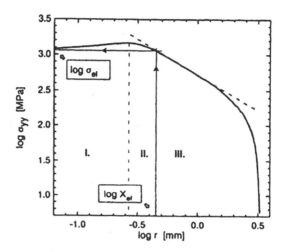

Figure 4. Stress distribution at notch root together with relative stress gradient versus distance from notch tip for fine carbides CrMoV rotor steel. Definition of the effective distance as the distance of minimum relative stress gradient.

Table II gives a comparison between grain size, effective distance and notch plastic zone size.

TABLE II. Comparison between grain size, effective distance and notch plastic zone size.

Grain size	Effective distance	Plastic zone size
0.015 mm	0.380 mm	0.630 mm

We can notice that effective distance is larger than grain size but smaller than plastic zone. Following Beremin's model [4], the probability of fracture P_r is proportional to the fracture process volume through Weibull's analysis. This fracture process volume is precisely the plastic zone.

$$P_r = 1 - \exp\left[-\int_V \left(\frac{\sigma}{\sigma_{n0}} \right)^{m_w} \frac{dV}{V_0} \right] \qquad (3)$$

308

Where V_O is the reference volume and is usually determined by the corresponding finite mesh element used in FEM. σ_{no} is reference parameter and m_w the Weibull 's modulus. These parameters are considered as material constants. Expression of the Weibull's law versus the effective volume is given by:

$$P_r = 1 - \exp\left[-\left(\frac{\sigma_{max}}{\sigma_{n0}}\right)^{m_w} \frac{V_{eff}^W}{V_0}\right] \qquad (4)$$

The effective volume in the Weibull's meaning V_{eff}^W is:

$$V_{eff}^W = \left[\int_V \left(\frac{\sigma}{\sigma_{max}}\right)^{m_w} dV\right] \qquad (5)$$

σ_{max} is the maximum of the stress distribution. Using experimental and computed results showing the cylindrical shape of the plastic zone, relationship (6) can be modified for a two dimensional approach to give the effective distance according to Weibull meaning and compared to the effective distance.

$$X_{eff}^W = \left[-\int_V \left(\frac{\sigma}{\sigma_{n0}}\right)^{m_w} dx\right] \qquad (6)$$

For this a polynomial description of the stress distribution at notch tip has been used. The Weibull's modulus m_w has got the value of $m_w = 12.5$ according to a recent work by Dlouhy et al [5].

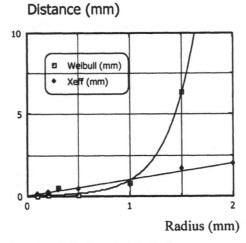

Figure 5. Comparison of variation of effective and Weibull's distances versus notch radius.

A linear evolution of the effective distance with notch radius has been noticed. This distance is generally of the same order of magnitude than the ρ value. The effective distance in the Weibull's meaning increases exponentially with notch radius. For large values of ρ this distance has no physical meaning and can be considered as a fitted parameter to satisfy equation 4.

The effective stress is defined as the average of the weighted stress inside the fracture process zone:

$$\sigma_{eff} = \frac{1}{X_{eff}} \int_0^{X_{eff}} \sigma_{ij} dx \tag{7}$$

For this material the mean value of the effective stress is 1223 MPa which can be compared to the average maximum local stress at fracture σ_{max} of 1310 MPa.

In the classical Ritchie, Knott and Rice [6] local fracture criterion, cleavage stress is generally taken as the yield stress at 0 Kelvin or computed by the way of local tensile stress. Comparisons of effective stress with cleavage stress and Weibull's stress are made in Table 3

TABLEIII. Comparisons of effective stress with cleavage stress and Weibull's stress.

Effective stress	Cleavage stress	Weibull's stress
1223 MPa	1106 MPa	3450 MPa

3. Notch Effects on the Brittle-Ductile Transition, Notch Sensitivity

Some materials like ferritic steels exhibit a brittle to ductile transition for fracture toughness represented by the critical notch stress intensity factor this transition is promoted by increasing temperature or decreasing loading rate. This effect can be explained by the fact that plasticity is necessary to initiate fracture process and is a thermal activated process.

A third physical parameter influences this transition: the notch radius or in another way the level of stress triaxiality. In agreement with different experimental observations on relationship of fracture toughness $K_{\rho c}$ versus the alpha root of the notch radius, it has been seen that under a critical notch radius ρ_c, fracture toughness is independent of ρ. Similarly, the presence of a ductile plateau is independent of the notch radius. In the transition regime, K_{Ic} is a linear function of the square root of the notch radius. A similar transition curve can be drawn to more generally represent the notch effect on fracture toughness. However to take into account any notch geometry, the critical notch stress intensity factor is plotted versus the notch radius at the power α (figure 5). The notch sensitivity m can be represented by the slope of the transition line.

$$m = \tan g(\beta) \tag{8}$$

310

If the test temperature decreases, the material becomes more and more brittle and the notch sensitivity increases, i.e. notches are more dangerous at low temperature (Figure 6). Similar facts are obtained when increasing loading rate (figure 7).

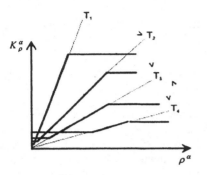 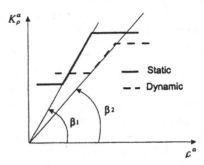

Figure 6. Evolution of the critical notch stress intensity factor versus the notch radius at the power α for three different temperatures.

Figure 7. Evolution of the critical notch stress intensity factor versus the notch radius at the power α for two loading rate conditions.

4. Notch Sensitivity of Static Fracture Toughness of a Steel

The critical energy parameter J_0 at initiation is defined according to Turner [8] as proportional to the specific energy for initiation U_i/Bb_0

$$J_0 = \eta \frac{U_i}{Bb_0}$$ (9)

Where η is a parameter, B is the specimen thickness and b_0 is the initial ligament size. η is defined from the formula of the energy parameter J according to Eq.(9). It is considered as equivalent of J integral and consequently:

$$J = -\frac{1}{B} \frac{\partial U}{\partial a}$$ (10)

Then η can be written as follows:

$$\eta = \frac{b_0}{U} \frac{\partial U}{\partial a} = -(W - a) \frac{\partial Ln U}{\partial a}$$ (11)

η has been computed by finite element using stress strain behaviour as described by Ludwik's law. Figure 8 shows the evolution of η for a constant notch radius equal to 0.7 mm versus the relative notch depth a/W. Figure 9 shows the evolution of η for constant relative notch depth a/W = 0.5 versus notch radius. We can notice that η differs from the deep notch solution for elastic case (η = 2) and is sensitive to notch radius. These solutions for η have been used for the determination of J_{IC} fracture toughness.

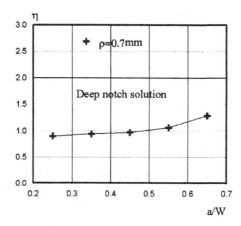

Figure 8. Evolution of η versus relative notch depth (a/W) for constant notch radius ρ=7mm (Charpy U notched specimen).

We notice that η increases with the relative depth a/W. The evolution of η as a function of notch root radius shows an absolute minimum whose abscissa called ρ_c is ranging between 0.75 and 1 mm (Figure 9). Similarly, we notice that for radius values below ρ_c, η decreases linearly with the increase of ρ. Beyond this critical abscissa η increases with ρ and becomes approximately constant for a radius ranging between 1.54 and 2 mm. The difference between η got for crack and that got for a notch with the same length can reach 36% of the last one, which is relatively important and justifies the present approach.

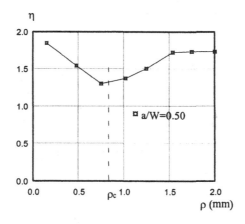

Figure 9: Evolution of η versus notch radius for constant relative notch depth a/W = 0.50 (Charpy U specimen).

The material under study is steel. Its mechanical properties are reported in table 4.

TABLE IV. Mechanical properties of a XC38 Steel.

Yield stress) Re (MPa)	Ultimate strength Rm (MPa)	Elongation A%	Ludwick law	Vickers Hardness	Fracture toughness J_{Ic} (KJ/m2)
304	430	30	88^n	137	798

A microanalysis of the material gives an atomic percentage of chemical elements: 0,24% of C.14% of Si, 0.53% of Mn and traces of Cr, Ni, and Mo. The present experiments used U-notched specimens of dimensions 25 x 25 x 150 mm obtained from a 30 x 30 mm plate, specimen length being parallel to the rolling direction. Different notches are introduced employing a wire-cut EDM (Electrical Discharge machine) and using different diameter of wires. The notch root radius was measured using a profile projector with an enlargement of x 50.

The experimental displacements were measured as the central deflection of the specimen in order to avoid all deformation influence of the machine testing columns on our results (taking into account the importance of loading stresses over 4 tons). We assume that the critical non-linear fracture energy was reached at maximum load. This critical value of the used energy is obtained as the area below load-displacement curve from zero to maximum load P_{max}. Figure 10 shows the variation of non linear fracture energy as a function of ρ a/W constant; we can notice that the energy absorbed until fracture increase non linearly with notch radius.

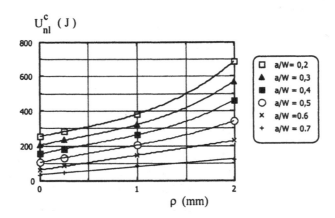

Figure 10. Evolution of the work done for fracture
versus notch radius for different values of ratio a/W.

In figure 11, we give the values of the apparent fracture toughness got by considering, in the expression (9) a critical state. In this relationship, work done $U_{nl}^c(a, \rho)$ and $\eta(a, \rho)$ factor are associated.

$$J_{Ic,App} = \frac{\eta(a,\rho).U_{nl}^c(a,\rho)}{Bb} \qquad (12$$

We can notice that radii less than 0.85 mm have no influence on fracture toughness. But, for a radius beyond 0.85 mm, $J_{Ic,App}$ increases linearly with ρ where $J_{Ic,App}$ unit is MJ/m^2, ρ in mm. The critical radius value ρ_c is $\rho = 0.85$ mm. This can be summarised as follows:

(a) plateau: $\rho < 0.85$ mm, $\qquad\qquad J_{Ic,app}$ = Constant= J_{IC},

(b) linear evolution: $\rho \quad 0.85$ mm, $J_{Ic,app} = 8.6.10^{-4}\,\rho$

We notice that the extrapolation of linear evolution passes through origin (0,0).

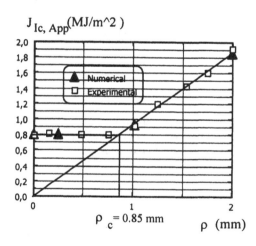

Figure 11. $J_{Ic,App}$ versus ρ. Comparison between experimental and numerical results using eq. 12.

5. Fracture Toughness from Charpy Impact Test in the Upper Shelf Plateau

A typical load-time diagram of ductile fracture obtained in the upper shelf plateau is presented in Fig. 12a. Load increases until a maximum value and decreases slowly. Initiation takes place just before the maximum load and is followed by ductile tearing.

314

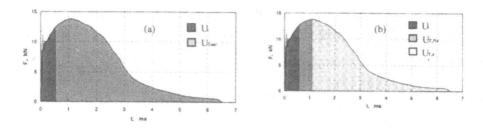

Figures 12 a and 12 b. Instrumented Charpy impact test force-time curve. Definition of energy for initiation, non-stable and stable tearing (Cast steel, Test temperature +20°C).

In the following diagram total fracture energy U_t can be separated into two parts: one for initiation U_i and one for tearing U_{Tear}

$$U_t = U_i + U_{Tear} \tag{13}$$

Tearing process can be also separated into two parts: transient tearing process from initiation to maximum load $U_{T,NSt}$ and stable tearing $U_{T,St}$ (Fig.12b). Stable tearing is governed by the fact that crack opening angle (COA) is assumed to be constant. Eq. (13) can be written as:

$$U_{Tear} = U_{T,Nst} + U_{T,st} \tag{14}$$

Therefore:

$$U_t = U_i + U_{T,Nst} + U_{T,st} \tag{15}$$

Only the absorbed energy at maximum load U_m and U_t are required as experimental input data, which can be obtained very easily and unambiguously from the force-time curve (Fig. 13) and then equation (15) can be rewrite:

$$U_t = U_m + U_{T,st} \tag{16}$$

Where:

$$U_m = U_i + U_{T,Nst} \tag{17}$$

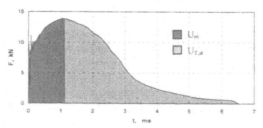

Figure 13 : Instrumented Charpy impact test force-time curve ; definition of total energy at maximum load and energy at stable tearing.

These different mechanisms of ductile failure can be represented on a J-Δa fracture resistance curve, (Figure 14). The energetic parameter J is plotted versus the crack extension Δa.

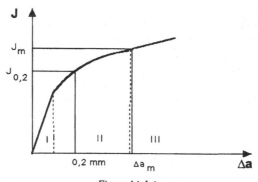

Figure 14. J-Δa curve

The J-Δa curves can be divided into 3 parts:

Part I:
initiation (by convention, a crack extension of $\Delta a_0 = 0,2$ mm is considered as an initiation). The J values reach initiation value $J_{0,2}$.

$$J(\Delta a) = m\,R_c\,\Delta a \qquad\qquad \text{For } \Delta a < \Delta a_0 \qquad (18)$$

m is a parameter and R_c is the flow stress.

Part II:
transient tearing process where energetic parameter J is a power function of crack extension

$$J(\Delta a) = R(\Delta a)^p \qquad\qquad \text{For } \Delta a_0 < \Delta a < \Delta a_m \qquad (19)$$

Δa_m is crack extension at maximum load

Part III:
stable tearing process.

During this process the following assumptions are used to compute the energy for stable tearing $U_{T,S}$: crack opening angle is constant, crack opening displacement is obtained assuming a fixed rotational center position; and bending moment is computed from an elastic-plastic material behaviour, strain hardening being taken into account using the flow stress R_c. This model has been proposed by Schindler [9], for $\Delta a > \Delta a_m$ it leads to:

$$J(\Delta a) = J_{\max} + s_2 \cdot \left[(\Delta a - \Delta a_m) - \frac{(\Delta a - \Delta a_m)^2}{2b_0} \right] \qquad\qquad \text{For } \Delta a > \Delta a_m \qquad (20)$$

with

$$s_2 = \frac{2\eta(U_t - U_m)}{B(b_0 - \Delta a_m)} \tag{21}$$

This model of fracture resistance curve leads to a three parameters model with parameters: p, R and Δa_m. Obtention of these three parameters is achieved only through values of the absorbed energy at maximum load U_m and total energy for fracture U_t. Detection of initiation is not necessary using this procedure.

The investigated steel material is a cast ferritic steel used for nuclear waste containers. The chemical composition and the mechanical properties are given in Table 5 and in Table 6 respectively.

TABLE V. Chemical composition of a cast steel.

Weigth %	C	Si	Mn	Cr	Mo	Ni	Cu
Cast Steel	0, 09	0,37	1, 18	0,12	0,03	0,29	0,29

TABLE VI. Mechanical properties of a cast steel.

Yield stress (MPa)	Ultimate strength (MPa)	A%	K_{CV} +20°C daJ / cm^2
375	478	31,7	8

We assume that the material is strain hardening and obeys the following stress-strain law:

$$\sigma = K\varepsilon^n \tag{22}$$

Where K is the strain-hardening coefficient and n the strain-hardening exponent (K = 737 MPa; n = 0.12). Charpy U notched specimens were used in the experiments with constant notch radius and different notch depths as given in Table 7 and with constant relative notch depth and different notch radii as given in Table 8.

TABLE VII. Notch depths and ligament sizes of the Charpy U notched specimens

ρ(mm)	0.7mm					
a(mm)	2	3	4	5	6	7
b_0=W-a (mm)	8	7	6	5	4	3

TABLE VIII: notch radii of the Charpy U

notched specimens

a/W	0.5						
ρ(mm)	0.13	0.25	0.4	0.7	1	2	2.5

Test temperatures have been chosen in the upper shelf region (+20°C) and in the lower transition region (-20°C) as it is shown in Figure 15.

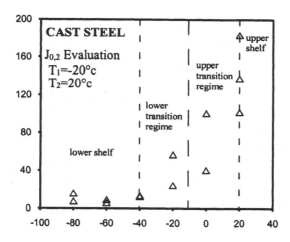

Figure 15. Charpy impact energy versus temperature for cast steel indicating the differentmode of fracture.

Despite a Large scatter on maximum and total energy, average values of $p = 0.85$ and $\Delta a_m = 0.79$ mm have been found. Evolution of $J_{0,2}$ versus notch radius is plotted in Figure 16. We can notice that for notch radius below $\rho_c = 1$mm the initiation toughness $J_{0,2i}$ has a constant value equal to 0.2 MJ/m^2. Notch sensitivity has been found to be equal to $\alpha_j = 100$ MJ/m^3.

318

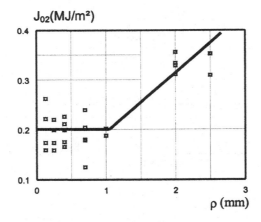

Figure 16. J$_{0,2}$ versus different values of Notch radius.

6. Notch Sensitivity in Mixed Mode Fracture

Notch sensitivity has been measured on a ring specimen made from a high strength steel (French standard 45 SCD 16).

Description of the specimen is given in figure 15. The ring has a small notch of 4 mm depth and notch radius varies in range [0.2 – 2 mm]. The notch plane is inclined with 3 different angle values from the loading direction in order to promote different modes of loading: $\beta = 0°$ mode I ; $0° < \beta < 33°$ mixed mode I + II and $\beta = 33°$ mode II

After heat treatment, material has the following mechanical properties.

Table IX: mechanical properties of 45 SCD 16 steel.

Yield stress	Ultimate strength	Elongation	K_{IC}
7 463 MPa	1 162 MPa	2,8 %	97 MPa√m

It has been noticed that critical load increases with inclination angle and also with notch radius when greater than a critical values $\rho_c = 0,75$ mm.

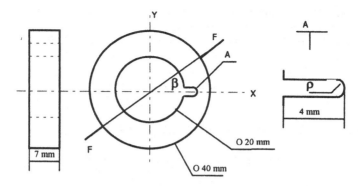

Figure 17. Geometry of ring specimen used for mixed mode I + II of fracture.

Fracture toughness has been determined using a local notch fracture criterion involving effective stress and effective distance.

Hoop stress has been computed by a finite elements method. Effective distance has been determined using a graphical procedure involving the minimum of the relative stress gradient.

Effective stress has been computed by averaging hoop stresses over effective distance. It can be noticed that according to Erdogan and Sih [7] criterion for mixed mode of fracture, effective stress can be expressed versus the value of mode I and mode II notch stress intensity factor.

$$\sigma_{ef} = \frac{1}{\left(2\pi X_{ef}\right)^a} \cos\frac{\theta}{2}\left[K_{\rho,I}\cos^2\frac{\theta}{2} - \frac{3}{2}K_{\rho,II}\sin\frac{\theta}{2}\right] \tag{23}$$

(Where θ is the crack bifurcation angle, $K_{\rho,I}$ mode I notch stress intensity factor and $K_{\rho,II}$ mode II notch stress intensity factor.).

It has been noticed that effective distance is related to notch radius by the following relationship.

$$X_{ef} = A'\rho^a \tag{24}$$

with a' (β) and A' (β) parameters which depend on inclination angle.

Similarly, effective stress is also related to notch radius and loading angle by:

$$\sigma_{ef}^c = \frac{B'}{\rho^{b'}} \tag{25}$$

with B' (β) and b' (β) parameters which are dependant of inclination angle.

Fracture toughness determined as the equivalent notch stress intensity factor has been plotted versus notch radius.

We can notice that the equivalent critical notch stress intensity factor is practically constant beyond a critical notch radius of $\rho_c \cong 0.70$.

$$\frac{K_{II}}{K_{IC}} = A\left(\frac{K_I}{K_{IC}}\right) + B\left(\frac{K_I}{K_{IC}}\right) + C \tag{26}$$

320

A new criterion for mixed mode brittle fracture of notched specimen is based on the criterion proposed by Erdogan and Sih [7]. For this criterion, fracture of cracked specimens occurs when the product of the critical hoop stress by the square root of the distance reaches a critical value

$$\sigma_{\theta\theta}^{c}\sqrt{2\pi r} = \frac{K_{IC}}{F_{\sigma}} \tag{27}$$

Where K_{Ic} is the fracture toughness and F_{σ} a geometrical factor.

For notched specimen, the criterion is modified as follow:

$$\sigma_{\theta\theta,ef}^{c}(2\pi r)^{\alpha} = K_{\rho,ef}^{c} \tag{28}$$

Where α is the pseudo singularity exponent. This formula uses the critical effective hoop stress and the effective critical notch stress intensity factor.
This criterion can be written as:

$$\cos\frac{\theta}{2}\left[K_{\rho,I}\sin\theta + K_{\rho,II}(3\cos\theta - 1)\right] = 0 \tag{29}$$

$$\cos\frac{\theta}{2}\left[K_{\rho,I}\left(\cos\frac{\theta}{2}\right)^{2} - \frac{3}{2}K_{\rho,II}\sin\theta\right] = K_{\rho,I}^{c} \tag{30}$$

The following assumptions are generally used for steels $K_{Ic} = K_{IIC}$, consequently:

$$K_{\rho,I}^{c} = K_{\rho,II}^{c} \tag{31}$$

The fracture criterion can be rewrite:

$$\cos\frac{\theta}{2}\left[\frac{K_{\rho,I}}{K_{\rho,I}^{c}}\sin\theta + \frac{K_{\rho,II}}{K_{\rho,I}}(3\cos\theta - 1)\right] = 0 \tag{32}$$

$$\cos\frac{\theta}{2}\left[\frac{K_{\rho,I}}{K_{\rho,I}^{c}}\left(\cos\frac{\theta}{2}\right)^{2} - \frac{3}{2}\frac{K_{\rho,II}}{K_{\rho,I}^{c}}\sin\theta\right] = 1 \tag{33}$$

$$\frac{K_{\rho,I}}{K_{\rho,I}^{c}} = f\left(\frac{K_{\rho,II}}{K_{\rho,I}^{c}}\right)$$

Equation can be plotted in plane.

$$\frac{K_{\rho,II}}{K_{\rho,I}^{c}} = A\left(\frac{K_{\rho,I}}{K_{\rho,I}^{c}}\right)^{2} + B\left(\frac{K_{\rho,I}}{K_{\rho,I}^{c}}\right) + c \tag{34}$$

With $A = -0.247\rho - 0.6$; $B = 0.35\rho - 0.1$; $C = 0.1\rho + 0.86$

This curve is called fracture intrinsic curve. We can notice that for $K_{\rho,I}/K_{\rho,I}^c = 0.6$ notch radius has no influence on this intrinsic curve. In mode II and for a crack we found that:

$$K_{\rho,II}^c = 0.86 K_{\rho,I}^c \qquad (35)$$

Value given by Erdogan and Sih criterion. Notch sensitivity appears only, for the range of explored notch radius and for this steel only if the bimodality ratio is more than 1.2.

$$\frac{K_{\rho,II}}{K_{\rho,I}} = 1.2 \qquad (36)$$

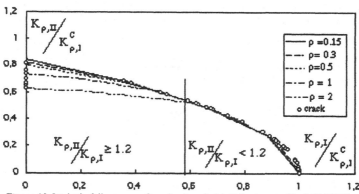

Figure 18. Intrinsic failure curve for mixed mode I + II fracture on Steel 45 SCD6.

7. Conclusions

Notch effect is difference between limit stress and critical stress for elastic or elastoplastic fracture. This definition does not take into account the geometrical effect of ligament area reduction.

Notch effects is related to stress rising over an effective fracture process volume. A semi-local fracture criterion involves 2 parameters: an effective distance and an effective stress.

Notch affects leads to a transition between brittle and ductile fracture. In the transition zone, fracture toughness versus notch radius at some power exhibits a linear evolution. The slope of this linear evolution is called notch sensitivity.

Notch sensitivity occurs by additional plastic work for fracture. Assuming that fracture toughness is proportional to the total work done for fracture, a linear relationship between fracture toughness and notch radius can be founded if notch radius exceeds a critical value.

This linear relationship remains if we considered fracture resistance to tearing.

Notch sensitivity can be also defined for mixed mode of fracture and can be presented as a modification of fracture intrinsic curve.

References

1. Yokobori, Y.and Konosu, S.(1977) Effects of Ferrite Grain size, notch acuity and notch length on Brittle fracture stress of Notched Specimens of low Carbon Steel, *Engineering Fracture Mechanics* Vol 9, 839-847

2. Krassowsky, A. and Pluvinage, G.(1993) Structural parameters governing fracture toughness of Engineering Materials *Physico Chemical Mechanics of Materials* vol 29, N°3, 106-113

3. Chacrone, A., Azari, Z., Bouami, D. and Pluvinage, G., Plastic Zone Size And Effective Distance at Notch Root

4. Beremin, F., M.(1983) *Metall. Transactions*, Vol 14 A , 2287-2296

5. Dlouhy, I, Chlup, Z and Holzman, M. (2001) Local Characteristics of (Brittle) Failure assessed from Charpy Type specimens *Notch effects in fatigue and Fracture Ed Pluvinage, G., and Gjonaj, M., Kluwer Academic publishers*, 127-146.

6. Ritchie, R., Knott, S. and Rice, J.(1973) *Journal of Mechanic Physics of Solids*, Vol 21, 395-410

7. Erdogan,F and Sih, G.C.(1963) On the crack extension in plates under plane loading and transverseshear. *Journal of basic engineering*, Vol 85,519-527

8. Turner, C.E. (1979) Methods for post-yield fracture safety assessment, *Post-Yield Fracture safety*, 23-210

9. Schindler, H. J. and Veidt, M. (1988) Fracture toughness evaluation from instrumend sub-size Charpy-type tests, *Small Specimen Test Techniques*, ASTM STP 1329, 48-61

QUANTIFICATION OF NOTCH EFFECTS – IN BRITTLE-DUCTILE TRANSITION

L. TÓTH
Bay Zoltán Institute of Logistics and Production Systems, Miskolc-Tapolca, Iglói u.2. H-3519, Hungary

Abstract: In the presents paper notch effects on brittle to ductile transition behaviour of steels and its description has been followed. Based on experimentally measured data for mild steel and data from literature local parameters, the notch stress, strain and energy intensity factors, respectively have been introduced arising from the global stress, strain or deformation energy fields (gradients) and characteristic distances. The relationships between the notch intensity factors and the notch radius in $1/K$ vs. $\sqrt{\rho}$ co-ordinates are linear for stress and strain fields and non-linear (parabolic) for deformation energy. The brittle to ductile transition sensitivity of materials can be characterised by the slope of the curves obtained. The experimental results performed at -196 °C on notched tensile specimens made of mild steel with different grain size are in full agreement with presented concept and they can be regarded as an experimental verification of the proposed approach.

Keywords: notch effect, ductile to brittle transition, mild steel, notch sensitivity factor, strain intensity factor, energy intensity factor, notch root effect, fracture toughness

1. Introduction

The reliability, service lifetime and functional properties of engineering components and structures are predetermined, among others, by external and internal local stress and/or strain concentrators. The internal stress/strain concentrators are predominantly caused by microstructural inhomogeneities present in materials, the external ones usually by notches connected, more or less, with a component and structure design. It is not possible to design the components without considering the effects of these stress and strain concentrators. The material behaviour in (local) areas affected by notches as stress/strain concentrators is controlled mainly by the notch geometry (notch depth, including angle and notch root radius), loading conditions (loading rate and test temperature effects) and the local material response on the stress - strain fields below the notch root.

From the engineering point of view the so-called "notch effect" is very important both in brittle transition behaviour and in fatigue failure of structural elements because 90 % of the failures are initiated but the stress-strain concentrators. Considering the statistical data that each country loses approximately 4 % of the GDP due to failures, the

I. Dlouhý (ed.), Transferability of Fracture Mechanical Characteristics, 323–336.
© 2002 *Kluwer Academic Publishers. Printed in the Netherlands.*

"notch effect in engineering practice" seems to be very important issue from the economical point of view as well.

The aim of the present contribution is to summarise the main ideas of notch effect quantification on fracture resistance of materials, steels with transition behaviour in particular. The basic definition for the notch stress, strain and energy intensity factors are here introduced based on published data and own experiments as well.

2. Notch Stress, Strain and Energy Intensity Concept

The notched round specimens tested in tension may have different fracture surfaces morphologics. From the *macroscopic* point of view they may be totally ductile, or brittle or they may have a brittle to ductile transition character. It is clear that the ratio of ductile and brittle areas depends on factors influenced by the notch geometry, testing conditions (temperature and loading rate) and material behaviour. These fracture surfaces are schematically illustrated in *Figure 1*. From *microscopic* point of view we can find a ductile fracture morphology at the notch vicinity even at the most dangerous loading conditions (i.e. at high loading rate, and at low temperatures). This ductile fracture morphology can be also realised at the notches with a possible highest notch acquity (i.e. in the case of crack). Obviously these areas could have only a magnitude of grain size. Considering only the *macroscopic* approach the following question can arise: *How can be the brittle-ductile behaviour of steels of notched elements characterised and how can the effect of loading conditions (temperature and loading rate) and the notch geometry be taken into account?*

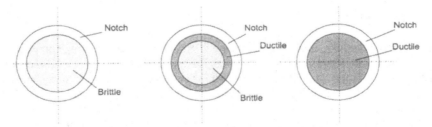

Figure 1. The possible fracture surfaces of notched specimens tested in tensile

It is obvious that the ductile, brittle or ductile-brittle transition behaviour of steels depends on the local conditions at the notch influenced areas. These local conditions will be created by the external (loading rate, temperature and notch geometry) and internal parameters of materials which is the most simple self-organised system. It means that if a high value of energy can be stored at the notch vicinity before fracture than the material behaves brittle during fracture because the crack will start with a higher rate. If the measure of energy concentration is not enough high than the crack will start with a smaller rate and the fracture surface will have either totally ductile or ductile-brittle character. They may schematically be characterised by the local stress, strains or absorbed energy distribution as it is shown in *Figure 2*.

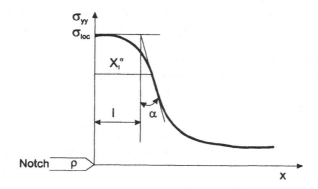

Figure 2. The stress distribution at the area affected by notch (scheme)

Figure 2 shows that the *local conditions* at the notch vicinity can be characterised by three parameters, i.e. by the local maximum stress (σ_{loc}), by the length of the plateau of the maximum stress (l) and by the gradient of stress (α). Both of the last two parameters may be summarised by a single length parameter denoted by X_I^σ (where the index I denotes the loading mode, i.e. the loading perpendicular to the fracture surface). Using these parameters the measure of stress concentration in the area affected by notch can be characterised by the *notch stress intensity factor* that can be stressed in the form of

$$K_I^\sigma = \sigma_{loc} \sqrt{\pi X_I^\sigma} \qquad (1)$$

The measure of strain concentration in the area effected by notch can be characterised by the *notch strain intensity factor* (K_I^ε) and the measure of absorbed energy concentration by the *notch energy intensity factor* (K_I^w), i.e.

$$K_I^\varepsilon = \varepsilon_{loc} \sqrt{\pi} X_I^\varepsilon \qquad (2)$$

$$K_I^w = w_{loc} \sqrt{\pi} X_I^w \qquad (3)$$

where ε_{loc} and w_{loc} are the local strain or energy concentrations on the surface, the X_I^ε and X_I^w are the characteristic distances respectively. The local stress (σ_{loc}), strain (ε_{loc}) or energy (w_{loc}) is proportional to the effective stress, strain or energy concentration factor.

Taking a notched cylindrical specimen with a crack (i.e. notch radii is app. equal to zero) and loaded by a given value tensile force the value of K_I^σ (in an elastic media) is equal to infinity, because the local stress is also infinity (the stress field has an $1/\sqrt{r}$ type singularity, where r is the distance from the crack front). If only the notch radii increases

326

in the above mentioned and loaded specimen then the notch stress intensity factor decreases. Because the local stress concentration factor (K_t) at the notch vicinity in elastic media is proportional to the $1/\sqrt{\rho}$, where ρ is the notch radii, it can be predicted that a linear relationship exists between $1/K_I^\sigma$ and $\sqrt{\rho}$ as it is illustrated by the straight line starting at the origin in *Figure 3*. It means that the reciprocate value of the measure of stress or strain concentration in the area affected by notch is proportional to the value of $\sqrt{\rho}$.

Considering the fracture surface morphology it could be brittle, transition or totally ductile, respectively. It is obvious that the fracture parameters are different. We can also say that if the measure of stress concentration is enough high in the notch influenced area of the material than the material is totally brittle. This situation can be characterised by the *plane strain fracture toughness* (K_{Ic}) which has no geometry dependence, i.e. it is not influenced by size and geometry of specimen. If the fracture surface is not totally brittle, or it is totally ductile than the fracture parameter is some kind of *fracture toughness* (K_c or σ_c) which has a geometry dependence (i.e. its value depends on the size and geometry of specimen and the notch).

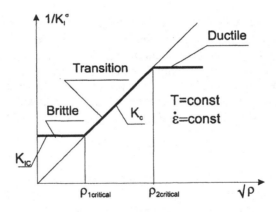

Figure 3. Relationship between measure of stress concentration in the area affected by notch i.e. between the notch stress intensity factor and other fracture parameters

This is perfectly in agreement with the experimental observations, i.e. there exists a small notch radius by which the brittle fracture of materials can be reached. This notch radius is denoted by ρ_{1crit}. In this case the fracture parameter is the plain strain fracture toughness which is constant.

The suggested model can be validated by the experimental data which can be found in the literature. In the *Figure 4.* the fracture toughness parameters of H-11 modified stainless steel is presented measured on 1 inch wide edge notched specimens with different notch radius [1]. The *Figure 5.* shows the fracture parameter of the same steel measured on *3 inch wide edge notched specimens* and the *Figure 6* shows the same fracture parameter measured on *3 inch wide central slotted specimens*.

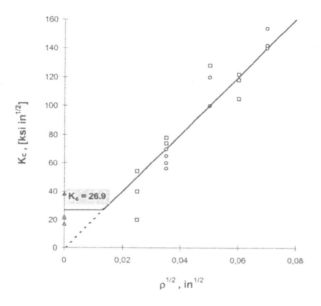

Figure 4. The apparent fracture toughness of H-11 modified stainless steel measured on 1 inch *wide single notched specimens* [1].

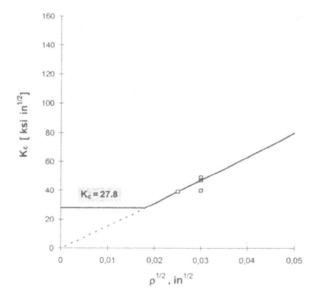

Figure 5. The apparent fracture toughness of H-11 modified stainless steel measured on 3 inch *wide edge notched specimens* [1].

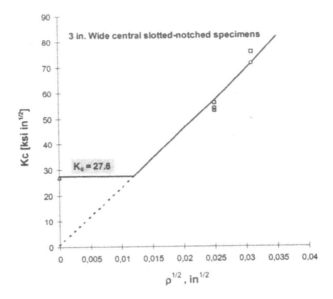

Figure 6. The apparent fracture toughness of H-11 modified stainless steel measured on 3 inch *wide central notched specimens* [1].

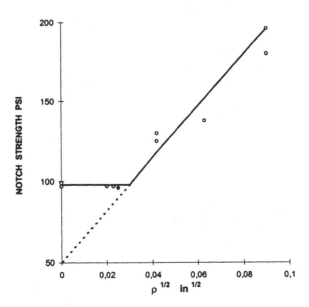

Figure 7. The fracture parameter (notch strength) of B 120 VCA alloy measured on notched cylindrical specimens with different notch radius [2].

In addition, the *Figure 7*. shows the notch strength of B 120 VCA alloy measured on notched cylindrical specimens with outer diameter of D=0,5 inch and diameter below the notch root d/D = 0,707 and notch angle 60° [2].

The *Figure 8*. summarises the J_{lc} values of glasses with different sulphur content (S=0,007, 0,017 and 0,034wt%, respectively) measured on notched specimens with different notch radius.

Considering the *Figures 4-6* the following conclusions can be drawn:

- the apparent fracture toughness vs. $\sqrt{\rho}$ has a linear character the straight line of which has the cross-section with the origin,
- the apparent fracture toughness is constant up to a given notch radius (which has been denoted by ρ_{1crit}) and it has no geometry dependence, i.e. its value does not depend on the specimen geometry.

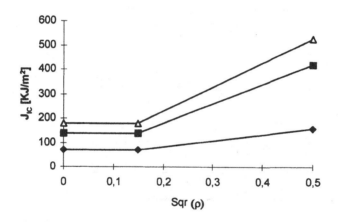

Figure 8. The J_{ic} values of glasses with different sulphur content (0,007-0,034wt%) measured on notched specimens with different notch radius.

These findings are in full agreement with the proposed model. Considering the results summarised in the *Figure 7*, where the notch strength (as an other fracture parameter) is plotted being measured on cylindrical notched specimens with different notch radius, it can be also concluded that the fracture parameter vs. $\sqrt{\rho}$ has a linear character. The straight lines fitted through the experimental data have the common cross session in the origin. The results summarised in the *Figure 8* for glasses are also in full agreement with the above mentioned knowledge.

If only the notch radius increases, the fracture has brittle-ductile transition character and upon reaching an upper limit of notch radius, for instance ρ_{2crit} (*Figure 2*), the fracture is totally ductile. At these ranges the values of fracture parameter depends on the ratio of the ductile and brittle morphology measured on fracture surface i.e. the brittle tendency of materials can be characterised by the slope of the straight line as it is shown in *Figure 9*.

If the value of slope (tan β_1) is small then the material is brittle, because the total brittle fracture can occur on a specimen even with larger notch radius. In contrast, if the slope is high (tan β_2) then the material is generally ductile, because the brittle fracture can only be found for a very small notch radius i.e. on a specimen with high notch acuity.

The brittle-ductile transition sensitivity of materials i.e. the slope (tan β_i) of $1/K_I^\sigma$ vs. $\sqrt{\rho}$ curve depends on

- type of materials,
- testing temperature and
- loading rate.

The last two parameters will be analysed in the next paragraphs.

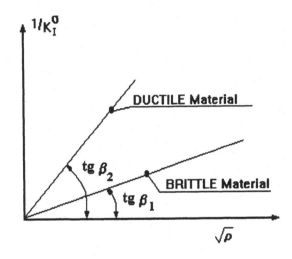

Figure 9. The brittle to ductile transition sensitivity of materials characterised by the slope (tan β_i) of $1/K_I^\sigma$ vs. $\sqrt{\rho}$

2.1 THE EFFECT OF TEMPERATURE

If the test temperature decreases, then the material becomes more and more brittle due to low temperature embrittlement. It means that the value of (tan β_1) decreases as it is demonstrated in *Figure 9*. It can be also formulated, that by decreasing the test temperature the notch sensitivity increases, i.e. notches are more dangerous at low test and operation temperatures. The mentioned details of course are in good agreement with experiences.

2.2 THE LOADING RATE EFFECT

The loading rate effect on brittle-ductile transition sensitivity of a given material is illustrated in *Figure 10*. The quasi-static loading condition is characterised by continuous line. With increasing loading rate the material becomes more and more brittle and its yield stress increases, i.e. the perfectly brittle fracture can be found on a specimen with higher value of the notch radius, consequently tan β_4 < tan β_3 (dashed line).

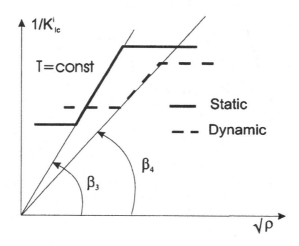

Figure 10. The loading rate effect characterisation on brittle -ductile transition sensitivity of a selected material by the slope (tan β_i) and $1/K_1$ vs. $\sqrt{\rho}$ curve

2.3 THE LOADING RATE AND TEMPERATURE EFFECTS

The same tendency can be observed if the testing temperature decreases. These are summarised in *Figure 11*.

2.4 NOTCH RADIUS EFFECT ON THE NOTCH SENSITIVITY OF MATERIALS

It is also well known, and experimentally verified fact, that with increasing notch radius the brittle-ductile transition temperature also increases (supposing the other experimental conditions are the same). This fact directly follows from the proposed concept.

Following the *Figure 12*. if the test is performed on a specimen with a notch radius of ρ_1, then at T_1 temperature the fracture is ductile. At a lower temperature T_2 the fracture has a ductile-brittle transition character. If the temperature is lower than T_3 i.e. T_4, then the tested material is totally brittle. If the notch radius is higher i.e. $\rho_2 > \rho_1$, then in accordance with *Figure 11*. the brittle-ductile transition has a lower value.

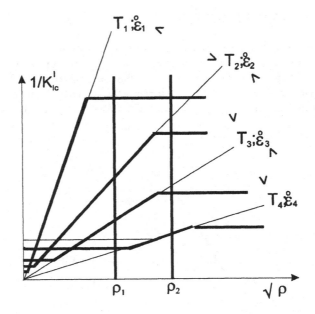

Figure 11. The loading rate and temperature effect characterisation on brittle-ductile transition sensitivity of a selected material by the slope of $1/K_{ic}^{i}$ vs. $\sqrt{\rho}$ curve

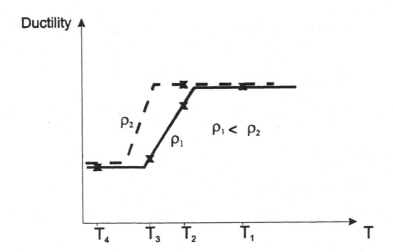

Figure 12. Notch radius effect on brittle-ductile transition

3. Effect of Grain Size on Brittle-Ductile Sensitivity of Mild Steel

It is well known that by increasing the grain size the brittle-ductile transition sensitivity
increases too. From the above mentioned approach it follows, that the slope (tan β_i) of
$1/K_I^\sigma$ vs. $\sqrt{\rho}$ curve increases with decreasing of grain size. This statement is fully
supported by the tensile test results performed by Yokobori, Y., Konosu S. at -196 °C,
as published in [6].

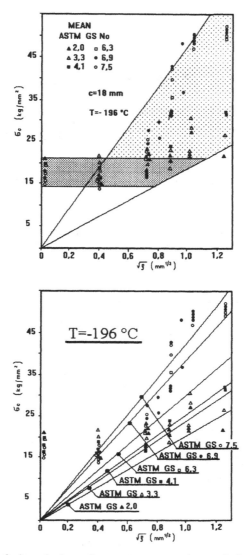

Figure 13. Effect of the ferrite grain size on fracture stress measured on notched specimens at -196 °C, low
carbon steel. a) for all range b) for a given grain size

In accordance with *Figure 13*. the values of slopes (tan $\beta_i \equiv$ m) of $1/K_I^\sigma$ vs. $\sqrt{\rho}$ curves for different grain diameter (d) are plotted in *Figure 14*.

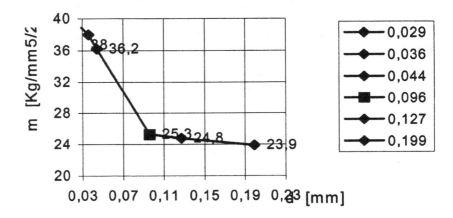

Figure 14. The value of slope (tan $\beta_i \equiv$ m) of $1/K_I^\sigma$ vs. $\sqrt{\rho}$ curves for different grain diameter (d)

Figure 14. exactly shows that the grain size has a great influence on brittle-ductile sensitivity only in a given range of grain sizes i.e. ASTM GS N 2.0-4.1. If the grain size is high enough then its influence is not remarkable i.e. in the range of ASTM GS 4.1-7.5.

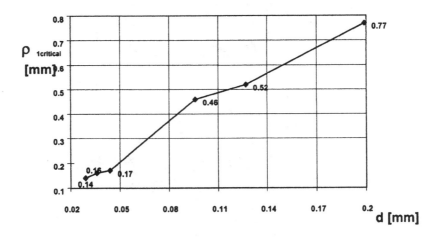

Figure 15. Critical notch radii (ρ_{1crit}) defined in *Figure 4*. vs. grain diameter (d) for mild steel tested notched specimens at -196 °C.

In accordance with *Figure 3*. the values of notch radii belongs to the beginning of brittle to ductile transition (ρ_{1crit}) for different grain size (d) are plotted and *Figure 15*.

Figures 14. and 15. are in full agreement with the presented concept and they can be regarded as an experimental verification of the proposed approach for description of brittle-ductile behaviour of notched structural components.

4. Summary

The elastic-plastic deformation state of the area affected by notch (the measure of stress, strain or energy concentration in the materials as the most simple selforganised system) can be characterised by local parameters such as notch stress, strain or energy intensity factors.

The local parameters, the notch stress, strain and energy intensity factors are expressed by:

$$K_I^\sigma = \sigma_{glob} \, K^\sigma \sqrt{X_\sigma} = \sigma_{loc} \sqrt{X_\sigma}$$
$$K_I^\varepsilon = \varepsilon_{glob} \, K^\varepsilon \sqrt{X_\varepsilon} = \varepsilon_{loc} \sqrt{X_\varepsilon}$$
$$K_I^w = W_{glob} \, K^w \sqrt{X_w} = W_{loc} \sqrt{X_w}$$

where the global stress, strain or deformation energy fields are characterised by, σ_{glob} ε_{glob} and W_{glob}; the local fields at the notch tip are expressed by the local, K^σ, K^ε and K^w concentration factors and the distribution of these fields by characteristic distances, X_σ, X_ε or X_w respectively.

The relationships between the notch intensity factors (K_I^i) and the notch radius (ρ) in $1/K_I^i$ vs. $\sqrt{\rho}$ co-ordinate system are linear for stress and strain ($i=\sigma$ or ε) and non-linear (parabolic) for deformation energy.

The brittle-ductile transition sensitivity of materials can be characterised by the slope of $1/K_I^i$ vs. $\sqrt{\rho}$ curve (i= stress or strain).

The experimental results performed on notched tensile specimens made of mild steel with different grain size at -196 °C are in full agreement with presented concept and they can be regarded as an experimental verification of the proposed approach.

Acknowledgement

The author gratefully acknowledge the support for this work provided by the Hungarian National Foundation for Science (OTKA, Contract No T-15601).

References

1. ASTM's Materials Research and Standards, Vol. 1, No 11. November 1961. p. 877-885. Fracture Testing of High-Strength Sheet Materials. Third Report of a Special ASTM Committee.
2. ASTM's Materials Research and Standards, Vol. 2, No 3. March 1962. p. 196-203. Screeing Test for High-Strength Alloys Using Sharply Notched Cylindrical Specimens. Fourth Report of a Special ASTM Committee.

3. Tóth L., Gouair H., Azari Z., Pluvinage G.: Notch effect on Brittle-Ductile transition. General Approach. (in Hungarian). GÉP.1994. 1994/4. p.3-8.
4. L.Tóth., G. Pluvinage, H. Gouair, Z. Azari.: The Notch Stress, Strain and Energy Concept. FRACTURE'94. University of Witwatersrand, Johannesburg. 23-24 November. Ed. M.N. James. p. 180-189.
5. Tóth L.: Approach of the Brittle-Ductile Transition by the Concept of Critical Notch Stress, Strain or Energy Intensity Factor. Proc. of the 1st Workshop on " Influence of Local Stress and Strain Concentrators on the Reliability and Safety of Structures" Miskolc. 1995. p.61-69.
6. Yokobori, Y., Konosu S.: Effects of Ferrite Grain Size, Notch Acuity and Notch Length on Brittle Fracture Stress of Notched Specimens of Low Carbon Steel. Engineering Fracture Mechanics, 1977. Vol. 9. pp.839-847.

BRITTLE MIXED MODE FRACTURE I+II: EMANATING FROM NOTCHES -EQUIVALENT NOTCH STRESS INTENSITY FACTOR -H

EL MINOR[1], M. LOUAH[1], Z. AZARI[2], G. PLUVINAGE[2], and A. KIFANI[3].

[1] LaMAT, ENSET, BP 6207 R.I, Morocco.
[2] L. F. M. Université de Metz - ENIM, Ile de saulcy, 57045 Metz Cedex 01, France.
[3] L. M. M., Ufr MFS, faculté des sciences, Agdal - Rabat, Morocco.

Abstract: In the present paper, the crack initiation in mixed mode fracture I+II has been studied using notched circular ring specimens. A new criteria in brittle mixed mode fracture I+II, based on notch tangential stress and the volumetric approach has been developed. The critical value of the equivalent notch intensity factor has been considered as fracture toughness in mixed mode fracture I+II.

Keywords: brittle mixed mode fracture I+II, notch effect, angled crack problem, tangential stress criterions, effective tangential distance, relative tangential stress gradient, weight function, equivalent notch stress intensity factor, fracture toughness.

1. Introduction

First studies of fracture mechanics have been achieved on pre-cracked specimens loaded according to elementary fracture mode I, because it is considered as the most dangerous fracture mode. In the reality some situations drive to simultaneous presence of fracture mode I and II (opening mode and shear mode): e.g. welded joint, marine structures, weld tubular and rolling mill.

In this case, no standardised test procedure exists; there is a method similar to compared tension specimen. Different test conditions have been used for most of the establishing experiments made until now and numerous types of specimen have been proposed in recent years for investigating fracture toughness. Many mixed mode fracture criterion, measuring fatigue threshold values and determining of the crack growth laws applicable to multi-axial stress situations were tested. Some of these specimens have been presented and critically analysed by Richard [1]. Recent investigations [2, 3] showed that the pre-cracked circular ring specimens were well suited for mixed mode experiments.

However the used pre-cracked specimen's (ring shape) ones presents two inconveniences:

(i) These specimens are regarded as time consuming and expensive.

I. Dlouhý (ed.), Transferability of Fracture Mechanical Characteristics, 337–350.
© 2002 Kluwer Academic Publishers. Printed in the Netherlands.

(ii) For very brittle materials as ceramics and high strength steels it is practically impossible to pre-crack the specimens and the application of notched specimen is preferred.

The analytical solution to determine the stress intensity factors, for notch tip distribution is Creager's solution [4] at the present time. This one consists in adding a geometrical correction factor to Irwin's solution [5]. This method is based on the assumption that the characteristic distance (or process volume diameter) is equal to r/2 (ρ : notch radius). But this procedure has a limited applicability because of structures up dimension.

It has been proposed recently [6, 7] to characterise fracture conditions for a notched specimen by using the actual stress gradient at the notch root. This stress gradient can be characterising by a relationship different from the crack tip stress gradient. This method has been used in the present work to determine the fracture resistance in applied mixed mode fracture (I+II) using notched circular ring specimens. The toughness for high strength steel (45SCD6) was defined by a critical equivalent notch stress intensity factor. This approach is classified under generic stress of the Notch Fractures Mechanics [16].

2. Material and Specimen

In this part we propose to check the validity of this fracture criterion using experimental results obtained from notched specimens subjected to compression loading. The specimen is made of high strength steel and the stress distribution has been computed using the finite element method.

The material studied is a high strength steel named 45CDS6 according to French standard. Mechanical properties are listed in TABLE I. The microanalysis of the material gives: 0.45% of C, 1.60% of Si, 0.60% of Mn, 0.60% of Cr and 0.25% of Mo. The chemical composition is given in atomic percentage.

TABLE I. Mechanical properties

M. V. (Kg/m^3)	v	A%	E (MPa)	Rp^{0,2} (MPa)	Rm (MPa)	K_{IC} (MPa m$^{0.5}$)
7800	0,28	2.8%	210065	1463	1662	97

Tests are performed using U-notched circular ring specimens *Figure 1,* with external radius R_e = 20 mm, internal radius R_i = 10 mm, thickness B = 7 mm and notch length a = 4 mm. Different notch radius are introduced using a wire-cut EDM (Electrical Discharge Machine) and using wires of different diameter. Notch root radius was measured using a profile projector.

2.1. DEFINITION OF DIFFERENCES MODE FRACTURE

$\beta = 0°$	Mode I [8] ;
$\beta = 33°$	Mode II [2] et [3] ;
$0° < \beta < 33°$	Mixed mode fracture I+II;

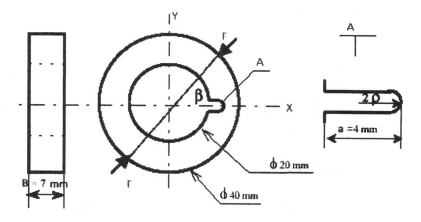

Figure 1. U-notched circular ring specimen

The *Figure 2* shows the mechanical testing using to notched specimens subjected to compression loading.

Figure 2. Mechanical testing

3. Experiments and Results

3.1. CRITICAL LOAD P_C

Figure 3 shows the experimental critical load P_c evolution versus notch radius ρ and inclination angle β ($\beta = 0°$ (mode I) and $\beta = 33°$(mode II)).

The *Figure 3* shows that the critical load P_c increases linearly whit notch radius ρ.

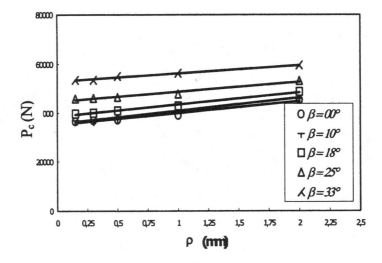

Figure 3. Critical load P_C versus notch radius ρ for different inclination angle β

3.2. BIFURCATION ANGLE.

Experimental values measured by optical microscope (see *Figure 4)* are compared to numerical values obtained from the maximum tangential stress criterion.

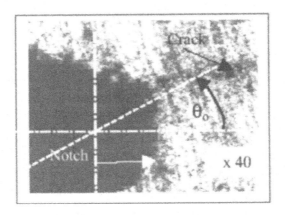

Figure 4. Example of bifurcation angle for $\rho = 1$ mm & $\beta = 18°$

3.3. MECHANISMS OF CRACK INITIATION: "VOLUMETRIC APPROACH".

According to an engineering approach, crack initiation exists whatever a high stress concentration is present and is defined as a short crack detectable with a magnification of x 50.

In the *Figure 5* can be seen that blunted notches give short crack mechanism, which means that crack growth rate is increased after a decreasing period.

Mechanisms of crack initiation have been described in detail. They consist of intrusions and extrusions mechanisms in pure ductile metals or dislocation pile-up on inclusions, decohesion of the matrix and finally crack initiation.

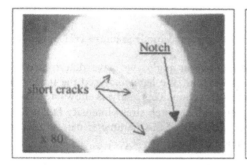

a) Short cracks mechanism

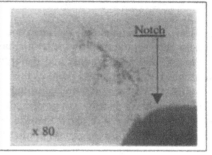

b) Evolution of micro-cracks

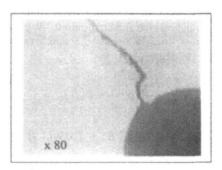

c) Crack growth

Figure 5. Fracture micro-mechanism

Two major elements indicate that the crack mechanisms need a physical volume to take place:
- The probability of crack initiation is proportional to the process volume where the probability to found an initiation site (inclusion) is assumed to be uniform.

- Crack resistance is influenced by the size of the specimen and relative stress gradient (i. e. the stress gradient divided by the stress value), which are dimensional parameters.

It has been proved that the crack mechanisms at notch root cannot be explained by the spot approach; i. e. the maximum local stress plays the important role. We consider that the effective stress range acting in the crack process volume plays this role.

In this volume the average stress is high enough to promote crack initiation, the relative stress gradient not to high to order these all points that are inside this volume called "effective tangential stress". The influence of relative stress gradient on the crack process was previously mentioned by Buch in [9].

4. Mixed Mode Fracture Criterion, Initiated from Notches: The Use of Equivalent Notch Stress Intensity Factor

This criterion assumes that the initiation crack in mixed mode fracture I+II emanating from notches is governed by the tangential stress.

For various notch radius and different inclination angles we have determined the maximum tangential stress point at the notch contour. We have analysed the tangential stress distribution at notch tip according to this direction. This analysis shows a "pseudo singularity" stress distribution governed by equivalent notch stress intensity factor K_{ep}. In the following we will show that the critical values of this parameter can be used to determine the fracture toughness in mixed mode fracture I+II.

4.1 MAXIMUM TANGENTIAL STRESS DIRECTION AT NOTCH TIP

Finite element calculations (CASTEM 2000) have been used to determine the maximum tangential stress on all points of the notch contour. We notice that the direction according to this stress varies linearly versus the notch radius. The following fitted relations give this variation (in degrees):

$$\theta_0 = A(\beta)\frac{\rho}{a} + B(\beta) \tag{1}$$

where:

$$A(\beta) = -0.0222\,\beta^2 + 1.4983\,\beta \tag{2}$$

$$B(\beta) = -0.0456\,\beta^2 + 3.6178\,\beta \tag{3}$$

In mode II ($\beta = 33°$) for $\rho = 0$: $\theta_0 = B\,(33°) = 70.39° \approx 70.5°$ (Sih et Erdogan [10]) $\approx 70.33°$ (Stroh [11]).

The numerical values according to the maximum tangential stress criterion (Eg. 1) are compared to experimental values measured by optical microscope in *Figure 6*. This figure shows that the angled crack θ_0 according to the maximum tangential stress criterion agree well with the experimental data and can be used to predict bifurcation angle for crack emanating from notches.

4.2. TANGENTIAL STRESS DISTRIBUTIONS AT NOTCH TIP.

In *Figure 7* we plot the evolution of the stress distributions at distance r according to maximum tangential stress direction θ_0 (for $\rho = 0.5$ mm, $\beta = 18°$).

The *Figure 7* shows that the tangential stress $\sigma_{\theta\theta}$ is distinctly superior to σ_{rr}, σ_{zz} and $\tau_{r\theta}$. This important difference is observed for various notch radius ρ and different inclination angles β.

We assume that the mixed mode fracture I+II, emanating from notch is governed by the tangential stress ($\sigma_{\theta\theta}$ plays the major role in crack process). In the following, we have studied this stress distribution according to maximum tangential stress direction.

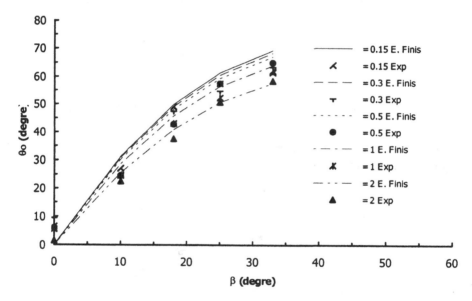

Note: Finis: finite elements
         Exp: experimental test

Figure 6. Direction of the maximum tangential stress θ_0 versus the inclination angle β

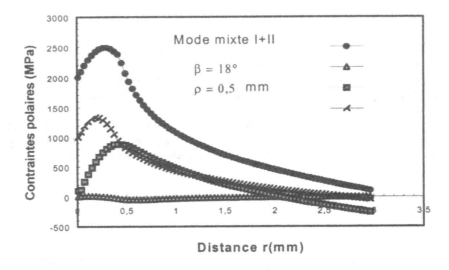

Figure. 7. Stress distributions $\sigma_{\theta\theta}$, σ_{rr}, σ_{zz} and $\tau_{r\theta}$ at notch tip

According to procedure described by [6, 7, 12] we plot the tangential stress distribution in a bi-logarithmic graph – see *Figure 8*.

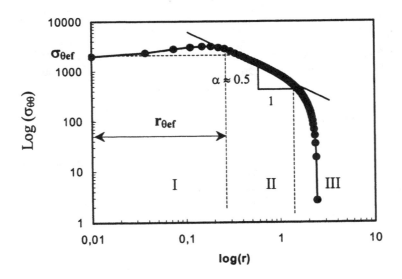

Figure 8. Distribution of tangential stress at notch tip

This distribution can be divided in three zones:

Zone I: High stress region.
Zone II: Pseudo singularity stress distribution governs by equivalent notch stress intensity factor.

$$\sigma_{\theta\theta} = \frac{K_{e\rho}}{(2.\pi.r)^{\frac{1}{2}}}$$ (4)

where:
$\sigma_{\theta\theta}$: tangentiel stress (MPa)
r: distance at notch tip (mm).
$K_{e\rho}$: Equivalent Notch stress intensity factor (MPa.m$^{1/2}$).

Zone III: far region.

4.3. EFFECTIVE TANGENTIAL DISTANCE $r_{\theta ef}$

The upper limit of the distribution "pseudo singularity" has the following co-ordinates ($r_{\theta ef}$, $\sigma_{\theta ef}$) where $r_{\theta ef}$ is the effective tangential distance and $\sigma_{\theta ef}$ the effective tangential stress.

By definition the effective tangential distance is the diameter of process volume assuming it has a cylindrical shape. A typical example of tangential stress distribution can be seen again in *Figure 9*. The relative tangential stress was plotted also versus distance r.

The relative tangential stress is defined as:

$$\chi = \frac{1}{\sigma_{\theta\theta}(r)} \cdot \frac{d\sigma_{\theta\theta}(r)}{dr}$$ (5)

The effective tangential distance can be determined using the following considerations:
• According to [13] the effective tangential distance is greater than the plastic zone diameter (Rp'θ_0)).
• The effective tangential distance is situated in the stressed region where the stress gradient is not too high.

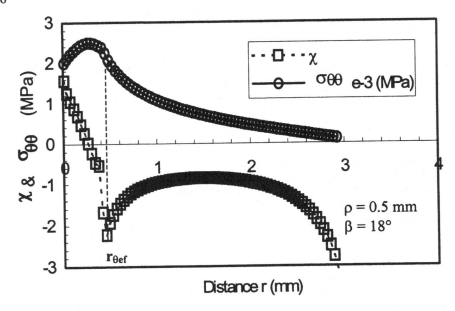

Figure 9. Tangential stress distribution at notch tip, relative tangential stress gradient versus the distance, definition of the effective tangential distance

4.4. WEIGHT FUNCTION

All the stressed point in the process volume plays a role in crack initiation emanating from notches. The role is different for each point and influenced by the distance between this stress and notch, and the stress gradient. We can define the weighted stresses, which take in to account these roles, as follows:

$$\sigma^*_{ij} = \sigma_{ij}\, \varphi(r,\, \chi) \qquad (6)$$

Where $\varphi(r,\, \chi)$ is the weight function. According to Weixing in [14] this function is defined by the relationship:

$$\varphi(r,\, \chi) = 1 - (r).\chi \qquad (7)$$

4.5. EFFECTIVE TANGENTIAL STRESS $\sigma_{\theta ef}$

The effective tangential stress is the stress corresponding to the distance on the stress distribution, is defined as follows:

$$\sigma_{\theta ef} = \frac{1}{r_{ef}} \cdot \int_{0}^{r_{ef}} \sigma_{\theta\theta} \cdot \varphi(r, \chi).dr \qquad (8)$$

It is the average value of the weighted tangential stress.

4.6. CRITICAL EQUIVALENT NOTCH STRESS INTENSITY FACTOR K $_{cep}$ AND FRACTURE TOUGHNESS K$_{IC}$

In this part, we determined critical equivalent notch stress intensity factor K$_{cep}$ in mode I, mode II and mixed mode fracture I+II. We compared its value to fracture toughness K$_{IC}$ according to Jones [8], who considered the notch as equivalent to a crack.

For circular ring specimens, Jones [8] expresses a fracture toughness K$_{IC}$ by the following relationship

$$K_{I}^{c} = 2,61. \frac{P_{C}}{B\sqrt{R_{o}}} \qquad (9)$$

where: $0,625 \leq (a+Ri)/Ro \leq 0,845$ and $Ri/Ro = 0,5$.

$a \geq 40 \% b$ (b = Ro-Ri)

2,61 correspond to the value of geometrical correction and P$_C$ critical load.

Fracture toughness K$_{IC}$ according to [8] and critical equivalent notch stress intensity factor K_{ep}^{c} have been determined from experimental and numerical results and reported respectively in *Figure 10* and *Figure 11* versus the notch radius ρ.

We notice that the critical equivalent notch stress intensity factor K_{ep}^{c} is practically constant and is independent on the notch radius and the inclination angle for a notch radius less than ρ$_c$. The critical value of this parameter is nearly equal to K^c$_{Ip}$ notch stress intensity factor in mode I fracture:

$$K^{c}_{ep} \approx K^{c}_{Ip} \approx K_{IC} \text{ (for } \rho < \rho_c = 0.7 \text{ mm}) \approx 97 \text{ MPa (m)}^{1/2} \qquad (10)$$

However for $\rho > \rho_c \approx 0.75$ mm, the fracture toughness K$_{IC}$, according to [8], exhibits a linear relationship with notch radius ρ it is not an intrinsic material characteristic.

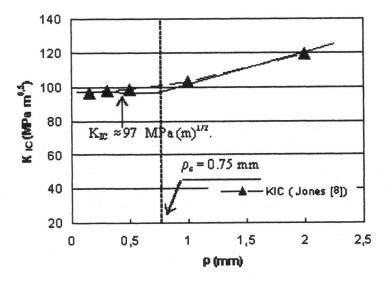

Figure 10. The influence of notch radius ρ on the fracture toughness (Jones) K_{IC}

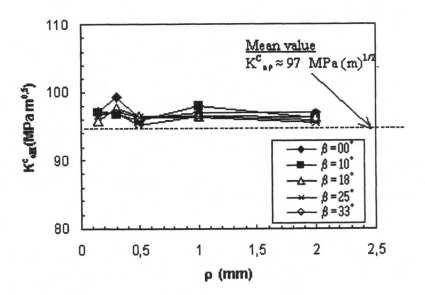

Figure 11. The influence of notch radius on the critical equivalent notch stress intensity factor K^c_{ep}

4.7. MIXED MODE I+II CRITERION: BASED ON THE EQUIVALENT NOTCH STRESS INTENSITY FACTOR $K^c_{e\rho}$

The tangential stress distribution *Figure 8* at notch tip can lead to two types of fracture criteria: global and local. In the case of notch, there is no stress singularity at crack tip, but a maximum stress followed by a pseudo singularity in which the stress distribution is governed by an equivalent notch stress intensity factor.

4.7.1. *Global fracture Criterion*
It is assumed that crack initiation occurs:
(i) in perpendicular direction to tangential stress when it reaches its maximum value,
(ii) when equivalent notch stress intensity factor $K^c_{e\rho}$ reaches a critical value:

$$K_{e\rho c} = \sigma^C_{eff} \cdot \left(2\pi . r^C_{eff} \right)^{0.5} = K^C_{l\rho} \tag{11}$$

4.7.2. *Local fracture criterion (volumetric approach)*
The local fracture criterion is based on the following consideration: for physical reason, fracture process needs a given volume called "effective volume". In this volume, the effective tangential stress can be considered as an average stress tangential weight which takes in to account the tangential stress distribution. This process volume can be described by the distance $r_{\theta ef}$, so called "effective tangential distance", considering that specimen thickness is constant and process volume is cylindrical.

The crack initiation is assumed to occur when the effective tangential stress $\sigma_{\theta ef}$ and the effective tangential distance reaches critical values.

5. Conclusions

Brittle fracture in mixed mode I + II emanating from notches needs like in a crack a physical volume for the elaboration the process of failure.

Using the effective tangential stress acting in the volume and its application in stress notch intensity factor seems to give good results.

Stress notch intensity factor application on brittle materials, where machining of sharp cracks is very difficult, is interesting for toughness determination.

With this parameter, linear fracture mechanics criteria can be taken as a particular notch fracture mechanics problem.

The results of this works and other [6, 12, 15] indicate that this approach gives a relative good description of notch effect.

References

1. H. A. Richard (1989) Specimens for investigating biaxial fracture and fatigue processes; Biaxial and Multiaxial fatigue, *Mechanical Engineering Publications*, 217-229.
2. T. Tamine, C. Chehimi, T. Boukharouba et G. Pluvinage (1996) Crack initiation in pure shear mode II, ISSN 0556-171Xn, *Problems of Strength*, Sp. Publisher.

350

3. T. Tamine (1994) Amorçage de fissures par fatigue-contact, *Thèse de Doctorat*, Université de Metz, France.

4. M. Creager et P. C. Paris (1967) Elastic field equations for blunt cracks with refrence to stress corrosion cracking; *Int. Jour. of Fract. Mech.* **3**, 247-252.

5. G. R. Irwin (1957) Analysis of stresses and strains near the end of crack traversing a plat; *Journal of Applied Mechanics*, Transactions of the A. S. M. E.

6. G. Pluvinage (1997) Rupture et fatigue amorcée à partir d'entailles – Application du facteur d'intensité d'entaille, *Revue française de mécanique*-1, 53-61.

7. G. Pluvinage (1999) Notch effect and effective stress in high cycle fatigue, présenté par le laboratoire de fiabilité mécanique, Université de Metz.

8. Jones (1974), *Eng. Fract. Mech.* **6**, 435-446

9. A. Buch (1974) Analytical approach to size and notch size effects in fatigue of aircraft material specimens., *Mat. Sc. and Eng.* **15**, 75-85.

10. G. C. Sih et F. Erdogan (1963) On the crack extension in plates under plane loading and transverse shear; *Journal if basic engineering.*

11. Stroh (1972) La rupture des matériaux D. Francois and L. Joly ; Masson and C.

12. N. Kadi (1997) Rupture par fatigue des arbres entaillés et clavetés, *Rapport de D. E. A.*, Laboratoire de Fiabilité Mécanique de l'Université de Metz.

13. H. El Minor Rupture fragile en mode mixte amorcée à partir d'entaille, Thèse de Doctorat, Université de Rabat, Maroc (en cour).

14. Y. Weixing (1995) Stress field intensity approach for predicting fatigue life, *Inter. Jour. of Frac.* **17**, N°4, 245-251,.

15. Qylafku, Z. Azari, N. Kadi, M. G.jonaj, G. Pluvinage (1999) Application of a new model proposal for the fatigue life prediction on notches and key-seats, *International Journal of Fatigue*, Elsevier, Volume 21/8, 753-760, U. K.

16. A. Nyoungue, H. Gouair, G. Pluvinage, Z. Azari (1996) Glass fracture Toughness from critical Notch Stress intensity Factor Concept, *Journal Problems strength of material*, 99-106, Ukraine.

SAFETY OF THE STEAM GENERATOR COVER: A PSA CASE STUDY

S. VEJVODA
Institute of Applied Mechanics Brno, Ltd., Veveří 95, 611 39 Brno, Czech Republic

D. NOVÁK, Z. KERŠNER, B. TEPLÝ
Brno University of Technology, Faculty of Civil Engineering, Institute of Structural Mechanics, Veveří 95, 662 37 Brno, Czech Republic

Abstract: The loss of functionality of a steam generator cover is investigated in probabilistic terms and its safety measures are assessed. The goal is to make the inspection strategies for NPPs more realistic and efficient. Mathematical description of the gradual damage of the stud bolt until its rupture is based on the three stages: origin of a surface defect such as a crack or pitting; growth of defect before conditions stress corrosion cracking; growth of defect under stress corrosion cracking. The results of probabilistic safety assessment of the analyzed demountable connection of the 1ˢᵗ circuit collector of steam generator are presented.

Keywords: Probability, failure, fracture mechanics, limit loads, nuclear power plant, plasticity, uncertainty, random variable, probabilistic safety assessment, reliability, risk assessment, structural integrity.

1. Introduction

The use of probabilistic techniques to assess the risk posed by activities in industry has significantly increased over the last decade. The relevant term used in the nuclear industry is the probabilistic safety assessment (PSA). The techniques used in the PSA involve applications of risk assessment procedure to determine:
(i)   how a system may fail and lead to hazardous scenarios;
(ii)   how likely are these scenarios to occur;
(iii)   what are the consequences of these scenarios.
This study is focused on the failure of the steam generator cover addressing issues (i) and (ii). The loss of functionality of the steam generator cover is investigated in probabilistic terms, and its safety measures are assessed with a goal to make the inspection strategies for the nuclear power plant (NPP) more realistic and efficient. In other words, the structural integrity is defined as the probability of a system satisfactorily performing its specified function under all stated conditions. In dealing with such problems the probability of failure is commonly utilized. It is generally defined [1] by using function g (.) of relevant load and resistance parameters, called

351

I. Dlouhý (ed.), Transferability of Fracture Mechanical Characteristics, 351–366.
© 2002 *Kluwer Academic Publishers. Printed in the Netherlands.*

basic random variables X_i: $g(X_1, X_2, ..., X_N) = g(X)$. The failure surface of this limit state of interest can than be defined as $g(X) = 0$. This is the boundary between the safe and unsafe regions in the design parameter space. The failure occurs when $g(X) \le 0$. Therefore, the probability of failure p_f is given by the integral:

$$p_f = P(g(X) \le 0) = \int_{D_f} f_x(x_1, x_2, ..., x_n) \, dx_1, dx_2, ..., dx_N \qquad (1.1)$$

in which $f_x(x_1, x_2, ..., x_N)$ is the joint probability density function to $x_1, x_2, ..., x_N$ and the integration is performed over the failure region D_f in which $g(X, t) \le 0$. Symbol t reflects the fact that limit state function depends on time t (deterioration).

Numerical methods for evaluating necessary integral (1.1) achieved at present the level enabling routine applications and are well documented, e.g. [2]. Taking the advantage of (1.1) the risk R may be quantified rather simply by [3]:

$$R = p_f C_f, \qquad (1.2)$$

where C_f is the consequence of an accident (event, limit state) given either by economical terms, number of fatalities or other appropriate measures.

Long-term experience has made it possible to identify many sources of errors in ordinary structures. However, there is only a limited experience with the structures in nuclear power plants concerning real failures and one cannot be confident that all likely sources of errors have been identified. But we should realize that gross (human) errors in design and construction are seldom apparent in NPP structures because of the enforcement of strict quality assurance programs. Errors arising from natural random physical variability of material, geometrical, loading, technological and environmental parameters are therefore in focus of the present case study.

2. Deterministic Model of Bolt Rupture

2.1. WORKING CONDITIONS OF DEMOUNTABLE CONNECTION

The demountable connection of the 1st circuit collector of the steam generator (see *Figure 1*) creates a barrier between the primary and secondary circuits. It is necessary to know what probability of the stud bolt rupture is and what the loss of tightness probability in case of different non-destructive tests interval is. Both probabilities as subjective measures of reliability can be called failure probabilities. Non-destructive tests of heat-exchange pipes require the cover dismantling. All stud bolts and the thread in the collector tapped holes are checked by the non-destructive testing. Stud bolts with defects detected on the surface are exchanged. In the repetitive manhole packing the new sealing rings are used.

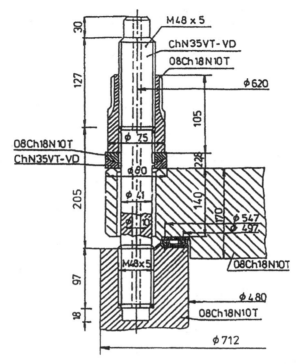

Figure 1. Scheme of the analysed demountable connection

Experimental measurements of stud bolts strain using strain gauges under service conditions during four years endorsed theoretical assumption that the axial force in stud bolts does not significantly change. So, fatigue is not a decisive damage mechanism for the bolt material. However, in case of the non-favorable environment influence, conditions, which are suitable for stress corrosion cracking, may arise. The process of stress corrosion cracking occurs at a sufficient level of stress, material sensitivity to the corrosive environment and at a sufficient inherence of chemical ingredients in the environment. The stress corrosion process will not occur when one of these factors is absent. In the above case, all three factors can be present. It is necessary to determine the allowable level, and subsequently, to provide probability analysis of the bolt rupture and the loss of tightness for the reached level. For the structure in question, material characteristics are given. Stress levels in stud bolts may decrease using a sealing made of expanded graphite instead of nickel gaskets and a tightening of nuts by means of a hydraulic pre-stressed device instead of their tightening through the torque moment. In case of using the hydraulic pre-stressed device, a bending stress of bolts will significantly decrease. When using a pre-stressed stud bolt, two pads with spherical surfaces that may mutually take suitable positions at a small thrust force. This may not happen in case of tightening by torque moment. The nut is set into recess in the upper pad.

In the probabilistic analysis, mechanical material characteristics are used according to stud bolt testimonials, supplied for the whole power station. Time before the defect initiation on a stud bolt surface is taken according to the results based on the non-destructive bolts testing under operational conditions.

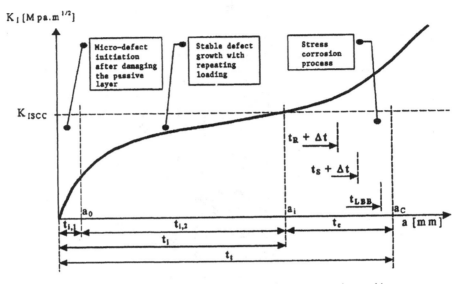

Figure 2. The process of material degradation under stress corrosion cracking

2.2. STAGES OF A STUD DEFECT ORIGIN AND GROWTH

Mathematical description of the gradual damage stud bolt until its rupture is based on the following stages [4], *Figure 2*:

1) *Stage 1* represents the origin of a surface microdefect, a defect such as a crack or pitting. Time of the 1st stage duration is indicated as t_{i1}. At this time, the microcrack will reach value $a_0 \sim 0,05$ mm.

2) *Stage 2* represents a defect growth caused by load in corrosion environment but excluding stress corrosion cracking conditions. The crack growth in comparison with the previous stage accelerated due to the cyclic loading. The bolt cross section is weakened, so the stress increased here. Time of stage 2 (duration) is indicated as t_{i2}, and at its end a defect depth will reach value a_i. These conditions may be described by stress intensity factor:

$$K_I = \sigma Y \sqrt{\pi a} = f(\sigma, a, Y) \qquad (2.1)$$

3) *Stage 3* represents a defect growth under stress corrosion cracking when stress intensity factor K_I reaches value K_{ISCC}, representing threshold value K_I. The crack growth continues by a mechanism of stress corrosion cracking until limit value a_L is reached. The stud bolt cross-section is weakened by the defect and is not capable of carrying the load induced by an axial force F and the bending

moment M_b. In the stud bolt cross section stressed in elastic-plastic state the value of stress $S_m = 0.5 (S_y + S_u)$ is reached. The time of duration of stage 3 is indicated as t_c.

4) *Total time* of the stud bolt operation its damage:

$$t_t = t_{i1} + t_{i2} + t_c \geq (t_R \text{ or } t_s \text{ or } t_{LBB}) + \Delta t \qquad (2.2)$$

In the probability analysis (section 4), time t_t is related to periods of non-destructive tests being 3, 4, 5, 6, 7 and 8 years.

2.3. STAGE 1, THE ORIGIN OF A SURFACE DEFECT

The time of the origin of a surface defect may be determined from relation [4], [5]:

$$t_{i1} = C_{t1} [Cl^-]^{-m_1} (V_{cor} - V_c)^{-m_2} \qquad [\text{ppm; mV; mm/s}] \qquad (2.3)$$

2.4. THE CRACK GROWTH DURING STAGES 2

It is assumed that cracks in stud bolts will grow perpendicularly to the maximum stress direction, i.e. to its axis, and will be of shape, shown in *Figure 3*. The defect will grow from value a_o to value a_i during the time t_{i2}:

$$a_i - a_o = v_i t_{i2} + \sum_{j=1}^{k} f_j \, t_j \, C \, (\Delta K_I)_j^m \,, \qquad (2.4)$$

where:

$$t_{i2} = C_{t2} [Cl^-]^{-m_1} (V_{cor} - V_c)^{-m_2} \qquad [\text{ppm; mV; mm/s}] \qquad (2.5)$$

The type of the cycle j has frequency f_j. The time $t_{i2} = t_1 + t_2 + \ldots + t_j + \ldots + t_k$, where an index k is equal to number of cycle types.

2.5. THE CRACK GROWTH DURING STAGE 3

Bolts almost work under nominal conditions when forces change insignificantly, $K_I = const.$ Relation determines this speed of a crack growth:

$$\frac{da}{dt_c} = C_c K_I^{m_c} = C_c [b_0 + b_1 a + b_2 a^2]^{m_c} \qquad [\text{mm/s}] \qquad (2.6)$$

The time the crack growth under corrosion cracking is determined from relation:

$$t_c = \int_{a_o}^{a_L} \frac{da}{C_c [b_0 + b_1 a + b_2 a^2]^{m_c}} \qquad [\text{s; mm}] \qquad (2.7)$$

Characteristics C, C_c, C_{t1}, C_{t2}, m, m_c, m_1, m_2 have to be determined by tests of specimens for a given material and environment.

For nominal operation conditions, when a stud bolt is stressed by an average stress in cross section $\sigma_{m,n}$ and by bending stress $\sigma_{b,n}$ the stress intensity factor K_I is possible to analyse by the finite element method (FEM) for different depths of defect, for example from a = 1 mm to 15 mm. Program SYSTUS [10] was used. Then it is possible a curve, for example the second order parabola, to put though these values of K_I as function of the depth a:

$$K_I = b_0 + b_1 a + b_2 a^2 \qquad (2.8)$$

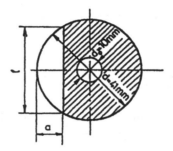

Figure 3. Crosssection of the weakened stud bolt by the crack

The range of the K_I is equal:

$$\Delta K_I = K_{I,max} - K_{I,min} \qquad (2.9)$$

Values of parameters b_0, b_1, b_2 are changed in dependence on the force F and bending moment M_b. Then the time t_c has to be determined as a sum of the t_{ci} calculated from relation (2.7) for the k groups of the F and M_b:

$$t_c = \sum_{i=1}^{k} t_{ci} \qquad (2.10)$$

2.6. FORCE RELATIONS IN DEMOUNTABLE CONNECTION AT A BOLT RUPTURE

The experiment on the model in scale 1:1 proved, that after the rupture of every stud bolt the forces in neighboring stud bolts will increase and that the cover is rigid and remains flat during a gradually rupture of stud bolts. It is assumed that bolts 1, 2 and 20 (*Figure 4*) will break as the last from all n = 20 stud bolts. Using of these assumptions it is possible to write the relation of the force F_i in the stud bolt i on the changes of forces ΔF_1, ΔF_2 and ΔF_{20} in the stud bolts 1, 2 and 20. The force F_{si} was in the stud bolt i before the rupture of the stud bolt j. The stud bolts j, j+1, ..., n-1, n were broken to this moment. Then [6]:

$$F_i = F_{si} + \Delta F_i + \frac{x_i}{x_2} (0.5\,\Delta F_2 + 0.5\,\Delta F_{20} - \Delta F_1) + \frac{y_i}{y_2} (\Delta F_2 - \Delta F_{20}) \quad (2.11)$$

The unknown values ΔF_1, ΔF_2 and ΔF_{20} are calculated from three conditions of equilibrium $\Sigma F = 0$, $\Sigma M_x = 0$, $\Sigma M_y = 0$ for every case of the rupture stud bolts j, j+1, ..., n-1, n.

The specific pressure action on the sealing changes about $\Delta q(\varphi)$ value from the specific pressure before the rupture of the stud bolt j [6]:

$$\Delta q(\varphi) = -\frac{k_t}{2\pi R_t b} \frac{R_t - 3R_w}{R_t} F_{sj} - \frac{3}{4} \frac{R_w k_t}{R_t^2 b} F_{sj} \left| \sin\left[\frac{\varphi}{2} + \frac{(11-j)\pi}{20}\right]\right| \quad (2.12)$$

where:

$$k_t = \lambda_s / (\lambda_s + \lambda_t) \quad (2.13)$$

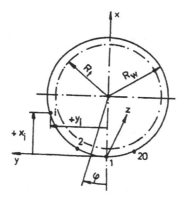

Figure 4. Location of the stud bolts of radius R_w and the seal of radius R_t, coordinates x, y, z = 0 of the 20 stud bolts, 1, 2, ...,i, ...20

3. Reliability Analysis

3.1. GENERAL REMARKS

Deterministic analysis of computational model described above cannot provide realistic results concerning the assessment of reliability. There are many uncertainties related to the input parameters of the problem and they have to be modeled as random variables. Then the reliability of the cover can be expressed in probabilistic way: The probability that an event (failure) occurs (or does not occur) is assessed. In case of the steam generator cover, such an event represents e.g. the breaking of a stud bolt, the loss of the functionality of the cover (leakage from the primary to the secondary system of the circuit) or the break off of the complete cover. Generally, an event is described by an appropriate limit state function constructed using the computational model. Within the framework of reliability analysis based on Monte Carlo type simulation technique, the limit state function should be evaluated many times with different realizations of random variables, and consequently the statistical, sensitivity and probability information may be gained.

3.2. LIMIT STATE FUNCTIONS

The general form of the limit state function used in reliability engineering was already shown. Limit state condition requires (safe state):

$$g(\mathbf{X}) > 0 \tag{3.1}$$

A case of negative limit state function represents failure condition, then unreliability of the system can be expressed by a theoretical failure probability:

$$p_f = P\left(g(\mathbf{X}) \le 0\right) \tag{3.2}$$

Such a failure probability is the quantitative measure of unreliability resulting from uncertainties involved in the problem. The measure of reliability simply is:

$$p_s = 1 - p_f \tag{3.3}$$

The basic task of our study is to assess the possible loss of functionality of the cover – a leakage occurrence. The aim of reliability analysis is to estimate the probability of such a loss of functionality. Limit state function for such a situation can be expressed in the form:

$$g(\mathbf{X}, t) = q_{min}\left(\mathbf{X}, t\right) - q_{lim} \tag{3.4}$$

where q_{min} is the minimum actual pressure on the seal between bolts, and q_{lim} is the limit (allowable) minimum pressure. The function is defined for a specific time – deterministic parameter t.

Alternatively, for the determination of the probability of the breaking of N_p bolts (and less), $N_p = 1, 2, \ldots, 20$, the limit state function can be defined in the form:

$$g(\mathbf{X}, t) = N_a\left(\mathbf{X}, t\right) - N_p - c \tag{3.5}$$

where N_a represents the actual number of broken bolts in the actual simulation. In case $N_p \ge N_a$, the limit state function is negative, which is the indication of a failure. The arbitrary constant $(0 < c < 1)$ was introduced in limit state function in order to make the function negative for $N_p = N_a$. Note that this probability of the breaking of a certain number of bolts can be solved in the following alternatives:

- Probability of the braking of N_p stud bolts resulting in simultaneous loss of the seal function of the cover.
- Probability of the breaking of N_p bolts while the cover remains in function.
- Probability of the breaking of N_p bolts – no attention is paid to the function of the cover.

Together four different limit state functions are introduced (3.4), (3.5) – see Par. 4.1. Note that q_{min} in (3.4) and N_a in (3.5) are functions of many geometrical and material parameters, as described in section 2, the majority of them are of random nature.

They have to be considered as random variables when assessing the safety of the constructional member in different operational situation of the NPP and when making the inspection or repair strategy.

3.3. RANDOM VARIABLES

In probabilistic calculations, the definition of the set of random variables is the primary task. It consists of two steps: Firstly, the choices of the parameters, which will be considered stochastic and will be simulated as random variables, thus forming vector **X**. Secondly, every random variable X should be described by the theoretical model of probability distribution function. Such a function is in many cases (two-parametric distributions) characterized by a particular type of distribution simply described by mean value $\mu(X)$ and standard deviation $\sigma(X)$.

If statistical data are available for all the variables, the statistical characteristics can be estimated, and a theoretical model of probability distribution function can be assigned by well-known procedures of mathematical statistics. In most cases, there is scanty data available. The statistical properties will then have to be estimated based on different assumptions. The following sources for the estimation (also for deterministic parameters involved) were used: In site measurements, information published in literature, experience and opinions of experts, engineering judgment and intuition. A synthesis of all pieces of information resulted in the estimation of limit values, x_{high}, x_{low}, above or below which the relevant variable will not fall with a certain probability. For the normal probability distribution it is well known that there is a 95% probability that a variable will have a value situated between $\mu - 2\sigma$ and $\mu + 2\sigma$. Based on this assumption, it is thus possible to estimate the mean value and the standard deviation:

$$\mu(X) = \frac{1}{2}\left(x_{high} + x_{low}\right), \qquad \sigma(X) = \frac{1}{4}\left(x_{high} - x_{low}\right) \qquad (3.6)$$

For a log-normal distribution, the corresponding formulae are:

$$\mu(X) = \sqrt{x_{high} \cdot x_{low}}, \qquad COV(X) = \frac{1}{4}\ln\left(x_{high}/x_{low}\right) \qquad (3.7)$$

where COV(X) is coefficient of variation of variable X.

The choice between the normal, log-normal and other distribution will depend on the physical nature of the random variable. Log-normal distribution is to be preferred for variables low coefficient of variation (COV $\leq$ 0.10), the difference between normal and log-normal probability distribution function is for practical purposes negligible.

The above assumptions resulted in a choice of the type of distribution, the mean value and the standard deviation of the random variables involved in (3.4) or (3.5). Together 78 random variables X_1, X_2, ..., X_{78} considered here are listed in TABLE I. The mean values and standard deviations adopted are those values, which were rated as providing the best estimates of the type indicated above. The normal probability distribution was considered to model the majority of parameters. The statistical correlation among random variables was not considered here due to the lack of sufficient information.

For the purpose of this pilot study, a rather high number of input parameters were considered as random. Let us note that some of them will influence the scatter of limit state functions in a dominant way. It could be a task of the sensitivity analysis (see e.g. [9]) to distinguish such parameters with the aim to make to future reliability computations more simple by considering variables exhibiting low sensitivity as those that are deterministic.

3.4. FAILURE PROBABILITY ESTIMATION BY IMPORTANCE SAMPLING

The utilization of classical Monte Carlo simulation (simple random sampling) is impossible for very low failure probabilities and computationally intensive limit state function. Our problem belongs to this category – the step-by-step breaking of bolts and the calculation of crack lengths due to the stress corrosion represent iteration algorithms (see 2.2 to 2.6) used in every simulation. Also the probabilities we face here are rather small. This is the reason why the so-called advanced simulation techniques (e.g. importance sampling, adaptive sampling, directional sampling, etc.) have to be applied [1, 7, 8]. These techniques concentrate simulation in the failure region around the surface $g(\mathbf{X}) = 0$. They enable us to estimate also very low probabilities quite efficiently using only e.g. a few thousands of simulations. For our limit state functions, we utilized importance sampling around mean values, which represents a rather conservative and stable method for the problem with many random variables involved [1].

During the simulation process, the bolts have to be broken one by one so that the limit state surface $g(\mathbf{X}) = 0$ may be achieved. For the problem of loss of the structural function, the number of broken bolts necessary to reach this stage was approximately from 1 to 6.

3.5. SUMMARY OF BASIC STEPS OF PROBABILISTIC ANALYSIS

The deterministic computational model described in section 2 is an iterative one (step-by-step breaking off of stud bolts). In addition, Monte Carlo type simulation used for the failure probability assessment requires repetitive calculations.

The algorithm for the calculation of the limit state function in focus can be briefly summarized as follows:

- Simulation of realizations of vector $\mathbf{X}$.
- Calculation of actual and limit forces in bolts. The influence of corrosion under stress is taken into account here.
- Conditions for corrosion fracture of bolts are checked individually for every bolt (based on comparison of stress intensity factor and fracture toughness).
- If above stated conditions are satisfied (as suitable for fracture corrosion initiation) two values are calculated: Maximum length of crack and actual crack length for time given.
- Determination of the bolt (out of the 20 existing ones), which will be broken first. It will be controlled by crack length and forces in bolt calculated in previous steps.
- Calculation of limit state function selected ((3.4) or (3.5)). In case that $g(\mathbf{X}) < 0$, there is a contribution to failure probability, and next simulation is performed.

– In opposite case, i.e. $g(\mathbf{X}) > 0$, the redistribution of forces should be calculated; then the process of step-by-step breaking off bolts continues.

TABLE I. Input random variables

	Variable	Unit	Symbol	Mean	COV	PDF	Note
1	Force in bolt	kN	F_s	305.96 (343.8)	0.058	log-normal	Nickel
				245.37 (283.9)	0.073	log-normal	Expanded graphite
2-21	Correction of force in bolt	-	F_{sc}	1.0	0.038	normal	Torque moment
				1.005	0.045	normal	Hydraulic device
22	Bending stress	MPa	σ_b	163.83 (175.0)	0.358	log-normal	Torque moment
				95.39 (76.2)	0.172	log-normal	Hydraulic device
23-42	Correction of bending stress	-	σ_{bc}	0.96	0.047	normal	
43	Tensile stress	MPa	S_u	877.0	0.042	Gumbel (min)	Set of data
44	Yield strength	MPa	S_y	513.69	0.067	gamma	Set of data
45	Radius of shank	mm	R	20.5	0.001	normal	
46	Radius of shank axis hole	mm	r	5.0	0.005	normal	
47	Radius of bolt circle	mm	R_w	310.0	0.001	normal	
48	Pressure in primary circuit	MPa	p_I	13.65 (12.3)	0.09	log-normal	
49	Pressure in secondary circuit	MPa	p_{II}	3.36 (4.7)	0.26	log-normal	
50	Testing pressure	MPa	p_h	16.33 (16.34)	0.002	log-normal	Control
51	Diameter of inside sealing ring	mm	$D_{tm,1}$	497.0	0.0002	normal	
52	Diameter of outside sealing ring	mm	$D_{tm,2}$	547.0	0.0002	normal	
53	Proportion of seal yielding	-	K_t	0.989	0.002	normal	Nickel
53	Proportion of seal yielding	-	K_t	0.986	0.002	normal	Expanded graphite
54	Characteristics of sealing	-	m	2.5	0.012	normal	Nickel only
55	Width of sealing	mm	b	7.2	0.014	normal	Nickel
55	Width of sealing	mm	b	32.5 (34.5)	0.038	normal	Expanded graphite
56	Model uncertainty factor	-	Φ	1.0	0.15	normal	
57	Fracture toughness	MPa.$m^{1/2}$	K_{ISCC}	71.06 (86.0)	0.176	log-normal	
58	Initiation period of crack growth	years	t_i	6.38	0.25	Gumbel (min)	Set of data
59-78	Correction of initiation period	-	$t_{i,c}$	1.0	0.025	normal	

3.6. SUMMARY OF BASIC STEPS OF PROBABILISTIC ANALYSIS

The deterministic computational model described in section 2 is an iterative one (step-by-step breaking off of stud bolts). In addition, Monte Carlo type simulation used for the failure probability assessment requires repetitive calculations.

The algorithm for the calculation of the limit state function in focus can be briefly summarized as follows:

- Simulation of realizations of vector X.
- Calculation of actual and limit forces in bolts. The influence of corrosion under stress is taken into account here.
- Conditions for corrosion fracture of bolts are checked individually for every bolt (based on comparison of stress intensity factor and fracture toughness).
- If above stated conditions are satisfied (as suitable for fracture corrosion initiation) two values are calculated: Maximum length of crack and actual crack length for time given.
- Determination of the bolt (out of the 20 existing ones), which will be broken first. It will be controlled by crack length and forces in bolt calculated in previous steps.
- Calculation of limit state function selected ((3.4) or (3.5)). In case that $g(X) < 0$, there is a contribution to failure probability, and next simulation is performed.
- In opposite case, i.e. $g(X) > 0$, the redistribution of forces should be calculated; then the process of step-by-step breaking off bolts continues.

4. Numerical Results

4.1. PROBABILITY OF THE LOSS OF COVER FUNCTIONALITY

In order to calculate failure probability, a higher number of simulations had to be used to achieve acceptable error of estimation ($250000 \div 500000$ simulations). All random variables summarized in TABLE I have been used. Probabilities were determined for different combinations of the following alternatives:

- Tightening of bolts by torque moment (K)
- Tightening by a hydraulic pre-stressed device (H)
- A seal made of expanded graphite (G)
- A seal made of nickel gaskets (N)

The results of reliability analysis are shown in *Figure 5*. It is obvious that differences are not very significant, nevertheless an alternative HG (hydraulic device, expanded graphite) appeared to be most reliable.

4.2. PROBABILITY OF STUD BOLT BREAKING

The probabilities of these alternatives were calculated without paying attention on the functionality loss of the whole cover. $N_p = 1$ was considered in limit state function (3.5). Achieved results are presented in *Figure 6*. One can see that the alternative KN resulted in the largest failure probabilities during the whole time specified. In this case, the failure probability of alternative HG dropped much more comparing to the previous case.

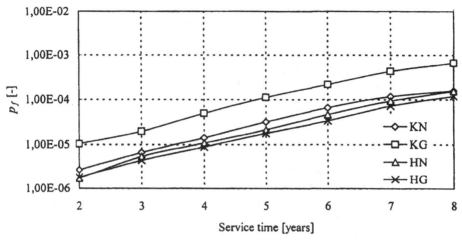

Figure 5. Probability of the loss of cover functionality vs. service time

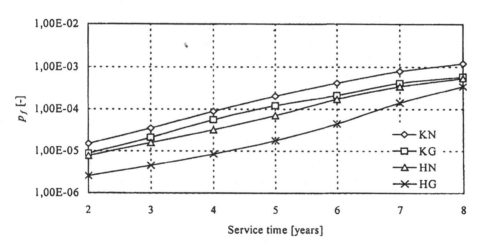

Figure 6. Probability of stud bolt breaking vs. service time

5. Summary

- The probabilistic safety assessment approach described above represents a suitable tool for the prediction of reliability decrease in the course of time, thus making it possible to make the planning of inspection and repair strategies more effective. The minimum protection system integrity required to reduce the risk to a tolerable level may be assessed.

- The presented probabilistic integrity study has shown the better alternative of the tightening process and used kind of a seal at respecting of both cover functionality tightening loss and breaking of a stud bolt.
- The results indicate that no significant decrease of reliability was detected prolonging the inspection period from 4 to e.g. 5 years.
- Other suitable gaskets have to be analysed, for example a grooved gasket with expanded graphite sheet on its both surfaces.
- A rather serious task is to reliably choose the set of basic random variables and their statistical parameters. A more detailed experimental study is generally needed in this context.

Nomenclature

a, a_o, a_i, a_c	-	depth of a crack (a_o – microdefect at time t_{i1}; a_i – origin of stress corrosion conditions; a_c – critical value), mm
b	-	width of seal, mm
b_o, b_1, b_2	-	coefficients, dimensionless
C	-	coefficient, crack growth at cyclic loading, dimension given by (2.4)
C_c	-	coefficient, crack growth at stress corrosive conditions, dimension given by (2.6)
C_{t1}, C_{t2}	-	coefficient, time of a crack growth at corrosive conditions, dimension given by (2.3) and (2.6)
COV (X)	-	coefficient of variation of random variable X
d	-	diameter of the stud bolt shank, mm
d_o	-	diameter of the shank axis hole, mm
$\Delta q(\varphi)$	-	specific pressure acting on the sealing exchanges, N/mm
f	-	frequency, Hz
F_i, F_j	-	tensile force (I, j – number of stud bolts), N
ΔF	-	change of the tensile force in a stud bolt, N
F_s	-	tensile force in a stud bolt before its breaking, N
φ	-	angle determined position of the stud bolt, radian
g(.)	-	limit state function
k_t	-	coefficient of seal yielding proportion, dimensionless
K_I	-	stress intensity factor, MPa $m^{1/2}$
K_{Imax}, K_{Imin}	-	max. and min. value of the K_I cycle, MPa $m^{1/2}$
ΔK_I	-	range of stress intensity factor, MPa $m^{1/2}$
K_{ISCC}	-	threshold value of K_I for given stress corrosion conditions, MPa $m^{1/2}$
l	-	length of crack, mm
λ_s	-	yielding of stud bolt, mm/N
λ_t	-	yielding of seal, mm/N
m	-	exponent, crack growth at cyclic loading, dimensionless
m_1, m_2, m_c	-	exponent, time of crack growth at corrosive conditions, dimensionless
$\mu (X)$	-	mean value of random variable X
M_b	-	bending moment, kNm
N_a, N_p	-	number of broken bolts, dimensionless
p_f	-	probability of failure, dimensionless

q_{min} (.)	-	minimal actual pressure on the seal, MPa
q_{lim}	-	limit (allowable) minimum pressure on the seal, MPa
R_t	-	radius of sealing ring, mm
R_w	-	radius of bolt circle, mm
S_m	-	flow stress (average value of the S_y and S_u), MPa
S_u	-	ultimate strength, MPa
S_y	-	yield stress, MPa
σ	-	stress, MPa
σ_b	-	bending stress, MPa
σ_m	-	average stress, MPa
σ_n	-	nominal stress, MPa
σ (X)	-	standard deviation of random variable X
t	-	time, s
Δt	-	safety time reserve, s
t_c	-	time of crack growth under stress corrosion cracking conditions, s
t_{i1}	-	time when crack achieves depth a_o, s
t_{i2}	-	time when crack achieves depth a_i, s
t_{LBB}, t_R t_s	-	time to: LBB – leak before break; R - repair of a defect; s – end of service, s
t_t	-	total time of the stud bold operation until ist damage, s
V_c	-	critical potential, mV
V_{cor}	-	corrosive potential, mV
v_i	-	speed of the crack growth, mm/s
X	-	vector of random variables X
x, y, z	-	coordinates, mm
Y	-	geometry function

Acknowledgements

The work on the present paper has partially been sponsored by the Czech Ministry of Educations, Project No. J22/98:261100007 and by the Czech Grant Agency, Grant No. 103/00/0093.

References

1. Schuëller, G.I. and Stix, R. (1987): A Critical Appraisal of Methods to Determine Failure Probabilities, J. Struct. Safety, Vol. 4, No. 4, 293-309.
2. Schuëller, G.I. (1998): Structural reliability – Recent advances, Proc. of Conf. on Structural Safety and Reliability ICOSSAR'97, Balkema, Rotterdam, 3-35.
3. Melchers, R.E. and Stewart, M.G. (1993): Probabilistic Risk and Hazard Assessment, Proc. Conf. on Probabilistic Risk and Hazard Assessment, Newcastle N.S.W., 243-252.
4. Vejvoda, S. (2001): Assessment of resistance of structures against stress corrosion cracking. Pros. of Conf. KOROZE 2001, Brno University of Technology, (In Czech).
5. Matocha, K. and Wozniak, J. (1994): Assessment of fatigue and brittle crack characteristics of 440 MW WWER steam generator and pressurize materials. Report of VÍTKOVICE company, IME, No. CD-51/94.

6.   Vejvoda, S. (1998): Backgrounds of probability analysis for demountable collector connection of the 1st circuit of PG VVER 440 MW CEZ-EDU. Conclusions to probability analysis. IAM Brno Report, arch. No. 2640/98, (In Czech).

7.   Bucher, C.G. (1995): Adaptive Sampling – An Iterative Fast Monte-Carlo Procedure, J. Struct. Safety, Vol. 5, 119-126.

8.   RCP – Reliability Consulting Programs (1995): STRUREL: A Structural Reliability Analysis Program System, COMREL & SYSREL Users Manual. RCP Consult, München.

9.   Novák, D., Teplý, B. and Shiraishi, N. (1993): Sensitivity analysis of structures: A review. Proceedings of Conference CIVIL COM´93, Edinburgh, 201-207.

10. SYSTUS version 233, Users Manual Framasoft+CSI, Framatome group.

INDEX